11/87

D0080745

Modern Elementary Particle Physics

Modern Elementary Particle Physics

Gordon Kane
Harrison M. Randall Laboratory of Physics
University of Michigan, Ann Arbor

Addison-Wesley Publishing Company, Inc.
The Advanced Book Program
Redwood City, California • Menlo Park, California • Reading, Massachusetts
New York • Amsterdam • Don Mills, Ontario • Sydney
Bonn • Madrid • Singapore • Tokyo • Bogotá • Santiago
San Juan • Wokingham, United Kingdom

Publisher: *Allan M. Wylde*
Editorial Coordinator: *Pearline Randall*
Electronic Book Production: *Mona Zeftel*
Production Administrator: *Karen Lynne Garrison*
Promotions Manager: *Celina Gonzales*

Library of Congress Cataloging-in-Publication Data

Kane, Gordon.
    Modern elementary particle physics : quarks, leptons, and their
interactions / Gordon Kane.
        p.   cm.
    Includes bibliographies and index.
    ISBN 0-201-11749-5 : $39.95
    1. Particles (Nuclear physics)   2. Quarks.   3. Leptons (Nuclear
physics)   4. Nuclear reactions.   I. Title
    QC793.2.K36 1987
    539.7'21—dc19   87-18628
                                                                         CIP

This book was composed by Mark Schutze using the T_EX typesetting language on a VAX II/780 computer. Camera ready output from an Apple Laser Writer.

ABCDEFGHIJK-HA-8987
0-201-11749-5

To the experimenters and theorists who discovered the Standard Model, and to the scientists who build the accelerators and detectors that will take us further.

Preface

All life is a struggle in the dark.........This dread and darkness of the mind cannot be dispelled by the sunbeams, the shining shafts of day, but only by an understanding of the outward form and inner workings of nature.

And now to business. I will explain...

Lucretius, "On the Nature of the Universe"
Translation of R. E. Latham, Penguin Books

A few years ago I was asked to teach our University of Michigan undergraduate course in particle physics. Soon after agreeing, I found that there was no book available at the undergraduate level which presented particle physics as the successful theory of quarks, leptons, and their interactions that it has become. Fifteen years ago there was no theory of weak or strong interactions; no possibility that a fundamental set of constituents had been identified; no way to calculate or explain a variety of results. Today there is a theory of strong, weak, and electromagnetic interactions, with the latter two unified. There have been several extraordinary experimental discoveries and there are no experimental results that appear to fall outside of the framework of that theory. The so–called Standard Model of particle physics that accomplishes all of that is now widely tested in a variety of ways, and it is here to stay. It is expected that deviations from the Standard Model will someday be found, as clues to improved understanding and to new physics, but the Standard Model will describe physics on the scale where strong, weak, and electromagnetic interactions are important.

Given that development, it seemed important to have a presentation of the Standard Model that could be used in an undergraduate course for physics majors. In addition, it seemed as important to have a book that any scientist who understood the necessary background material could read in order to learn about the developments in particle physics. Those developments should become

vii

a part of the education of anyone interested in what mankind has learned about the basic constituents of matter and the forces of nature. After teaching the course once, I was convinced that an introductory course in quantum theory and normal undergraduate courses in mechanics and electromagnetism were the minimal background. With those tools it is possible to obtain a good, generally quantitative, understanding of modern particle physics. Physics books on electrodynamics normally begin with Maxwell's equations and explore their consequences, rather than proceeding with the historical development of the subject. I believe the Standard Model should be taught the same way—by writing down the basic form of the theory and working out the consequences.

It quickly became clear that a good deal of the subject matter that had been very important in the historical development of particle physics was not essential for describing its present status. Many areas, such as parity violation, the hadron spectrum and the flavor $SU(3)$ symmetry, elastic scattering and total cross sections, and more, are of interest to professional particle physicists but not to those who want to understand only the fundamentals of the field today. So, with apologies to the people whose work is not covered, I have treated the Standard Model deductively rather than historically when that seemed suitable, and suggested places to read history. I have left out a great deal that is already available in other books. In most cases, I give references where a reader who wants to explore some area further can begin. Since historical treatments of the past two decades have already appeared, and more are coming, I have not discussed the history and the credit at all, rather than doing so superficially.

A second major goal emerged as I taught the course again. Although particle physics has clearly reached a plateau where many of the historical goals of physics have been achieved, no one feels that its development is complete. There are many open questions: why the theory takes the form it does, why there are some particles and not others, what is the physical origin of mass, and so on. Once a reader has understood the basic structure of modern particle physics, it is a small extension to add a framework for understanding why some directions of frontier research are emphasized most, and from what directions progress is thought likely to come. There are no guarantees, of course, but now that there is a theory it is possible to evaluate ideas in ways that were not available before. For example, one of the most crucial problems of the Standard Model is to find the Higgs boson or to establish limits on what mass it might have. Because the Standard Model is available, it is possible to enumerate all ways the Higgs boson might be produced, what would be required to detect it if it were produced, *etc.*

Further, because accelerators and detectors have become so expensive, we know long in advance which experimental facilities will be built. It is essentially impossible for a major facility to turn on before 1994 unless it is already planned. So the directions from which data will come in the next decade are rather well known now. Consequently, this book also attempts to give the reader the information needed to understand what most particle physicists will be doing for the next decade, and why, in so far as it is connected with extending the Standard Model experimentally and theoretically. Of course there will also be theoretical efforts in directions we will not consider because they are beyond the scope of this book, such as quantum gravity or superstrings, and there will be accelerator and detector developments.

It was necessary to decide what fraction of the course and the book should be spent on detailed calculations. Clearly the reader should learn to understand the qualitative numerical structure of the Standard Model. On the other hand, only active particle physicists need to be able to calculate precisely, and good books have appeared at the graduate level. I have taken an intermediate path, using the notation of quantum field theory, which is compact and easily understandable, but not calculating with it. Some simple approximate calculational procedures are developed which allow almost all decay rates and cross sections to be calculated to within about a factor of two, and especially allow one to keep track of their dependence on the important physical quantities. Thus, the reader learns to estimate in a controlled way many of the main rates necessary to understand the Standard Model and its tests.

The book can be thought of as being divided into three parts: the Standard Model, some advanced topics in the Standard Model, and a few topics beyond the Standard Model. The main treatment of the Standard Model is subdivided into several sections. Chapter 1 and 2 compose the first section. They are surveys meant to prepare the reader for the systematic treatment, which begins in Chapter 3. Chapter 1 describes the particles and the forces as we think of them today. This material will be unfamiliar to many readers, and much of it is only explained as the book unfolds, but a preview of the material has been helpful for many students. Chapter 2 presents and ties together a variety of material necessary for proceeding with the main purpose of the book. It will be useful for some readers, though perhaps frustrating for others, depending on how their backgrounds fit what is presented and on how they find the heuristic arguments that are given. Probably some readers should read Chapters 1 and 2 initially and then return to them after some exposure to the rest of the book. Others can skip Chapter 1 and 2 entirely.

The explanation of the Standard Model begins in Chapter 3. The development proceeds with massless fermions and bosons through Chapter 7, including what is needed about the Dirac equation and its solutions in Chapter 5. The Higgs mechanism is presented in Chapter 8. The formal development concludes in Chapter 9 with a derivation of the W and Z widths.

The third section covers, in Chapters 10–19, a variety of tests and predictions and properties of the Standard Model. I have found that this portion of the book, through Chapter 19, constitutes most of a one–semester advanced undergraduate course, with time to include about three topics from the rest of the text.

Chapters 20–25 contain short treatments of several important but somewhat more advanced Standard Model topics. I try to include parts of Chapters 20 and 21 in a one semester course.

The third part of the book provides a few possibilities for what might happen beyond the Standard Model. I have restricted the subjects to those which fit naturally into the gauge theory framework of the Standard Model, and have treated them basically as applications of the Standard Model techniques. The calculation of $\sin^2 \theta_w$ in Chapter 27, or the derivation of the photino cross section in matter in terms of the structure function F_2, are good exercises in Standard Model physics even though they are used in the context of a hypothetical new theory.

Finally, several Appendices make the book more self–contained in various ways. About sixty homework problems are included, more than enough for a one–semester course. Most chapters (particularly Chapter 1) end with some guide to additional historical, pedagogical, or technical information for readers who wish to pursue the subject further.

In general, the guiding principle I followed in deciding what to include, how detailed a discussion should be, what should be derived, and so on, was to teach interested readers to understand the structure of modern particle physics, not to train them to become particle physicists. The main purposes of the book then, are to provide what is needed for people (with some scientific background) who are eager to understand the extraordinary progress in particle physics over the past two decades, to appreciate the beauty of the Standard Model, and to have a framework within which to appreciate the developments yet to come.

* * * *

I am grateful to many people for the encouragement and enthusiasm that helped convince me to write this book. The comments and questions of the students in Physics 468, and their desire to understand modern particle physics, were very important. I appreciate comments on the content of the book or assistance from Joel Primak, Frank Paige, Rudi Thun, Dennis Hegyi, Jay Chapman, Greg Snow, Bob Cahn, Bob Tschirhart, Tim Jones, and especially from Jean-Marie Frére, Marc Ross, and Chien-Peng Yuan who provided many valuable suggestions. The figures were drawn by Les Thurston. The enthusiasm and assistance of my editor, Allan Wylde, were important to me. The contribution of Mark Schutze, who took the course and produced the manuscript intelligently and effectively, was essential. Finally, I am very grateful for all the valuable support as well as suggestions about the manuscript from my wife, Lois.

Contents

The Standard Model

1   Survey   1
   1.1   Introduction   1
   1.2   The Framework   3
   1.3   The Forces   5
   1.4   The Particles   7
   1.5   Natural Units   11
   Problems   12
   Suggestions for Further Study   13

2.   Relativistic Notation, Lagrangians,
   Currents, and Interactions   15
   2.1   Some Relativistic Notation   16
   2.2   Lagrangians   17
   2.3   Lagrangians in Particle Physics   19
   2.4   The Real Scalar Field   19
   2.5   Sources and Currents in Non-Relativistic
       Quantum Theory   22
   2.6   Complex Scalars, Conserved Currents,
       and Noether's Theorem   23
   2.7   Interactions   27
   2.8   Summary of the Lagrangians   30
   2.9   Feynman Rules   32
   Problems   33

Suggestions for Further Study   34

3.   Gauge Invariance   35

    3.1   Gauge Invariance in Classical Electromagnetism   35

    3.2   Gauge Invariance in Quantum Theory   36

    3.3   Covariant Derivatives   38

    Problems   41

    Suggestions for Further Study   42

4.   Non-Abelian Gauge Theories   43

    4.1   Strong Isospin, an Internal Space   43

    4.2   Non–Abelian Gauge Theories   46

    4.3   Non–Abelian Gauge Theories for Quarks
          and Leptons   47

    Problems   52

    Suggestions for Further Study   53

5.   Dirac Notation for Spin   55

    5.1   The Dirac Equation   56

    5.2   Massless Fermions   57

    5.3   Fermions with Mass $\neq 0$ 57

    5.4   The γ–matrices   58

    5.5   Currents   60

    5.6   Free Particle Solutions   62

    5.7   Particles and Antiparticles   64

    5.8   Left–handed and Right–handed Fermions   65

    5.9   Useful Relations   66

    5.10 The Dirac Lagrangian   68

    Problems   68

    Suggestions for Further Study   69

6.   The Standard Model Lagrangian   71

    6.1   Labeling the Quark and Lepton States   73

    6.2   The Quark and Lepton Lagrangian   76

    6.3   Gauging the Global Symmetries   78

    Problems   79

	15.2	Color–Singlet Hadrons	179
	15.3	Quantum Numbers of Mesons and Baryons	181
	15.4	Comments and Perspective	183
		Problems	184
		Suggestions for Further Study	184
16.		Light Mesons, Baryons, and Strong Isospin	185
	16.1	The $L = 0$ Meson States	186
	16.2	The $L = 0$ Baryon States	187
	16.3	Decays and Transitions	188
	16.4	The Origin of Strong Isospin Invariance	189
		Suggestions for Further Study	190
17.		Heavy Quarks (c, b)	191
	17.1	Some Charmonium Properties	193
	17.2	The Charmonium Spectrum	195
	17.3	Charmed Mesons	196
	17.4	More Leptons and Quarks	199
	17.5	The b quark	199
		Problems	199
		Suggestions for Further Study	200
18.		Deep Inelastic Scattering and Structure Functions	201
	18.1	Deep Inelastic Scattering	201
	18.2	Analysis of the Parton Model	204
	18.3	The Structure Functions	205
		Problems	210
		Suggestions for Further Study	211
19.		e^+e^- Colliders and Test of the Standard Model	213
	19.1	Are quarks, leptons, and gluons point–like?	215
	19.2	The ratio $R = \sigma/\sigma_{\text{point}}$	216
	19.3	The τ and Heavy Leptons	218
	19.4	Observation of Gluons	220
		Problems	222
		Suggestions for Further Study	222

Advanced Topics in the Standard Model

20. Coupling Strengths Depend on the
     Momentum Transfer   223
        20.1   QED   224
        20.2   QCD   230
        Problems   233

21. Production and Detection of a Higgs Boson   235
        21.1   Higgs Couplings   237
        21.2   Higgs Decays   237
        21.3   Ways to Search for Higgs Bosons   240
        21.4   Large M_H 244
        21.5   Comments   246
        Problems   247
        Suggestions for Further Study   248

22. Quark (and Lepton) Mixing Angles   249
        Problems   254

23. Quark and Hadron Masses   255

24. CP Violation   257
        Problems   260
        Suggestions for Further Study   260

25. Why the t–Quark and the τ–Neutrino Must Exist   261
        25.1   Forward–Backward Asymmetries   261
        25.2 b–Quark Decays   263
        25.3   The τ–Neutrino   264
        25.4   The Mass of the t–Quark   265
        Problems   265
        Suggestions for Further Study   266

Applications of the Standard Model
to Questions Beyond the Standard Model

26. Open Questions   267

27.   Grand Unification   271

       27.1   Unifying Quarks and Leptons;
               Electric Charge and the Number of Colors   271

       27.2   Unification of Forces   273

       27.3   Calculation of $\sin^2 \theta_w$ 276

       27.4   Proton Decay   280

       27.5   The Baryon Asymmetry   283

       Suggestions for Further Study   284

28.   Supersymmetry   285

       28.1   Production and Detection
               of Supersymmetric Partners   288

       28.2   The Lightest Supersymmetric Particle and
               Dark Matter   291

       Problems   293

       Suggestions for Further Study   293

29.   Neutrino Masses?   295

       29.1   If $m_\nu \neq 0$; Neutrino Oscillations   296

       29.2   Solar Neutrinos   300

       29.3   Measurement of m_ν in Decays   301

       29.4   Expectations for m_ν 302

       Suggestions for Further Study   303

APPENDICES

A.   Angular Momentum and Spin and $SU(2)$ 305

B.   Some Group Theory   311

       B.1   The $SO(n)$ Groups   313

       B.2   The $SU(n)$ Groups   314

       B.3 $SU(2)$ and Physics   316

       B.4 $SU(3)$ 317

       B.5   Abelian and Non-Abelian Groups   318

C.   Some Relativistic Kinematics   319

D.   The Point Cross Section   323

E.   When Are Our Approximations Not Valid?   325

F.   Lagrangians and Symmetries;
     The Euler–Lagrange Equations   327
        Problems   330

Bibliography   331

Index   337

CHAPTER 1

Survey

1.1   Introduction

Remarkably, the field of particle physics is completely different today from what it was fifteen years ago; what most particle physicists work on and think about today has essentially no resemblance to what they learned in graduate school. Quarks and leptons are the fundamental objects; they interact via the exchange of gauge bosons. The forces that significantly affect them are the unified electroweak force, whose gauge bosons are the photon and the $W^{\pm}$ and Z° bosons, and the strong force. The theory of the strong force is called quantum chromodynamics (QCD); the gauge bosons of the strong force are the (eight) gluons. (The new terms here will be defined later, as the new physics appears in context.)

In another sense there is great continuity. The theory fully incorporates special relativity. And there has been a continuous development of relativistic quantum field theory from its inception over fifty years ago. Theorists have learned how to deal with difficult problems such as mass and renormalization. There has been steady and extraordinary progress in particle physics, both in understanding quantum field theory and in learning what to include in the Lagrangian; no revolution has occurred. The theories which describe the particles and their interactions seem to be gauge theories, a special class of quantum theories where there is an invariance principle that necessarily implies the existence of interactions mediated by gauge bosons. In gauge theories the interaction Lagrangian is, in a sense, inevitable rather than being introduced in an *ad hoc* way as in quantum theory.

Although technical work in a relativistic quantum gauge field theory can be very difficult, the basic formulation of the theory is accessible to anyone having an undergraduate knowledge of classical mechanics and electrodynamics, plus an introduction to quantum mechanics including spin and angular momentum. Though the theory is formulated in a fully relativistic way, the student has to learn mainly notation in order to make simple calculations and estimates. (I hope that the reader will find the present structure of particle physics beautiful and compelling, and thus feel challenged to learn more of the advanced quantum theory needed in order to grasp the particle theory at a deeper level.) Our goals in this book will be (a) to understand the way the theory of quarks and leptons and their interactions (called the "Standard Model") is formulated, (b) to learn to calculate or approximately estimate a number of predictions of the theory, both to see how it works and to learn the techniques, and to understand the tests of the theory and why it is believed to actually describe nature, and (c) to have a framework within which future major research efforts can probably be understood. To avoid misunderstanding, it should be emphasized that although the Standard Model is called a "Model" it is in fact as fully a mathematical theory as there has ever been in the history of science.

Most of the book will deal with the established Standard Model. One part of the Standard Model, concerning the physics associated with the Higgs mechanism and Higgs bosons, is poorly understood (although technically satisfactory); we will examine it in some detail. Although the Standard Model describes all known experiments and particle interactions, there are many questions that can be raised about the values of the parameters it depends on and about the form it takes. Most workers hope for and expect a number of future developments to help answer such questions; we will survey some of these at the end, and see how they fit into and extend the Standard Model.

Next we will briefly describe the structure the theory takes and introduce the quarks and leptons, the gauge bosons, and the forces. The purpose of such an introductory survey is to give an overview so it is more clear where we are heading, particularly later on when we are deep into a long series of arguments or calculations. The material covered in the rest of this chapter is developed in detail in the first two–thirds of the book.

1.2   The Framework

Recall the way in which force enters in Newton's laws. $F = ma$ is used to compute the motion of an object, given *any* force F on the object. And specific classical forces have been discovered, such as gravity with $F = G_N mM/r^2$, Coulomb's law with $F = KqQ/r^2$, etc. Hamilton's or Lagrange's equations are equivalent to $F = ma$ in a different formalism.

In quantum theory there is an analogous structure. The Schrödinger equation, $H\Psi = i\frac{\partial \Psi}{\partial t}$, is like $F = ma$. It holds for any Hamiltonian. Specific forces lead to specific Hamiltonians.

In particle physics it will help to keep these distinctions in mind. We will briefly review (Chapter 2 and Chapter 9) the formalism, analogous to $F = ma$ or the Schrödinger equation, that allows one to start with any Lagrangian and to compute "the motion", *i.e.* to compute cross sections and decay rates. In practice that means extracting Feynman rules to write matrix elements, and converting the matrix elements into transition probabilities. The procedure here is the standard one of a relativistic quantum theory. Although many new insights have been gained in this area in the past two decades, most of them are more technical and beyond the level of this book, though some will be covered.

The specific Lagrangians for the electroweak force and the strong force are what is particularly new about the Standard Model. Today not only the electromagnetic force, but also the weak force and the strong force are known. In addition, the electromagnetic and weak forces are unified in a certain sense, and in an elegant way that is promising for future progress in particle physics. These Lagrangians are somewhat more complicated than $G_N mM/r$, and it will take a significant part of the book to write them down and learn to use them for simple calculations.

To say it differently, the Hamiltonians that describe all the known interactions of particles have been found in recent years, and the main goal of this book is to describe them. There has also been much progress in understanding the structure of the quantum field theory equivalent of the Schrödinger equation; we will not cover much of that, partly for reasons of space but mainly because it is considerably more difficult technically.

The combination of quantum theory and relativity leads to the introduction of quantum fields and of associated particles. To see intuitively why that must occur, suppose various particles can interact with one another, and you give one particle a push. The forces, due to that particle, that act on nearby particles cannot produce instantaneous changes in their motions, since no signal can travel faster than the speed of light. Instead, as with electromagnetism and gravity, we say the pushed particle is the source of various fields which carry energy, and perhaps other quantum numbers, through the surrounding space; eventually the fields interact with other particles.

Because of the quantum theory, the energy (and perhaps other quantum numbers) is carried by discrete quanta, which become identified with the particles transmitting the force. Thus in a quantum field theory, the elementary particle interactions are interpreted in terms of exchanges of (some of the) particles themselves.

The theories called gauge theories are a special class of quantum field theories where there is an invariance principle that necessarily requires the existence of interactions among the particles. When we speak of gauge forces, as we will extensively later, we will mean forces which respect a gauge symmetry, and in addition, forces whose strengths are proportional to a "charge" of some kind. This is familiar for electrodynamics, where the fine structure constant α measures the strength of the electromagnetic force. Just as for electromagnetism the charge both measures the strength of the interaction ($\alpha = e^2/4\pi\hbar c \simeq 1/137$), and gives the amount of charge (particles have charges of 0, $\pm\frac{1}{3}e$, $\pm\frac{2}{3}e$, $\pm 1e$, ...), for other forces new charges arise which play both these roles. In particle physics the words "force" and "interaction" are used essentially interchangeably.

The basic view, then, of particle interactions is as shown in Figure 1.1.

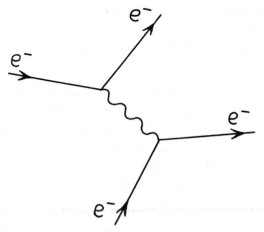

Figure 1.1
Basic view of a particle interaction.

For electrodynamics, a charged particle emits a photon and recoils; the photon is absorbed by another charged particle, which changes its motion as a consequence. Such diagrams can be useful pictures of what is occurring. They are more than that. When a set of rules (the Feynman rules of the theory) is given to convert each diagram into a matrix element, and to calculate transition probabilities, the theory can be summarized in its diagrams (assuming situations where perturbative calculations are relevant). We will deduce the Feynman rules of the Standard Model. In the case of electromagnetism, the matrix element in Figure 1.1, in the nonrelativistic limit, gives Coulomb's law.

Table 1.1
The Known Forces

Forces	1970 Real Theory?	now	soon?
Gravity	yes but classical	unchanged	?
Electromagnetism	yes (QED)	unified into a single real theory (electroweak)	single "grand unified theory"?
Weak Interaction	no		
Strong Interaction	no	yes (QCD)	

1.3 The Forces

The progress in particle physics, both theoretical and experimental, in the past two decades has been remarkable. Table 1.1 illustrates how the understanding of the theory has improved. Around 1970, the situation was as shown on the left–hand side of the table. The theory of general relativity was fully satisfactory but unconnected with the other forces, and a quantum theory of gravity had not yet been constructed. That is unchanged today, though various approaches to constructing a quantum theory of gravity and to unifying gravity with other forces are being pursued actively, and there is considerable optimism that progress will come soon. By "Real Theory?" we mean "Does a Lagrangian Quantum Field Theory exist in which all observables are finite?" Indeed, although the theory of quarks and leptons and their interactions is called the "Standard Model," for historical reasons, it should be emphasized that it is not a model in the usual sense that word is used in physics, but a complete quantum field theory.

In 1970 there was no theory of weak interactions. Weak interactions are a class of interactions that are observed to exist and that would not occur if only gravitational and electromagnetic interactions existed; an example is neutron $\beta-$decay ($n \rightarrow pe\overline{\nu}$). They play a critical role in the process of generation of energy in the sun, and in the building up of heavy elements. Life on earth could not exist in the absence of the weak interactions (or if any of the other known forces were missing, for that matter). They are called "weak" because the typical time scale on which they occur is much longer (of order 10^{-13} sec when appropriately defined) than for electromagnetic processes (of order 10^{-19} sec when similarly defined). Now there is a theory of weak interactions, and further, the weak and electromagnetic interactions have been unified into one force.

Similarly, in 1970 there was no theory of strong interactions. That some "nuclear" force existed had been known since the 1930's or longer, since a nucleus containing several protons would hold together in spite of their electrical repulsion—consequently, another (attractive) force, stronger than electromagnetism, must exist. Since the force was strong, it was expected that perturbative calculations would not apply to observable phenomena. It is all the more astonishing that today it is believed that a real theory of strong interactions exists and is experimentally checked in a variety of ways. It is quarks that undergo the strong interactions. Hadrons (protons, neutrons, pions, etc.) are formed from quarks. The force between quarks is called a "color" force—we will see later how to describe it. [This "color" is not related to the colors we see.] The quarks carry the color charge, and combine to make color neutral hadrons, just as electrically charged electrons and nuclei combine to make electrically neutral atoms. And just as a residual electric field outside of neutral atoms causes them to combine into molecules, the residual color field outside of protons and neutrons is the nuclear force that forms nuclei. Many other hadrons would form nuclei if they lived long enough, but they are unstable because of the strong or weak forces. The theory of the color force is called quantum chromodynamics (QCD).

The success of past unifications (electricity and magnetism into electromagnetism, electromagnetism and weak interactions into the electroweak theory) has encouraged people to try to unify the electroweak theory and QCD into one ("grand unified") theory. There is some theoretical progress in this direction, and candidate theories exist; we will briefly explore grand unification in Chapter 27 to apply some of what has been learned, to define some of the open questions in particle physics, and to see what some answers might look like.

1.4   The Particles

Understanding the structure of the natural universe can be thought of as (at least) a three-part problem. We have to identify the basic particles that are the constituents of matter, know what forces the particles feel, and know how to calculate the behavior of the particles given the forces. [In addition, some properties could be determined by otherwise accidental initial or boundary conditions.] We have mentioned the forces and how to calculate them. The particles can be put into two categories, matter particles and gauge bosons (some newer ideas, that we will briefly mention in Chapter 28, attempt to relate them). The matter particles are the quarks and the leptons.

A quark can be defined as a fermion that carries the color charge of QCD, while a lepton is a fermion with no color charge. Both have spin $\frac{1}{2}$. So far there are known to be six kinds of quarks (called six quark "flavors") and six kinds of leptons (six lepton "flavors"). It is not understood why there are six, nor whether more will be found as machines become available to look at higher energies. The quarks are called, for historical reasons: up, down, strange, charmed, bottom, top. They are denoted by the first letters of their names. As we will see later, they naturally fall into doublets (called "families" or "generations"):

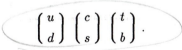

$$\begin{pmatrix} u \\ d \end{pmatrix} \begin{pmatrix} c \\ s \end{pmatrix} \begin{pmatrix} t \\ b \end{pmatrix}.$$

The top row has electric charge $q = \frac{2}{3}e$ and the bottom row has $q = -\frac{1}{3}e$, where e is the magnitude of the electron's electric charge. Each quark flavor comes in three colors. The quarks all carry another quantum number called baryon number, B. Quarks have $B = 1/3$; protons and neutrons have $B = 1$. Baryon number is observed experimentally to be conserved to a very good accuracy, but no reason is known why it should be absolutely conserved.

Five of the quarks have been studied experimentally for some time. The t-quark must exist, given the requirements of the electroweak theory and the measured properties of the b-quark (we will understand this statement in Chapter 25 as an example of the way the theory works), but its mass is not yet measured.

Any colored particle, according to what is currently believed (but not yet rigorously proved) to be a consequence of QCD, is permanently bound inside a colorless hadron. That does not at all mean that quarks do not exist in the same sense as electrons exist, as we will discuss later. But it does make quantitative determination of quark masses a subtle point when the quark masses are smaller than the typical hadronic masses of about 1 GeV/c^2. As a result, it is normal to report quark masses in two ways, which we will label "free"–quark masses and constituent–quark masses, as in Table 1.2. The "about" in the table reflects variations in how the mass is defined as well as the actual determination. Why the constituent masses are larger than the "free" ones, contrary to our naïve experience with planets or atoms, and some other aspects of masses, will be discussed in Chapter 23. The values of the masses are not calculated or understood, they are simply measured. For historical reasons, what we have called "free"-quark masses are usually called "current algebra" masses.

Table 1.2
Quark Masses

	"free"–quark		constituent–quark	
d about	15	MeV/c^2	330	MeV/c^2
u about	7	MeV/c^2	330	MeV/c^2
s about	200	MeV/c^2	500	MeV/c^2
c about	1.3 GeV/c^2		1.5 GeV/c^2	
b about	4.8 GeV/c^2		5 GeV/c^2	
t about	??		??	

The flavors of leptons are also arranged in three families of doublets,

$$\begin{pmatrix} \nu_e \\ e \end{pmatrix} \begin{pmatrix} \nu_\mu \\ \mu \end{pmatrix} \begin{pmatrix} \nu_\tau \\ \tau \end{pmatrix}.$$

The electron (e), muon (μ), and tau (τ) have electric charge $-e$, while each has its own neutrino of electric charge zero. As far as is known, the separate lepton types do not undergo transitions into one another. A lepton number can be defined for each family; it is observed experimentally to be conserved, though fundamental reasons why it should be absolutely conserved are not known. The neutrino masses are not measured yet; they are consistent with zero, but most theorists do not expect them to be zero. The present limits are given in Table 1.3.

Table 1.3
Lepton Masses

ν_e	$< \; 46 \quad$ eV/c^2
ν_μ	$< \; 0.25$ MeV/c^2
ν_τ	$< \; 70 \quad$ MeV/c^2
e	0.51 MeV/c^2
μ	105.6 MeV/c^2
τ	1784.2 ± 3.2 MeV/c^2

As with the quarks, the lepton masses are not understood. The ν_τ has not been directly observed, but (just as was remarked for the t-quark; see Chapter 25) given the requirements of the electroweak theory and the properties of the τ, it must exist. Indirect cosmological arguments imply that the sum of all the neutrino masses is less than about 100 eV/c^2.

Table 1.4
Forces and Gauge Bosons

Force	acts on	transmitted by
gravity	all massive particles	graviton (massless, spin-2)
electromagnetism	all electrically charged particles	photon (γ) (massless, spin-1)
weak interaction	quarks, leptons, electroweak gauge bosons	$W^\pm$, Z° (heavy, spin-1)
strong interaction (QCD)	all colored particles (quarks and gluons)	eight gluons (g) (massless, spin-1)

The quarks and leptons are the basic particles of matter. In addition, there are the particles which transmit the forces. They are all bosons (integral spin). Table 1.4 summarizes the forces, what they act on, and the bosons that transmit them. Since the quantum field theory that allows us to calculate the behavior of the particles is a gauge theory, they are called "gauge bosons." As we will discuss in detail later, gauge bosons were expected to be massless, and most of them are; that the $W^{\pm}$ and Z° were heavy held up the development of the theory for some time. We will study the mass problem in detail.

Gravitons interact too weakly to be detected singly; their existence and properties are inferred from the structure of the theory, in the same way that quantum electrodynamics leads to a photon. The other gauge bosons all have been discovered. The photon is familiar. The gluons were predicted to exist, and were observed at the electron-positron collider in Hamburg, PETRA, in 1979, with the expected properties. The $W^{\pm}$ and Z° were predicted by the theory and were observed at the proton–antiproton collider at CERN in 1983, also with the expected properties ($M_W \simeq 82$ GeV and $M_Z \simeq 94$ GeV). The prediction and discovery of gluons and the $W^{\pm}$ and Z° rank among the major intellectual achievements of mankind.

An interesting thing to keep in mind is that the known universe is made of u-quarks, d-quarks, electrons, ν_e , and the gauge bosons. All the other quarks and leptons have been made at accelerators (and occasionally by collisions of energetic cosmic rays in the earth's atmosphere), and existed at an early stage of the universe, but are very short-lived and play no known role in the universe today.

In addition to the quarks, leptons and gauge bosons, it turns out that one more class of particles is needed to make a consistent theory of particle masses and interactions, the spin-zero or scalar bosons called Higgs bosons. The electroweak theory requires one electrically neutral Higgs boson, but more could exist. We will study them extensively later. Detecting them experimentally and understanding their properties is the major remaining gap in the electroweak theory, and should be considered the central problem in particle physics.

As was already briefly mentioned, hadrons (protons, neutrons, pions, ...) are built up from quarks bound by gluons. The QCD force between particles with color charge binds them into hadrons. The residual color force outside color neutral hadrons is the nuclear force, that binds stable hadrons into nuclei. The electrically charged nuclei and stable electrically charged leptons (only the electron) are bound into atoms by the electromagnetic force, mediated by photons. The residual electromagnetic force outside electrically neutral atoms binds them into molecules. Thus is the hierarchy of structures in nature built.

Finally, each particle has an antiparticle. The antiparticles have opposite values of electric charge, color charge, and flavor, from the particles, but the same mass and spin. We will sometimes denote the antiparticles just by giving the charge labels: e^- and e^+ for electron and positron, π^+ and π^- for the positive and negative pions, W^+ and W^- for the electroweak gauge bosons, which are antiparticles of each other, and p^+ and p^- for the proton and antiproton, etc. Sometimes we will use a bar over a letter for the antiparticle, as $\bar{p}$ for antiproton, and $\bar{e}$ for positron, $\bar{f}$ for antifermion where fermion can denote any of the quarks or leptons. Some particles such as the photon or π^0 are their own antiparticle. For our purposes, antiparticles are just particles, having some properties (such as mass) that are the same as those of some other particles.

1.5   Natural Units

It is customary, and convenient, to use what are called natural units—the more quantities which can be set to unity, the simpler the formulas will look, and the more readable they will be. Normally only $\hbar$ and c are eliminated, $\hbar = c = 1$.

Then energy (mc^2), momentum (mc), and mass (m) will all appear as mass, and will normally be stated in GeV (so if translation is required, GeV $\rightarrow$ GeV/c or GeV/c^2). The actual expressions for $\hbar$ and c provide the conversion factors,

$$\hbar = 6.6 \times 10^{-25} \text{GeV} \cdot \sec \tag{1.1}$$

so

$$1 \text{ GeV} = \frac{1}{6.6 \times 10^{-25}} \sec^{-1} \tag{1.2}$$

and

$$c = 3 \times 10^{10} \text{ cm}/\sec \tag{1.3}$$

so

$$1 \sec = 3 \times 10^{10} \text{ cm}. \tag{1.4}$$

Combining these gives

$$10^{-13} \, cm = 1 \text{ fermi} \approx 5 \text{ GeV}^{-1} \tag{1.5}$$

so

$$1 \text{ mb} \equiv 10^{-27} \text{ cm}^2 = 2.56 \cdot \text{GeV}^{-2}, \tag{1.6}$$

or

$$1 \text{ GeV}^{-2} = 0.39 \times 10^{-28} \text{ cm}^2.$$

We will often use (1.2) and (1.6) to convert widths to lifetimes and to express cross sections in cm^2.

In natural units we can write, for example,

$$m_e = m_e c \simeq \frac{1}{2} \text{MeV} \tag{1.7}$$

$$= m_e c^2 \simeq \frac{1}{2} \text{MeV} \tag{1.8}$$

$$= \frac{1}{\hbar / m_e c} \simeq \frac{1}{4} \times 10^{11} \text{ cm}^{-1} \tag{1.9}$$

$$= \frac{1}{\hbar / m_e c^2} \simeq \frac{1}{1.3} \times 10^{21} \text{ sec}^{-1}. \tag{1.10}$$

Problems

1.1: Why do you believe that atoms exist? Nuclei? Electrons? Could you convince someone who lived two hundred years ago? A high school student today? A philosopher?

1.2: How many fundamental forces of nature were there considered to be (a) before Newton, (b) around 1750, (c) around 1880, (d) around 1950?

1.3: What is the Hamiltonian for the hydrogen atom? Where did it come from?

1.4: Plot the quark and lepton masses in some way which suggests regularities; *e.g.* family number versus mass. Guess what the t quark mass might be. If there is a fourth family, what masses might you predict for the members?

1.5: What nuclei have about the same masses as the fundamental $W^{\pm}$ and Z^0 particles?

1.6: If a cross section is calculated to be $\sigma = 10^{-3}/M_W^2$, what is it in the usual units for a cross section, cm^2? If a lifetime is $\tau = 1$ GeV^{-1}, what is it in seconds?

1.7: What is Newton's constant G_N in natural units?

Suggestions for Further Study

Readers are likely to want further information of one sort or another. After most chapters a few suggestions for additional sources or references will be provided. They are not intended to be complete, but they do suggest where the reader might begin to look.

Some will want more technical treatments than we have provided. The books which are most like this one in spirit, but designed to be at graduate student level for people who are or will be active in particle physics, are those by Halzen and Martin, and Quigg. Quigg's book also contains extensive, detailed, annotated references to sources and suggestions for further study. The book by Perkins is more a survey and less pedagogical, also largely at the graduate level. The approach of Gottfried and Weisskopf, in two volumes, is quite different, ranging from the principles of quantum theory through the Standard Model and beyond, with few formulas or derivations.

Currently, the most complete graduate level particle physics and field theory text is the book by Cheng and Li. Introductory graduate level field theory books have recently appeared by Mandl and Shaw, and by Ryder. The structure of gauge field theories is well described in an introductory way by Aitchison and Hey. The book by DeWit and Smith (Volume 1 available so far) will be a very useful field theory and particle physics text.

The question of measuring quark masses is subtle and technical. We will remark on it briefly in Chapter 23. The interested reader could consult J. Gasser and H. Leutwyler (from which the numbers we quote are taken), or Weinberg.

As I stated in the preface, I am assuming that the reader wants the Standard Model explained here, and will go elsewhere to learn the history. Undergraduate students are certainly more interested in the ideas, and many scientists feel they have had some exposure to the history (some are old enough to have lived through it) but that they have much less access to the ideas and tests of the theory. Such views have led to the ahistorical approach I have, with some concern, taken. Essentially all of the people who made the discoveries and constructed the ideas that became the Standard Model are still active workers, deserving of great credit. I urge the reader to also study some of the recent history of particle physics. Fortunately, three books are available that contain a great deal of the history, from three different points of view. One is by Pickering; he has written a "sociological history of particle physics." The history is very professional and well documented, and the particle physics well described; I do not endorse the sociological and philosophical aspects of the book. Another, by Crease and Mann, is a pleasantly anecdotal and reliable historical account of the development of the Standard Model. The third is complementary to the present book; Cahn and Goldhaber have taken a pedagogical historical approach to the discoveries that are the basis and the tests of the Standard Model. In addition to these, the final seventy–five pages of Pais' book gives a personal historical view of the development of the Standard Model. All references quoted here and elsewhere in the book can be found in the bibliography.

One of the UA 1 experimenters, P. Watkins, has written "The Story of the W and Z", covering the experiment from the beginning through the major discoveries. Another recent historical survey has appeared, by Ne'eman and Kirsh. Finally, a masterful survey of modern particle physics for nonexperts has been written by Okun (1985).

Two books that were largely written before the Standard Model was fully developed, and which can be read to learn more about the topics of importance in the two decades before the early 1970's, are those by Perl and by Hughes.

The Nobel lectures of S. Glashow, A. Salam, and S. Weinberg can be read in Reviews of Modern Physics. A number of Scientific American articles have been published on the Standard Model. The most complete is by G. 't Hooft; others are mentioned in the appropriate chapters.

Unless otherwise stated, all data quoted in the book is from the "Review of Particle Properties" by the Particle Data Group, referred to as "Particle Data Tables". This valuable source of information is updated every two years and published as an issue of Physics Letters. Individual copies can be obtained from the Particle Data Group.

CHAPTER 2

Relativistic Notation, Lagrangians, Currents, and Interactions

In this chapter we will go through some of the background material we need to construct the full theory. Various ingredients and ways of thinking have converged to generate the present synthesis. Some of what is covered here is found in undergraduate courses in mechanics and modern physics, though perhaps with a different emphasis. We will use relativistic notation, since it greatly simplifies how equations look, and it makes many operations almost automatic; we will do few calculations which require more than an elementary skill with the relativistic notation.

The goal of this book is to explain the new physics of the Standard Model. To do that requires the use of a variety of physics that has been developed over the past six decades. In order to be self–contained, the needed background physics is described here or in later chapters or appendices. The present chapter may leave some readers feeling a little frustrated, since we survey several topics, but we often stop short of complete treatments. We rely mainly on heuristic arguments and analogies to develop a number of points of view that are used extensively in particle physics. All the subjects not covered in detail here can be found well treated in a variety of places, and the reader is encouraged to study them. For the most part, we emphasize only areas that are not well covered elsewhere. To understand this book the reader should be familiar with (but not necessarily extremely proficient with) the material of this chapter and Chapter 5.

2.1   Some Relativistic Notation

Let a^μ be a four–vector, with components

$$a^\mu = (a^0; a^1, a^2, a^3). \tag{2.1}$$

The most familiar four–vector is $x^\mu = (t; x, y, z) = (x^0; x^1, x^2, x^3)$. We distinguish between upper and lower indices,

$$a_\mu = (a_0; a_1, a_2, a_3) = (a^0; -a^1, -a^2, -a^3). \tag{2.2}$$

This defines a "metric tensor"

$$g_{\mu\nu} = \begin{pmatrix} 1 & & & \\ & -1 & & \\ & & -1 & \\ & & & -1 \end{pmatrix} \tag{2.3}$$

with zeros off the diagonal, and $a_\mu = g_{\mu\nu} a^\nu$. Repeated indices are always summed. Of course, $x^\mu = \left(t; \vec{x}\right)$, and $p^\mu = \left(E; \vec{p}\right)$.

The scalar product of two four–vectors is

$$a_\mu b^\mu = a^0 b^0 - a^1 b^1 - a^2 b^2 - a^3 b^3. \tag{2.4}$$

Appendix C gives some examples of relativistic kinematics.

We will frequently use derivatives in writing Lagrangians. They are defined as

$$\partial^\mu = \frac{\partial}{\partial x_\mu} = \left(\frac{\partial}{\partial t}; -\frac{\partial}{\partial x}, -\frac{\partial}{\partial y}, -\frac{\partial}{\partial z}\right) = (\partial^0; -\nabla) \tag{2.5}$$

$$\partial_\mu = \frac{\partial}{\partial x^\mu} = (\partial_0; \nabla). \tag{2.6}$$

Then from equation (2.4), noting ∂_μ is formed of components from x^μ (with an upper index),

$$\partial_\mu a^\mu = \frac{\partial a^0}{\partial t} + \nabla \cdot \vec{a}. \tag{2.7}$$

The volume elements are $d^3x = dx\,dy\,dz = dx^1 dx^2 dx^3$, and $d^4x = d^3x\,dt = d^3x\,dx^0$. Note the minus signs in equation (2.2); they give the relative minus sign in equation (2.4) so $a_\mu b^\mu = a^0 b^0 - \overrightarrow{a} \cdot \overrightarrow{b}$. If extensive calculations are undertaken, it is necessary to carefully track the signs. Since we will not do any extensive calculations, we only need to keep in mind that $\partial_\mu a^\mu = \partial a^0/\partial t + \nabla \cdot \overrightarrow{a}$ and $a_\mu b^\mu = a^0 b^0 - \overrightarrow{a} \cdot \overrightarrow{b}$ for essentially all our uses of this formalism. Since the time component does not change sign, we do not have to distinguish p^0 and p_0 for energies.

Finally, occasionally we will need

$$
\begin{aligned}
\partial_\mu \partial^\mu &= \frac{\partial^2}{\partial t^2} - \frac{\partial^2}{\partial x^2} - \frac{\partial^2}{\partial y^2} - \frac{\partial^2}{\partial z^2} \\
&= \partial_0^2 - \nabla^2 \\
&= \partial^\mu \partial_\mu \;.
\end{aligned}
\tag{2.8}
$$

2.2   Lagrangians

Classical mechanics can be expressed in terms of Lagrangians for point particles, or for continuous systems. The Lagrangian $\mathcal{L}$ is given by $T - V$, the kinetic energy minus the potential energy. The classical action is $S = \int_{t_1}^{t_2} dt \mathcal{L}$ and is taken to be a minimum. When S is minimized, the Euler–Lagrange equations result, and give rise to Newton's laws. The force enters as the derivative of the potential.

Similarly, classical electrodynamics can be written as a Lagrangian theory. Let us examine this case in some detail, since it is relatively familiar on the one hand, but is also rather similar to the way the subject is treated in particle physics. $\overrightarrow{E}$ and $\overrightarrow{B}$ are the classical electric and magnetic fields, and $\overrightarrow{A}$ and ϕ are the vector and scalar potentials, also classical fields. $\overrightarrow{J}$ and ρ are the current density and charge density. Using notation as in the previous section, and $\partial_j = \partial/\partial x^j$, we know we can write the field strengths in terms of the potentials as

$$
E_i = -\nabla V - \frac{\partial \overrightarrow{A}}{\partial t} = -\partial^i A^0 - \partial^0 A^i,
\tag{2.9}
$$

$$
\overrightarrow{B} = \nabla \times \overrightarrow{A}.
\tag{2.10}
$$

We can define a four–vector for the potentials,

$$
A^\mu = \left(V; \overrightarrow{A} \right) = \left(A^0; \overrightarrow{A} \right).
\tag{2.11}
$$

Similarly, the current four–vector is $J^\mu = \left(\rho;\ \overrightarrow{J}\right)$.

Then it is convenient to also define an antisymmetric tensor

$$F^{\mu\nu} = \partial^\mu A^\nu - \partial^\nu A^\mu \tag{2.12}$$

with components

$$F^{0i} = \partial^0 A^i - \partial^i A^0 = -E^i \tag{2.13}$$
$$F^{ij} = \partial^i A^j - \partial^j A^i = -\epsilon^{ijk} B^k. \tag{2.14}$$

Note that $F^{\mu\nu}$ is explicitly invariant under a transformation where A^ν changes by $\partial^\nu \chi$ since $F^{\mu\nu}$ changes by $\partial^\mu \partial^\nu \chi - \partial^\nu \partial^\mu \chi \equiv 0$ (this invariance is discussed in the next chapter).

The conventional Lagrangian (density) for electromagnetism is

$$\mathcal{L} = \frac{1}{2}\left(E^2 - B^2\right) - \rho V + \overrightarrow{J} \cdot \overrightarrow{A}. \tag{2.15}$$

Written in terms of $F^{\mu\nu}$, this is

$$\mathcal{L} = -\frac{1}{4} F_{\mu\nu} F^{\mu\nu} - J_\mu A^\mu. \tag{2.16}$$

The last two terms of equation (2.15) obviously give the last term of (2.16). The reader may want to show the first terms are equal. Given this Lagrangian, the Euler–Lagrange equations become Maxwell's equations. This is shown in Appendix F (partly in a homework problem), for interested readers. We could go through the calculation, but it takes us into more detail than we need, so we will not pursue it. The main point is that we could formulate the physics of the electromagnetic fields by writing a Lagrangian that is a function of the fields or the potentials (which are also classical fields). The first term in equation (2.16) is the kinetic energy term and the second is the interaction Lagrangian.

2.3   Lagrangians in Particle Physics

It has become conventional to formulate particle physics by giving the Lagrangian. From it, using the rules of quantum field theory, all observables can in principle be calculated.

The Lagrangian defines the theory. It is written in terms of the elementary particles of the theory. Any composite objects should appear as bound states that arise as solutions of the theory. For electrodynamics, the photon is the quantum of the electromagnetic field; it is represented by the vector potential field A^μ. The electron is represented by the fermion field ψ. The Lagrangian contains the fundamental interactions of the theory. For electrodynamics that is the conventional $\overrightarrow{J} \cdot \overrightarrow{A}$ interaction Hamiltonian, which becomes $J_\mu A^\mu$ relativistically.

More precisely, it is the potential energy parts of the Lagrangian that specify the theory. The kinetic energy parts are general and only depend on the spins of the particles. The potential energy parts specify the forces; we will often call them the interaction Lagrangian.

One of the main reasons why particle physics is formulated in terms of the Lagrangian is that $\mathcal{L}$ is a single function that determines the dynamics, and $\mathcal{L}$ must be a scalar in every relevant space, invariant under transformations, since the action is invariant. Making $\mathcal{L}$ invariant under Lorentz transformations guarantees that all predictions of the theory are Lorentz invariant.

Thus we will write the theory in terms of a Lagrangian. In a full field theory, the Lagrangian would be used in a variety of ways. For us it will only serve as a guide in writing the basic interactions, from which we will read off the Feynman rules of the theory.

2.4   The Real Scalar Field

In Section 2.1, we wrote the Lagrangian for the electromagnetic field. Appendix F shows how to go from that Lagrangian to Maxwell's equations. For much of what we do we will only need the much simpler Lagrangian of scalar (*i.e.* spinless) fields $\phi(x)$. The field $\phi(x)$ can be thought of as arising from a source in much the same way as the electomagnetic fields arise from charged particles; as for electromagnetism, we can consider the fields without concerning ourselves with the sources. We want to write the Lagrangian, analogous to the one for the electromagnetic fields, for a real scalar field $\phi(x)$ of mass m. The answer is

$$\mathcal{L} = \frac{1}{2}\left[\partial_\mu\phi\,\partial^\mu\phi - m^2\phi^2\right].\tag{2.17}$$

The factor of $1/2$ is a convention. In Appendix F, it is shown that this Lagrangian implies that ϕ satisfies a wave equation

$$\partial_\mu\partial^\mu\phi + m^2\phi = 0.\tag{2.18}$$

That is what we expect, since we can write

$$E^2 = \overrightarrow{p}^2 + m^2,\tag{2.19}$$

and $E = i\partial_0$, $\overrightarrow{p} = -i\nabla$ so $E^2 - \overrightarrow{p}^2 = -\partial_0^2 + \nabla^2 = -\partial_\mu\partial^\mu$. Thus every field that describes a particle of mass m should satisfy $-\partial_\mu\partial^\mu\phi = m^2\phi$. The first term in equation (2.17) is called the kinetic energy term, since it arises effectively from $\overrightarrow{p}^2$. The second term is called a mass term since it is proportional to m^2. Identifying the mass terms in Lagrangians will be important for us in interpreting the theory. There is no potential energy or interaction term in equation (2.17); it is written for a noninteracting field.

Note that what we have been calling a Lagrangian is usually a Lagrangian density, so the Lagrangian is really $\int d^4x\,\mathcal{L}(x,t)$. Since the context always makes clear whether a density is involved, it is conventional to just speak of the Lagrangian, whether or not a density is involved.

To help understand the formalism and make a few points, let us construct a field normalized to a single quantum of definite energy ω and momentum $\overrightarrow{k}$. We know that

$$\phi = A\,\cos(\overrightarrow{k}\cdot\overrightarrow{r} - \omega t)\tag{2.20}$$

is a solution of the wave equation (2.19) if $\omega^2 = k^2 + m^2$, and A is a normalization constant. The Hamiltonian associated with a Lagrangian is $H = \dot{\phi}\frac{\partial\mathcal{L}}{\partial\dot{\phi}} - \mathcal{L}$, and the total energy is $E = \int d^3x\,H$. Then since $\partial\mathcal{L}/\partial\dot{\phi} = \dot{\phi}$,

$$H = \frac{1}{2}\left(\dot{\phi}^2 + (\nabla\phi)^2 + m^2\phi^2\right)\tag{2.21}$$

so E is

$$E = \int\frac{1}{2}d^3x\left[\dot{\phi}^2 + (\nabla\phi)^2 + m^2\phi^2\right].\tag{2.22}$$

Substituting ϕ from equation (2.20) in equation (2.22) leads to

$$
\begin{aligned}
E &= \frac{1}{2}A^2 \int \left[(\omega^2 + k^2)\sin^2(\vec{k}\cdot\vec{r} - \omega t) + m^2 \cos^2(\vec{k}\cdot\vec{r} - \omega t) \right] d^3x \\
&= \frac{1}{2}A^2 \left[\frac{1}{2}(\omega^2 + k^2) + \frac{1}{2}m^2 \right] \\
&= \frac{1}{2}A^2\omega^2,
\end{aligned}
\tag{2.23}
$$

where we have replaced the integrals over $\sin^2(\vec{k}\cdot\vec{r} - \omega t)$ and $\cos^2(\vec{k}\cdot\vec{r} - \omega t)$ by their average values of $\frac{1}{2}$, and used $\omega^2 = k^2 + m^2$. Since we want a single quantum of energy $\hbar\omega$, we must choose $A = \sqrt{\frac{2}{\omega}}$, remembering $\hbar = 1$. So we have

$$
\phi = \frac{1}{\sqrt{2\omega}}\left[e^{i(\vec{k}\cdot\vec{r} - \omega t)} + e^{-i(\vec{k}\cdot\vec{r} - \omega t)} \right].
\tag{2.24}
$$

In a quantized theory we want to speak of the creation and destruction of particles, so we have written ϕ to facilitate that. If $|n_k\rangle$ is a state with n particles all of the same momentum $\vec{k}$ and energy ω at time t, we expect

$$
|n_k\rangle \propto e^{in\omega t}.
\tag{2.25}
$$

Then a state with one less particle will vary as

$$
|n_k - 1\rangle \propto e^{i(n-1)\omega t},
\tag{2.26}
$$

so we expect the destruction operator, a, to have the first term in equation (2.24) as a coefficient. Thus when we go from a field to an operator we expect

$$
\phi = \frac{1}{\sqrt{2\omega}}\left[a e^{i(\vec{k}\cdot\vec{r} - \omega t)} + a^\dagger e^{-i(\vec{k}\cdot\vec{r} - \omega t)} \right],
\tag{2.27}
$$

where $a^\dagger$ creates quanta associated with the field ϕ and a destroys them. We will not directly calculate with these operators. But keep in mind that every quantum field can create or destroy a particle. When it does, there is an associated factor that is the wave function of the particle. In a full derivation we would go through the quantization of the scalar field, just as one goes through the quantization of the electromagnetic field that leads to the interpretation of the photon as the quantum of the electromagnetic field. We have argued heuristically that ϕ must have a certain form and can be interpreted in terms of creation and destruction of spinless particles, with associated wave function factors.

2.5   Sources and Currents in Non-Relativistic Quantum Theory

Next we recall the idea of sources and currents in ordinary quantum theory. The Schrödinger equation is

$$2mi\frac{\partial \Psi}{\partial t} + \nabla^2 \Psi = 0 \ . \tag{2.28}$$

Multiply this by $i\Psi^*$, and add it to the equation obtained by multiplying the complex conjugate of equation (2.28) by $-i\Psi$. Then define

$$\rho = |\Psi|^2 \tag{2.29}$$

and

$$\overrightarrow{J} = -\frac{i}{2m}\left(\Psi^* \nabla \Psi - \Psi \nabla \Psi^*\right), \tag{2.29}$$

and the resulting equation is

$$\frac{\partial \rho}{\partial t} + \nabla \cdot \overrightarrow{J} = 0 \ . \tag{2.29}$$

For a free particle $\Psi = C \exp\left(i\overrightarrow{p} \cdot \overrightarrow{r} - i\omega t\right)$, so

$$\rho = |C|^2 \tag{2.30}$$

and

$$\overrightarrow{J} = \rho \overrightarrow{p}/m \ , \tag{2.31}$$

so it is customary to describe ρ as a probability density and $\overrightarrow{J}$ as a current density. In non-relativistic quantum theory this is not a particularly useful notion, but in a relativistic quantum field theory the generalizations have become very useful aids to thinking.

2.6   Complex Scalars, Conserved Currents, and Noether's Theorem

Some interesting physics emerges if we consider a system of two real scalar fields, ϕ_1 and ϕ_2, having the same mass m. Then we expect from equation (2.17) that

$$\mathcal{L} = \frac{1}{2}\left[\partial_\mu\phi_1\partial^\mu\phi_1 - m^2\phi_1^2\right] + \frac{1}{2}\left[\partial_\mu\phi_2\partial^\mu\phi_2 - m^2\phi_2^2\right] . \qquad (2.32)$$

We can combine ϕ_1 and ϕ_2 into a single complex scalar field ϕ by writing

$$\phi = (\phi_1 + i\phi_2)/\sqrt{2} , \qquad (2.33)$$

in which case

$$\phi^* = (\phi_1 - i\phi_2)/\sqrt{2} ; \qquad (2.34)$$

then $\mathcal{L}$ becomes

$$\mathcal{L} = \partial_\mu\phi^*\partial^\mu\phi - m^2\phi^*\phi. \qquad (2.35)$$

It is important to remember that ϕ, ϕ_1, and ϕ_2 all have the same mass m, so one is describing the same physics in different ways. The $\sqrt{2}$ is to have ϕ and ϕ^* normalized to unity if ϕ_1 and ϕ_2 are.

Now it is possible to make an observation which is significant and which will be fundamental to much of what we consider in this book. Nothing fixed the particular "direction" of ϕ_1 and ϕ_2. We could equally well have started with two fields ϕ_1' and ϕ_2' that were "rotated" in some sense by an angle α (α is a real constant),

$$\phi_1' = \phi_1\cos\alpha + \phi_2\sin\alpha , \qquad (2.36)$$

$$\phi_2' = -\phi_1\sin\alpha + \phi_2\cos\alpha . \qquad (2.37)$$

Then we can form $\phi' = (\phi_1' + i\phi_2')/\sqrt{2}$ and ϕ'^*,

$$\begin{aligned}
\phi' &= (\phi_1\cos\alpha + \phi_2\sin\alpha - i\phi_1\sin\alpha + i\phi_2\cos\alpha)/\sqrt{2} \\
&= \left(e^{-i\alpha}\phi_1 + ie^{-i\alpha}\phi_2\right)/\sqrt{2} \\
&= e^{-i\alpha}\phi
\end{aligned} \qquad (2.38)$$

and

$$\phi'^* = e^{i\alpha}\phi^*. \qquad (2.39)$$

There is clearly no change in $\mathcal{L}$ since it only depends on $\phi^*\phi$, so the physics is invariant under this transformation.

We will see that we can extract many instructive implications from this simple example. Whenever a physical system is invariant under a transformation, interesting results emerge. To proceed, assume the rotation "angle" α is infinitesimal to simplify the algebra (it is sufficient to consider infinitesimal transformations to get general results since a continuous transformation can be built out of infinitesimal ones). Then we can write

$$\phi' \approx (1 - i\alpha)\,\phi = \phi - i\alpha\phi \equiv \phi + \delta\phi \qquad (2.40)$$

so the change in ϕ is

$$\delta\phi = -i\alpha\phi\ . \qquad (2.41)$$

Similarly

$$\delta\phi^* = i\alpha\phi^*. \qquad (2.42)$$

Now we want to calculate the change in the Lagrangian as a result of this transformation. We know the answer is zero, but zero can be written in a very instructive form. The following derivation gives long equations but only depends on simple algebra and differentiation. Assume $\mathcal{L}$ depends on ϕ and $\partial^\mu\phi$. In general, for any variations $\delta\phi$ and $\delta\phi^*$,

$$\delta\mathcal{L} = \delta\phi\frac{\partial\mathcal{L}}{\partial\phi} + \left[\delta\left(\frac{\partial\phi}{\partial x_\mu}\right)\right]\frac{\partial\mathcal{L}}{\partial\left(\frac{\partial\phi}{\partial x_\mu}\right)} + (\phi \to \phi^*)\ . \qquad (2.43)$$

The second term can be written

$$(\delta(\partial^\mu\phi))\frac{\partial\mathcal{L}}{\partial(\partial^\mu\phi)} = \partial^\mu\left\{\delta\phi\frac{\partial\mathcal{L}}{\partial(\partial^\mu\phi)}\right\} - \delta\phi\left\{\partial^\mu\frac{\partial\mathcal{L}}{\partial(\partial^\mu\phi)}\right\} \qquad (2.44)$$

where the second term here is just to cancel the part of the first term where ∂^μ operates on the derivative of $\mathcal{L}$. Then the second term in equation (2.44) can be combined with the first term in equation (2.43) to give

$$\delta\mathcal{L} = \delta\phi\left\{\frac{\partial\mathcal{L}}{\partial\phi} - \partial^\mu\frac{\partial\mathcal{L}}{\partial(\partial^\mu\phi)}\right\} + (\phi \to \phi^*) + \partial^\mu\left\{\delta\phi\frac{\partial\mathcal{L}}{\partial(\partial^\mu\phi)} + \delta\phi^*\frac{\partial\mathcal{L}}{\partial(\partial^\mu\phi^*)}\right\}. \qquad (2.45)$$

For our case, the variations with respect to ϕ or ϕ^* or $\partial^\mu\phi$ or $\partial^\mu\phi^*$ are all independent, so

$$\frac{\partial \mathcal{L}}{\partial \phi} = - m^2 \phi^*$$

$$\frac{\partial \mathcal{L}}{\partial(\partial^\mu \phi)} = \partial_\mu \phi^*$$

$$\partial^\mu \left[\frac{\partial \mathcal{L}}{\partial(\partial^\mu \phi)} \right] = \partial^\mu \partial_\mu \phi^*$$

$$\frac{\partial \mathcal{L}}{\partial \phi} - \partial^\mu \frac{\partial \mathcal{L}}{\partial(\partial^\mu \phi)} = - m^2 \phi^* - \partial^\mu \partial_\mu \phi^* = 0 \ ,$$

the last equality following from equation (2.18). Thus the first term in equation (2.45), and the equivalent term for ϕ^*, drop out. [This is actually general, since the first term in brackets vanishes by the Euler–Lagrange equations, but our derivation does not require the reader to know that.] We finally obtain

$$\delta \mathcal{L} = \partial^\mu \left\{ \delta \phi \frac{\partial \mathcal{L}}{\partial(\partial^\mu \phi)} + (\phi \to \phi^*) \right\} . \tag{2.46}$$

Note that this is a general result, not dependent on the details of our transformation. The variation in $\mathcal{L}$ can be written as the derivative of the quantity in brackets. Since we know $\delta \mathcal{L} = 0$, we see that the quantity in brackets behaves like a conserved current, *i.e.* its four–divergence is zero.

Before we actually write the current, it is convenient to put in the forms of $\delta \phi$ and $\delta \phi^*$ so that the current is independent of the parameter(s) of the transformation, α. Using equations (2.41), (2.42) and (2.35) gives

$$\delta \mathcal{L} = \alpha \partial^\mu S_\mu \tag{2.47}$$

where

$$S_\mu = i \left(\phi \partial_\mu \phi^* - \phi^* \partial_\mu \phi \right) , \tag{2.48}$$

and

$$\delta \mathcal{L} = 0$$

requires

$$\partial^\mu S_\mu = 0 \ . \tag{2.49}$$

Several observations will help show the importance of this result.

(*i*) If we interchange ϕ and ϕ^*, S_μ changes sign. A relativistic theory has pairs of particles of the same mass and of opposite electric charge, *i.e.* antiparticles, which are just what we have been studying here. If ϕ corresponds to a particle of electric charge e, then ϕ^* corresponds to a particle of electric charge $-e$, and S_μ can be interpreted as a charge current density. Then equation (2.49) says that the change in the charge density $S_0(x)$ in some region is just equal to the current $\vec{S}(x)$ flowing out of the region. Thus the charge is locally conserved and can be used to label the states.

(*ii*) Nothing in the derivation requires the interpretation to be in terms of *electric* charge. We will see that particles have a number of "charges", and at least some of them can be related to conserved currents.

(*iii*) The transformation described by equations (2.38) and (2.39) is called a "gauge transformation of the first kind" or a "global gauge transformation". If the parameter α that described the transformation could vary with $\vec{x}$ or t, it would be a "gauge transformation of the second kind", or a "local gauge transformation". If the name had not arisen for historical reasons, we would probably say the theory was invariant under global or local phase transformations.

(*iv*) Note that equation (2.46) is very general. It is an example of a very basic property of quantum field theories, that whenever a physical system is invariant under some transformation it leads to conserved quantities. For continuous transformations it can be stated in the form of Noether's theorem: for a system described by a Lagrangian, any continuous symmetry which leaves invariant the action $\int \mathcal{L}dt$ leads to the existence of a conserved current S_μ, with $\partial^\mu S_\mu = 0$. It is always possible to define a charge

$$Q(t) = \int d^3x \, S_0(x) \,,$$

and the charge is conserved in the sense that $dQ/dt = 0$. Thus the charges mentioned in (i) and (ii) above are conserved quantities that characterize the properties of the particles.

That invariance under a transformation implies an associated conservation law should already be familiar in classical and quantum mechanics, with

rotational invariance	↔	angular momentum conservation
translational invariance	↔	linear momentum conservation
time translation invariance	↔	energy conservation.

A little more discussion of this physics is provided in Appendix F.

(v) Readers familiar with the neutral kaons can think of them as a physical realization of such a system. K_1 and K_2 are like ϕ_1 and ϕ_2; K° and $\overline{K}^\circ$ are like ϕ and ϕ^*. The charge is the strangeness, *i.e.* the number of strange quarks. For the kaons "charge" conservation is broken by doubly weak interactions which can convert $K \leftrightarrow \overline{K}$, and consequently introduces a small level splitting, the mass difference between the K_1 and K_2.

(vi) The analysis given here is a classical one. A similar analysis can be carried out in quantum theory, with generally the same results. Higher order radiative quantum corrections can give a result different from zero on the right hand side of equation (2.49) even when the classical result would be zero; such terms are called anomalies. Requiring that the equations contain no anomalies may be an important guide to determining the structure of the theory, particularly because anomalies sometimes vanish if the theory has certain symmetries; that is what happens in the Standard Model. Anomalies are currently an active field of research because of attempts to extend the Standard Model, but are technically beyond the scope of our level of study.

2.7   Interactions

So far we have only considered free fields, without asking about their source or how they interact. To understand the way fields and particles are thought of, it is helpful to recall some notions that are essentially the viewpoint described by Yukawa.

If we add to the Lagrangian (equation (2.17)) a part describing interactions, such as

$$\mathcal{L}_{\text{int}} = -\phi\rho(\overrightarrow{x}, t) , \tag{2.50}$$

then equation (2.18) is modified to have a source term,

$$\partial_\mu \partial^\mu \phi + m^2 \phi = \rho . \tag{2.51}$$

[This follows directly from the Euler–Lagrange equations.] By analogy with electrodynamics, we think of ρ as a source of the field ϕ. To study how the system behaves, we examine the simple case

$$\rho = g\delta(\overrightarrow{x}) ,$$

i.e. a time-independent point source at the origin, of strength g. Then we can "solve" the problem by a Fourier transform procedure.

Since ρ is not time dependent, equation (2.51) becomes

$$\left(-\nabla^2 + m^2\right)\phi = g\delta(\overrightarrow{x})\,. \tag{2.52}$$

Writing

$$\phi(\overrightarrow{x}) = \frac{1}{(2\pi)^{3/2}} \int d^3k\, e^{i\vec{k}\cdot\vec{x}}\widetilde{\phi}(\overrightarrow{k}) \tag{2.53}$$

and the inverse transform

$$\widetilde{\phi}(\overrightarrow{k}) = \frac{1}{(2\pi)^{3/2}} \int d^3x\, e^{-i\vec{k}\cdot\vec{x}}\phi(\overrightarrow{x}) \tag{2.54}$$

we get (since $\nabla^2 \to -\overrightarrow{k}^2$)

$$(\overrightarrow{k}^2 + m^2)\widetilde{\phi}(\overrightarrow{k}) = g/(2\pi)^{3/2}\,. \tag{2.55}$$

We can divide to obtain $\widetilde{\phi}$ and substitute back into equation (2.53) to get ϕ,

$$\phi(\overrightarrow{x}) = \frac{g}{(2\pi)^3} \int d^3k \frac{e^{i\vec{k}\cdot\vec{x}}}{\overrightarrow{k}^2 + m^2}\,. \tag{2.56}$$

[Note that if we had not taken a time–independent source, the denominator would have been $-k_0^2 + \overrightarrow{k}^2 + m^2 = m^2 - k^2$, where $k^2 = k_\mu k^\mu$. We will see that this denominator appears as a "propagator" whenever a particle is exchanged in an interaction.]

We can do the integral. Putting $\overrightarrow{k} \cdot \overrightarrow{x} = kr\cos\theta$ it becomes

$$\int_0^\infty \frac{k^2 dk}{k^2 + m^2} \int_0^{2\pi} d\phi \int_{-1}^1 d(\cos\theta)e^{ikr\cos\theta} =$$

$$= \frac{2\pi}{r} \int_0^\infty \frac{dk^2}{k^2 + m^2}\sin kr$$

$$= \frac{\pi}{ir} \int_0^\infty \frac{dk^2}{k^2 + m^2}\left(e^{ikr} - e^{-ikr}\right) \tag{2.57}$$

$$= \frac{\pi}{ir}\left[\int_0^\infty \frac{dk^2}{k^2 + m^2}e^{ikr} - \int_0^{-\infty} \frac{dk^2}{k^2 + m^2}e^{ikr}\right]$$

$$= \frac{\pi}{ir} \int_{-\infty}^\infty \frac{dk^2}{k^2 + m^2}e^{ikr}\,.$$

This can be evaluated as a contour integral. Closing the contour where Im $k > 0$ so the integral is convergent, and picking up a contribution from the residue at $k = im$, gives finally

$$\phi = \frac{g}{4\pi}\frac{e^{-mr}}{r} \; .$$

$$(2.58)$$

Yukawa identified ϕ as a meson field, with the nucleon as source, just as any electrically charged particle is the source of the electromagnetic field. And just as the effects of the electromagnetic field are transmitted by photons, the effects of the meson field should be transmitted by particles ("mesons"). When the particles have a mass m, the field is significant in size only out to a range of force $r \sim 1/m$ (in natural units), because of equation (2.58).

Now we want to extend this to see how one nucleon would interact with another by sensing its meson field. The interaction Hamiltonian between two nucleons, the second one described by $\rho_2(\vec{x})$, is

$$H = -\int d^3x \, \phi(\vec{x})\rho_2(\vec{x}) \; .$$

$$(2.59)$$

To make the result more symmetrical we can put ρ_1 back in ϕ,

$$\phi(\vec{x}) = \frac{1}{4\pi}\int d^3x' \, \rho_1(\vec{x}')\frac{e^{-m|\vec{x}-\vec{x}'|}}{|\vec{x}-\vec{x}'|} \; ,$$

$$(2.60)$$

which gives back equation (2.58) when $\rho_1(\vec{x}') = g\delta(\vec{x}')$. Then

$$H_{12} = -\frac{1}{4\pi}\int d^3x \, d^3x' \rho_1(\vec{x})\rho_2(\vec{x}')\frac{e^{-m|\vec{x}-\vec{x}'|}}{|\vec{x}-\vec{x}'|} \; .$$

$$(2.61)$$

That is, the potential can be written

$$V(r) = -\frac{1}{4\pi}\frac{e^{-mr}}{r} \; .$$

$$(2.62)$$

[Note the role of the mass, and note the analogy with electrostatics as $m \to 0$.] This result leads to the general interpretation, in a quantum field theory, that all interactions are due to the exchange of field quanta. The concepts of force and of interaction are used interchangeably. Equation (2.61) gives the interaction in position space. Usually we write matrix elements in momentum space. Then from equation (2.56)and the following remarks, we see that for a general situation the momentum space quantity representing the exchanged particle of mass m is

$$1/ \left(k^2 - m^2 \right) . \tag{2.63}$$

This is called a propagator, and will be used whenever we write the matrix element. The complete propagator also has a phase factor and a numerator that depends on the spin of the exchanged particle but for most of our calculations these can be considered as technical details that do not affect qualitative results.

2.8   Summary of the Lagrangians

Here we list in one place a few of the free particle Lagrangians that will be used later so they are easily accessible for identification of the signs, factors of 2, *etc.* Some of the equations and terms given here are only introduced in later chapters, but it is useful to include all of the information in one place. Our main use of these will be for identification of mass terms and interaction terms when they arise.

 a.   Real spin–zero field of mass m (scalar or pseudoscalar)

$$\mathcal{L} = \frac{1}{2} \left[(\partial_\mu \phi)(\partial^\mu \phi) - m^2 \phi^2 \right] \tag{2.64}$$

and ϕ satisfies

$$(\partial^\mu \partial_\mu + m^2)\phi = 0 . \tag{2.65}$$

 b.   Complex scalar (or pseudoscalar) field of mass m
         (or two real scalars of the same mass)

$$\mathcal{L} = \frac{1}{2} \left[(\partial_\mu \phi_1)(\partial^\mu \phi_1) - m^2 \phi_1^2 \right] + \frac{1}{2} \left[(\partial_\mu \phi_2)(\partial^\mu \phi_2) - m^2 \phi_2^2 \right] \tag{2.66}$$

$$= (\partial^\mu \phi)^*(\partial_\mu \phi) - m^2 \phi^* \phi$$

where

$$\phi(x) = (\phi_1(x) + i\phi_2(x))/\sqrt{2}, \tag{2.67}$$

and

$$\left(\partial^\mu \partial_\mu + m^2\right)\phi = \left(\partial^\mu \partial_\mu + m^2\right)\phi^* = 0 . \tag{2.68}$$

c. <u>Spin-$\frac{1}{2}$ fermion field of mass m</u>

At the end of Chapter 5 we will write

$$\mathcal{L} = \overline{\psi}(i\gamma^\mu \partial_\mu - m)\psi \tag{2.69}$$

and ψ satisfies the Dirac equation

$$(i\gamma^\mu \partial_\mu - m)\psi = 0 . \tag{2.70}$$

d. <u>Massive Abelian vector field</u>

If there were an Abelian field B^μ, like the electromagnetic field, but massive, all the equations of Section 2.3 would hold; in addition, to give a mass term in the wave equation, a term

$$\frac{1}{2}m^2 B^\mu B_\mu \tag{2.71}$$

is added to the Lagrangian. If we see a term $B^\mu B_\mu$ appear in a Lagrangian, we can identify its coefficient as $(\text{mass})^2$.

e. <u>Non–Abelian vector field</u>

For completeness, though we will not explicitly use it, we also note what happens if the gauge field is non–Abelian, such as is the case with the gluon or W fields. A reader who has not previously encountered such objects will understand the notation after completing Chapters 3, 4, and 6. The vector potential A^μ of electromagnetism will be generalized to a non–Abelian field that will now have an internal index, say a, so the non–Abelian vector potential can be written W_a^μ. For $SU(2)$ $a = 1$, 2, 3 and for $SU(3)$ $a = 1$, 2, $\cdots$, 8. Then we define (using W_a^μ as an example)

$$W_a^{\mu\nu} = \partial^\mu W_a^\nu - \partial^\nu W_a^\mu + g f_{abc} W_b^\mu W_c^\nu \ . \tag{2.72}$$

The f_{abc} are structure constants (see Appendix B); for $SU(2)$, $f_{abc} = \epsilon_{abc}$. The extra term is necessary to make the vector (in internal space) transform correctly under rotations (like equation (4.15)). Since we will not use these results, we will not derive the extra term in equation (2.72). This form of $W_a^{\mu\nu}$ is gauge invariant.

The associated Lagrangian is then just what is expected, analogous to the one for electromagnetism,

$$\mathcal{L} = -\frac{1}{4} W_a^{\mu\nu} W_{\mu\nu}^a + \frac{1}{2} m^2 W_a^\mu W_\mu^a \ . \tag{2.73}$$

As always, repeated indices are summed. Note that because of the term in equation (2.72) that is quadratic in W, the Lagrangian contains terms with three and four W's. Such terms must occur in a gauge invariant, Lorentz invariant theory. We will not use these terms, but they will be present in the full theory and are important in a complete treatment.

2.9   Feynman Rules

In practice, much of explaining the Standard Model and why it seems to describe nature amounts to writing the Feynman rules of the Standard Model. At the level of this book, the rules for fermion–boson interactions are the most important. Let us summarize here the basic arguments, all of which were motivated in the earlier sections of this chapter.

Consider the electromagnetic interaction. The interaction Lagrangian is (equation (2.16))

$$\begin{aligned} \mathcal{L}_{\text{int}} &= -J_\mu A^\mu \\ &= Q\overline{\psi}\gamma^\mu \psi A^\mu \end{aligned} \tag{2.74}$$

where we have inserted a current $J^\mu = -Q\overline{\psi}\gamma^\mu\psi$. Q is the electric charge, $\overline{\psi}$ and ψ represent final and initial electrons, and γ^μ is a factor such that $\overline{\psi}\gamma^\mu\psi$ is a four–vector; equation (2.74) will be explained in Chapter 5. Suppose we are describing an interaction where an electron of momentum p emits a photon of momentum k and recoils with momentum p'. For an electron, $Q = -e$. The wave functions are $\psi = u(p)e^{-ip \cdot x}$ and $\overline{\psi} = \overline{u}(p')e^{+ip' \cdot x}$, for the initial and final electrons, and $A_\mu = \epsilon_\mu e^{ik \cdot x}$ for the photon, where ϵ_μ is the polarization vector of the photon.

The factor that goes at the $ee\gamma$ vertex is then $\mathcal{L}_{\mathrm{int}}$ with the external wave functions removed, $-e\gamma^\mu$ in this case. Our rule for constructing the transition matrix element for any process is to (1) write the appropriate factor for any vertex, (2) put a factor $1/\left[Q^2 - m^2\right]$ for the propagator of any internal line of four–momentum Q and mass m, (3) multiply by external wave function factors $\overline{u}$ for final fermions, u for initial fermions, $\overline{u}$ for initial antifermions and u for final antifermions (Chapter 5 will contain further discussion of antiparticles, but the details will not be quantitatively important for us), 1 for scalar bosons, and ϵ_μ for vector bosons. The true rules for an arbitrary process are significantly more complicated than ours, involving phase factors (± 1, $\pm i$), spin factors for propagators, rules for loops, $etc.$ But a good semi–quantitative understanding of the Standard Model can be obtained from our approximate rules. Our rules are basically those for calculating the transition matrix element up to a phase factor in the Born approximation in quantum theory. That is, $M \simeq \langle f| V |i\rangle$, and the potential V is equivalent to the interaction Lagrangian.

Given the matrix element, in Chapter 9 we will learn how to calculate the observables, which are mainly decay widths and scattering cross sections. In the next chapter we begin the systematic treatment of Standard Model physics.

Problems

2.1: Given the Schrödinger current $\overrightarrow{J_{ca}} = \left(u_c^* \nabla u_a - u_a \nabla u_c^*\right)/2im$ and the transition density $\rho_{ca} = u_c^* u_a$, where u_c and u_a are wave functions:

    a) Write the current four vector J^μ.

    b) Verify that $\partial^\mu J_\mu = 0$.

The states u_a and u_c can be different.

Suggestions for Further Study

The standard review article on the modern way of dealing with space–time invariances of the Lagrangian is the review by E. L. Hill. Wentzel's quantum field theory book, although one of the oldest ones, begins at a level which is not too sophisticated, and provides a rather clear and explicit exposition of much of the material described in this chapter. The treatment of DeWit and Smith is a thorough and contemporary one. Chapter IC of Gottfried and Weisskopf overlaps significantly with this chapter.

CHAPTER 3

Gauge
Invariance

Where do Lagrangians or Hamiltonians come from? How do we know that a certain interaction should describe an actual physical system? Why is the electromagnetic interaction due to a massless spin–one particle, the photon, being exchanged between electrically charged objects? Within the framework of gauge theories, these questions have been answered, in a certain sense. Basically, if certain forms of matter exist and are to interact in a way consistent with quantum theory, then the structure of the interaction can be deduced. This is a great departure from the historical situation, where we are simply given the form of the interaction, which was itself guessed by clever physicists. Theories where the interaction is determined (because of the invariance of the theory under some local transformations) are called "gauge theories."

We will proceed by first looking at gauge transformations in classical physics, and then in quantum theory where the essential features are already present. Next we look at Abelian quantum field theories. The final stage is the one of interest for the real world, non–Abelian gauge theories.

3.1 Gauge Invariance in Classical Electromagnetism

In classical electrodynamics, the fields are related to the vector potential $\vec{A}$ by

$$\vec{B} = \nabla \times \vec{A}$$

$$\vec{E} = -\nabla V - \frac{\partial \vec{A}}{\partial t} .$$

$$(3.1)$$

If the transformations

$$\vec{A} \to \vec{A}\,' = \vec{A} + \nabla\chi \qquad\qquad (3.2)$$

$$V \to V' = V - \frac{\partial\chi}{\partial t} \qquad\qquad (3.3)$$

are carried out, with χ being appropriately differentiable but otherwise arbitrary, equations (3.1) are unchanged. It is convenient to combine $\vec{A}$ and V into a four-vector

$$A^\mu = (V; \vec{A}) \qquad\qquad (3.4)$$

so the transformations (3.2), (3.3) are

$$A^\mu \to A^{\mu\,\prime} = A^\mu - \partial^\mu\chi \,. \qquad\qquad (3.5)$$

Besides being a useful notation, this emphasizes that there is a connection between the transformations and that they are done simultaneously. These are called gauge transformations. Although the existence of such an invariance has been known for a long time, it was largely treated as a curiosity until the 1960's. Classical electrodynamics is seldom viewed as a gauge theory.

3.2   Gauge Invariance in Quantum Theory

The form that gauge invariance takes in quantum theory is quite different, and has led to the modern viewpoint. Since observables depend on $|\Psi|^2$, we can demand that the structure of the theory be invariant under

$$\Psi \to \Psi' = e^{-i\alpha}\,\Psi \qquad\qquad (3.6)$$

where α is a constant. This is called a global gauge transformation, since $\Psi(\vec{x},t)$ transforms the same way everywhere. That is, it should be possible to choose the phase of Ψ in an arbitrary way. Further, it should be possible to choose the phase of Ψ at each space-time point without affecting the theory—we ought to be able to fix our phase conventions here without regard for how they are chosen on the moon. Then the theory should be invariant under

$$\Psi(\vec{x},t) \to \Psi'(\vec{x},t) = e^{-i\chi(\vec{x},t)}\,\Psi(\vec{x},t) \,. \qquad\qquad (3.7)$$

This is called a local gauge transformation. If the name "gauge transformation" had not been used for historical reasons, these transformations would probably have been called "phase transformations."

A partial exception can occur when interference effects are measured, and intensities can become sensitive to phase *differences* of two interfering states. A nice recent description of these effects is given by Sakurai (1985).

A surprise occurs when we try to confirm that the Schrödinger equation ("the theory") is invariant under the transformation of equation (3.7). It is not! Consider a matter particle, which should be described by a wave function Ψ which satisfies

$$-\frac{1}{2m}\nabla^2\Psi(\overrightarrow{x},t) = i\frac{\partial\Psi(\overrightarrow{x},t)}{\partial t} \ . \tag{3.8}$$

If Ψ satisfies the Schrödinger equation, then Ψ' from equation (3.7) will not satisfy it, for a general $\chi(\overrightarrow{x},t)$, since the derivatives will not cancel.

For electrically charged particles we know the solution to this apparent puzzle. In the presence of an electromagnetic field we modify the Schödinger equation to be

$$\frac{1}{2m}\left(-i\nabla + e\overrightarrow{A}\right)^2\Psi = \left(i\frac{\partial}{\partial t} + eV\right)\Psi \tag{3.9}$$

where e is the magnitude of the electric charge of the electron. Then under the simultaneous transformations

$$\left.\begin{array}{rl} \Psi(\overrightarrow{x},t) \rightarrow & \Psi'(\overrightarrow{x},t) = e^{-i\chi}\,\Psi(\overrightarrow{x},t) \\[4pt] A \rightarrow & A' = A + \frac{1}{e}\nabla\chi \\[4pt] V \rightarrow & V' = V - \frac{1}{e}\frac{\partial\chi}{\partial t} \end{array}\right\} \tag{3.10}$$

we see that the form of (3.9) is unchanged if $(\Psi, \overrightarrow{A}, V) \rightarrow (\Psi', \overrightarrow{A}', V')$. [Without even writing it out this result is clear. In the bracket on the right hand side of equation (3.9) the $i\frac{\partial}{\partial t}$ brings down an extra term $-i^2\frac{\partial\chi}{\partial t}$, while eV picks up an extra $-\frac{\partial\chi}{\partial t}$, with e's and minus signs arranged so the extra terms just cancel. The same thing happens on the left hand side.] Note the relativistic connection is maintained, with $A^\mu \rightarrow A^{\mu\,\prime} = A^\mu - \partial^\mu\chi/e$.

We can reinterpret this standard result to say that the local phase invariance of the theory *requires* the presence of a field $A^\mu = (V; \overrightarrow{A})$. Since the field will have an expansion like equation (2.27) in terms of particle creation and destruction operators there must be an associated particle, and since the field is described by a four–vector it must be associated with a vector, *i.e.* spin–one, particle. Since the same effect occurs for any charged particle, the interaction of the new particle (which we interpret as being the photon) is the same with any charged particle, *i.e.* it is a universal interaction. Phase invariance of the theory for electrically charged particles requires that there must be a photon and an electromagnetic interaction of precisely the observed kind. Note however that the numerical value of e is undetermined.

In a certain sense the existence and form of the electromagnetic interaction has been derived. If a particle carries a charge and the theory is invariant under certain phase transformations, which are generally called gauge transformations, then associated fields (called gauge fields) and associated particles with spin-one (called gauge bosons) must exist. As we will see explicitly below, this allows us to write the associated interaction Lagrangians. What is not yet understood is under which gauge transformations the theory should be invariant. We will see that there are three known gauge transformations under which the theory is invariant, and three associated sets of particles—why these three and not others, or whether there are additional ones, is unknown.

Another way to word the interpretation of what we have observed is that we can not distinguish between the effects of a local change in phase convention and the effects of a new vector field. Perhaps that is not surprising given our experience with electromagnetism, where a local time dependent change in the potential V(*i.e.* $\frac{\partial \chi}{\partial t}$) can be compensated by an associated local change in the magnetic vector potential $\overrightarrow{A}$. By "compensated" we mean Maxwell's equations are unchanged. Such an effect requires that magnetic and electric fields are related.

3.3   Covariant Derivatives

By rewriting the above equations it is possible to put them in a nice form which makes their properties explicit and which is easily generalizable to the gauge invariant theories we will finally be interested in.

Define

$$\overrightarrow{\mathcal{D}} = -\nabla - ie\,\overrightarrow{A} \tag{3.11}$$

and

$$\mathcal{D}^\circ = \frac{\partial}{\partial t} - ieV \; . \tag{3.12}$$

Then the Schrödinger equation becomes

$$\frac{1}{2m}(i\overrightarrow{\mathcal{D}})^2\Psi = i\mathcal{D}^\circ\Psi \; . \tag{3.13}$$

Now, suppose we perform a local gauge transformation, *i.e.* use equation (3.10), and examine what happens to $\overrightarrow{\mathcal{D}} \Psi$.

$$
\begin{aligned}
-i\overrightarrow{\mathcal{D}}'\Psi' &= -i(-\nabla - ie\overrightarrow{A} - i\nabla\chi)e^{-i\chi}\Psi \\
&= -e^{-i\chi}(-\nabla - ie\overrightarrow{A})\Psi \\
&= e^{-i\chi}(-i\overrightarrow{\mathcal{D}}\Psi)
\end{aligned}
\tag{3.14}
$$

and similarly,

$$
\begin{aligned}
-i\mathcal{D}'^{\,\circ}\Psi' &= -i\left(\frac{\partial}{\partial t} - ieV + \frac{\partial\chi}{\partial t}\right)e^{-i\chi}\Psi \\
&= e^{-i\chi}\left(-i\mathcal{D}^{\circ}\Psi\right).
\end{aligned}
\tag{3.15}
$$

Let us also simplify the notation by combining $\overrightarrow{\mathcal{D}}$ and $\mathcal{D}^{\circ}$ into a four-vector,

$$
\mathcal{D}^{\mu} = (\mathcal{D}^{\circ}; \overrightarrow{\mathcal{D}}) .
\tag{3.16}
$$

$\mathcal{D}^{\mu}$ is called the "covariant derivative." Then equations (3.14) and (3.15) show that $\mathcal{D}^{\mu}\Psi$ transforms as a wave function if Ψ does, and that any equation written in terms of the covariant derivative will automatically be gauge invariant. Since $\mathcal{D}^{\mu}\Psi$ behaves like a wave function, $\mathcal{D}_{\mu}(\mathcal{D}^{\mu}\Psi)$ also behaves like a wave function under gauge transformations, so repeated applications of $\mathcal{D}^{\mu}$ still give a gauge invariant equation.

In the example of an electrically charged particle we knew how A^{μ} should change because of the historical role of the electromagnetic interaction. The arguments we have been using would be the same if some particles carried some other non-electromagnetic "charge," and we can rewrite the formalism a little to make it more general. Suppose we want the theory to be invariant under a transformation where particle states change as

$$
\Psi' = U\Psi
\tag{3.17}
$$

for some U. We want to define

$$
\mathcal{D}^{\mu} = \partial^{\mu} - igA^{\mu}
\tag{3.18}
$$

where A^{μ} represents the interacting field that has to be added to keep the theory invariant, but now we do not know how A^μ itself transforms. We also want

$$\mathcal{D}^\mu{}'\Psi' = U\left(\mathcal{D}^\mu\Psi\right) \tag{3.19}$$

and we can write this out,

$$\left(\partial^\mu - igA^\mu{}'\right)U\Psi = U\left(\partial^\mu - igA^\mu\right)\Psi .$$

This can be solved for $A^\mu{}'$:

$$
\begin{aligned}
-igA^\mu{}'U\Psi &= -\,\partial^\mu\left(U\Psi\right) + U\partial^\mu\Psi - igUA^\mu\Psi \\
&= -\left(\partial^\mu U\right)\Psi - igUA^\mu\Psi .
\end{aligned}
$$

Since each term acts on an arbitrary state Ψ, we can drop the Ψ and multiply from the right by U^{-1}, so

$$A^\mu{}' = -\frac{i}{g}\left(\partial^\mu U\right)U^{-1} + UA^\mu U^{-1}, \tag{3.20}$$

and for any U, we have found how A^μ must transform. The reader can verify that this gives the expected answer for $g = -e$ and $U = e^{-i\chi}$. Equation (3.20) is very general and stays valid if the U's are matrices in an internal space. The A^μ is also a matrix, and the order of factors cannot be interchanged. Consequently, we leave the last term in its present form, even though $UA^\mu U^{-1} = A^\mu$ if A^μ is not a matrix in some internal space.

Problems

3.1: If A_μ transforms according to equation (3.20), verify that for $U = e^{-i\chi}$, where χ is a function of $\vec{x}$ and t, the expected result is obtained for A'_μ .

3.2: Consider the gauge transformation $\psi' = e^{-i\chi}\psi$ and $A'_\mu = A_\mu - \frac{1}{g}\partial_\mu\chi$. If ψ satisfies the equation $\mathcal{D}^\mu\mathcal{D}_\mu\psi + m^2\psi = 0$, where $\mathcal{D}^\mu = \partial^\mu - ig A^\mu$, what equation does ψ' satisfy?

3.3: Suppose $\psi' = e^{i(\vec{k}\cdot\vec{r} - \omega t)}\psi$. (a) Does ψ' satisfy the free particle Schrödinger equation? (b) If ψ satisfies the Schrödinger equation with given potentials $\vec{A}$ and V, find the potentials for ψ' to satisfy the Schrödinger equation for a charged particle in the presence of an electromagnetic field. (c) Interpret the potentials obtained.

3.4: To obtain the electromagnetic interaction of a charged scalar, one can replace ∂^μ by $\mathcal{D}^\mu = \partial^\mu - ig A^\mu$ in equation (2.35). Show that if ϕ is a solution of the resulting equation with $A^\mu = (A^0; \vec{0})$, then ϕ^* is a solution with $A^0 \to -A^0$.

3.5: The gauge where $A^0 = 0$ and $\nabla \cdot \vec{A} = 0$ is called the Coulomb gauge. Suppose you start in an arbitrary gauge, where these equations are not satisfied. (a) Show that if $\chi(\vec{x}, t) = \int_{-\infty}^{t} A^0(\vec{x}, t')dt'$ then in the new gauge $A'^0 = 0$, and $\nabla \cdot \vec{A'}$ is given by some function $F(\vec{x}, t)$ that is in general not zero. (b) Now transform to $A''^\mu = (A''^0, \vec{A''})$ and show A''^μ satisfies the Coulomb gauge conditions if the new gauge function is

$$\chi' = \int d^3x' \frac{F(\vec{x}', t)}{4\pi|\vec{x} - \vec{x}'|} \; .$$

(Recall that: $\nabla^2 \left(\frac{1}{|\vec{x} - \vec{x}'|}\right) = -4\pi\delta(\vec{x} - \vec{x}')$.)

Suggestions for Further Study

The books by Aitchison and Hey, and by Moriyasu, are introductory treatments of gauge invariance. Sakurai (1985) has a brief but nice discussion of gauge
invariance in classical electrodynamics and quantum mechanics.

CHAPTER 4

Non-Abelian
Gauge
Theories

4.1   Strong Isospin, an Internal Space

Now we want to study the strong isospin symmetry. This is a symmetry of nucleons, pions, and other hadrons that plays an important role in the understanding of the physics of nuclei and of hadrons. It also had an important conceptual impact on the development of the ideas that led to modern gauge theories, and it is this aspect that is of most interest to us.

We will see later on that the weak isospin symmetry that we will study is the more fundamental one. Whenever we say "isospin" in the rest of the book we will mean the weak isospin.

Consider the neutron (n) and proton (p). Their masses are

$$m_n = 939.57 \text{ MeV}, \ m_p = 938.28 \text{ MeV} \qquad (4.1)$$

and differ by about 0.1%. No other particles have a similar mass. Both form nuclei, and both interact similarly. Why do we think of them as different? Well, obviously, the proton has an electric charge and the neutron does not. But strong interactions don't know about electric charge, and strong interactions are very strong compared to electromagnetic ones, so the electric charge should not matter much.

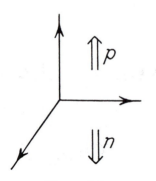

Figure 4.1
Strong Isospin Space.

This kind of reasoning has led to the idea that we should think of n, p as two states of the same thing, a nucleon N. The electric charge is then a label to distinguish the two states if that is needed. It is fruitful to imagine a particle space, called "strong isospin space," where the nucleon state points in some direction, as shown in Figure 4.1. If N points up it is a proton, if it points down it is a neutron, and in other directions it does not have a special name.

Now an important step is to assume that the theory that describes nucleon interactions is invariant under rotations in strong isospin space. That is, strong forces do not change if one makes rotations in this space. If this turns out to be true it can only be approximately true in nature, as electromagnetic forces will break the symmetry, but because they are only about 1% as strong as the strong force, this invariance might hold to a good approximation.

Next we want to put this in a more quantitative form. Since there are two nucleon states it is like spin-up and spin-down, so we try putting the proton and the neutron as states of a spin–like doublet, or $SU(2)$ doublet,

$$N = \begin{pmatrix} p \\ n \end{pmatrix}. \tag{4.2}$$

Can other hadrons be classified as states in $SU(2)$ multiplets? Yes! The pion, for example, has states $\pi^{\pm}$ and π°, with masses $m_{\pm} = 139.57$ MeV, $m_{\circ} = 134.96$ MeV, and can be represented as an isospin–one state,

$$\pi = \begin{pmatrix} \pi_1 \\ \pi_2 \\ \pi_3 \end{pmatrix}, \tag{4.3}$$

with charge states

$$\pi^{\pm} = (-\pi_1 \pm i\pi_2)/\sqrt{2}$$
$$\pi^{\circ} = \pi_3 . \tag{4.4}$$

The relation of charge states $\pi^{\pm}$ and π° to "Cartesian components" π_1, π_2, and π_3 is like that of the spherical harmonics Y_{lm} to Cartesian components x, y, and z. As for nucleons, the pion states have the same strong interactions, and differences in mass and in interactions of the typical size of electromagnetic effects. Later we will have W bosons in a similar weak isospin classification.

If the strong isospin notion is to be truly fruitful it must also hold for interactions. We can write an interaction Lagrangian to describe the most general pion-nucleon interaction. While perturbation theory and the lowest order matrix element of the Lagrangian may not work well in describing the experimental πN interaction, let us write it since the technique will be useful for W-boson interactions below. For notation, let $p^{\dagger}$ create a proton or destroy an antiproton, π^{+} destroy a π^{+} or create a π^{-}, and n destroy a neutron or create an antineutron, etc. Then clearly the most general three particle nucleon–conserving interaction Lagrangian is

$$\mathcal{L}_{\text{int}} = g_{pn} p^{\dagger} n \pi^{+} + g_{np} n^{\dagger} p \pi^{-} + g_{pp} p^{\dagger} p \pi^{\circ} + g_{nn} n^{\dagger} n \pi^{\circ}. \qquad (4.5)$$

This is not invariant under rotations in the isospin space unless certain relations hold among g's. For example, if we rotate $p \leftrightarrow n$, we have to require $g_{pp} = \pm g_{nn}$. How can we write an interaction Lagrangian which is invariant? Since π is a vector in isospin space, to make a scalar (invariant under rotations) we must make another vector from the nucleon. By analogy with spin, we know the answer. We form the vector (see Appendix A) $N^{\dagger} \vec{\tau} N$, where τ_i are the Pauli spin matrices. Then

$$\mathcal{L}_{\text{int}} = g \left(N^{\dagger} \vec{\tau} N \right) \cdot \vec{\pi} \qquad (4.6)$$

is clearly invariant under rotations since it is the scalar product of two vectors. To write it in the form of equation (4.5), we have

$$
\begin{aligned}
\vec{\tau} \cdot \vec{\pi} &= \tau_1 \pi_1 + \tau_2 \pi_2 + \tau_3 \pi_3 \\
&= \begin{pmatrix} 0 & 1 \\ 1 & 0 \end{pmatrix} \pi_1 + \begin{pmatrix} 0 & -i \\ i & 0 \end{pmatrix} \pi_2 + \begin{pmatrix} 1 & 0 \\ 0 & -1 \end{pmatrix} \pi_3 \\
&= \begin{pmatrix} \pi_3 & \pi_1 - i\pi_2 \\ \pi_1 + i\pi_2 & -\pi_3 \end{pmatrix} \qquad (4.7) \\
&= \begin{pmatrix} \pi^{\circ} & -\sqrt{2}\pi^{+} \\ -\sqrt{2}\pi^{-} & -\pi^{\circ} \end{pmatrix} \qquad (4.8)
\end{aligned}
$$

where equation (4.4) has been used in the last step. Then

$$
\begin{aligned}
N^\dagger \vec{\tau} \cdot \vec{\pi} N &= (\, p^\dagger \quad n^\dagger \,) \begin{pmatrix} \pi^0 & -\sqrt{2}\pi^+ \\ -\sqrt{2}\pi^- & -\pi^0 \end{pmatrix} \begin{pmatrix} p \\ n \end{pmatrix} \\
&= (\, p^\dagger \quad n^\dagger \,) \begin{pmatrix} \pi^0 p - \sqrt{2}\pi^+ n \\ -\sqrt{2}\pi^- p - \pi^0 n \end{pmatrix} \\
&= p^\dagger p \pi^0 - \sqrt{2} p^\dagger n \pi^+ - \sqrt{2} n^\dagger p \pi^- - n^\dagger n \pi^0 .
\end{aligned} \tag{4.9}
$$

Thus we see that the couplings g_{pn}, g_{np}, g_{pp}, g_{nn} must occur in the ratios $1 : 1 : -1/\sqrt{2} : 1/\sqrt{2}$ to have an interaction invariant under rotations in strong isospin space.

This technique of writing interactions invariant under rotations in internal spaces to obtain the form of the interaction will be used extensively in the following chapters. For weak isospin, it will be the W bosons rather than pions that have isospin one.

We will not go into details here about the ways that strong isospin has been checked experimentally, as they are covered elsewhere. Our interest has rather been in the conceptual aspects of strong isospin, as a historical and perhaps well known example of an internal symmetry of particles. Remarkably, as we will discuss later, that strong isospin is a good symmetry turns out to be essentially an accident rather than a fundamental feature of nature.

4.2   Non–Abelian Gauge Theories

Now we want to put together the ideas of internal spaces and of phase invariance. For the moment, we continue with the nucleon example. The proton and neutron are in an internal $SU(2)$ space, the strong isospin space. Then we can write a phase transformation where the change is expressed as an operator in isospin space,

$$
\begin{pmatrix} p' \\ n' \end{pmatrix} = e^{i\vec{\epsilon}\cdot\vec{\tau}/2} \begin{pmatrix} p \\ n \end{pmatrix} . \tag{4.10}
$$

The τ_i are the Pauli matrices, and ϵ_i are three parameters which specify the rotation from p and n to p' and n'. Remember that a function of a matrix is defined by its power series expansion, and because $\tau_i^2 = 1$, all powers of any Pauli matrix are either itself or the unit matrix; thus the exponential is simply expressed in terms of the Pauli matrices as demonstrated in Appendix A. Note that there is a new feature, in that the order of successive transformations matters, since the rotations do not commute. Formally, that is expressed by the commutator $[\tau_i, \tau_j] = 2i\epsilon_{ijk}\tau_k$. Whenever the order of transformations matters, they are called non–Abelian transformations.

We could equally well consider particles in a representation of any group, and demand invariance under the appropriate transformation. If particles a_1, a_2, and a_3 carry quantum numbers in an $SU(3)$ space, we could write

$$\begin{pmatrix} a_1' \\ a_2' \\ a_3' \end{pmatrix} = e^{i\vec{\alpha}\cdot\vec{\lambda}/2} \begin{pmatrix} a_1 \\ a_2 \\ a_3 \end{pmatrix} \tag{4.11}$$

where $\vec{\alpha} = (\alpha_1, \alpha_2, \ldots \alpha_8)$ are the eight rotation parameters, and λ_i $(i = 1, 2, \ldots 8)$ are the $SU(3)$ matrices analogous to the Pauli matrices; they are described in Appendix B. We will see that quarks have such a degree of freedom; it is called "color," a name chosen because some of its properties are analogous to those of real colors. In fact, however, it is a completely new property that particles possess, and has no direct connection to anything in classical physics or the everyday world.

At the present time there is no theoretical principle that tells us what internal spaces to examine. We will see that each internal space where particles carry nontrivial quantum numbers leads to an interaction between particles, mediated by a new set of gauge bosons. As far as is known, the complete set of spaces that lead to a description of all known experiments is the $SU(3)$ color space, and the $SU(2)$ and $U(1)$ electroweak spaces. These have been discovered empirically. We will examine the implications of demanding invariance under transformations in these spaces, but no one yet understands why it is these particular ones that apply and not others. Now that we have introduced all these ideas for the relatively familiar case of the proton and neutron, let us next go to the quarks and the leptons and their weak isospin.

4.3 Non–Abelian Gauge Theories for Quarks and Leptons

Suppose that the quarks and leptons can be put in representations ψ of a ("weak") isospin space, and that the theory should be invariant under transformations of the form of equation (4.10). Proceeding now as before with the $U(1)$ transformations of the previous chapter, we demand invariance under local transformations. That is, technically we make the parameters functions of space and time, $\epsilon_i(\vec{x}, t)$ or $\alpha_i(\vec{x}, t)$. That guarantees we can choose how we define the phase of the quark and lepton states at each space-time point, rather than having a choice here fix how the phase must be defined somewhere else or some time later. A theory with a local non-Abelian phase invariance is called a Yang–Mills gauge theory.

As before, no free particle can have an invariance under a non-Abelian gauge transformation, since the derivatives in the Schrödinger equation (or equivalent relativistic equation) will act on $\epsilon_i(\overrightarrow{x}, t)$. We are led again to define a covariant derivative. All the logic of Section 3.2 and Section 3.3 goes through here just as it did there. Instead of a function $\chi(\overrightarrow{x}, t)$ in the exponent of equation (3.7), we have a function $\overrightarrow{\epsilon} \cdot \overrightarrow{\tau}$ which transforms nontrivially under an $SU(2)$ group, or $\overrightarrow{\epsilon} \cdot \overrightarrow{\lambda}$ which transforms nontrivially under a $SU(3)$ group. The states ψ describe leptons or quarks.

To define $\mathcal{D}^\mu$, it is necessary for the $SU(2)$ case to introduce a set of three fields, each of which behaves as a four–vector under Lorentz transformations, in order that we can write a term that transforms as ∂^μ does. Before we needed one A^μ; now we need a W_i^μ for each τ_i. We can define

$$\mathcal{D}^\mu = \partial^\mu - ig_2 \frac{\overrightarrow{\tau}}{2} \cdot \overrightarrow{W^\mu}. \tag{4.12}$$

This is the generalization of the Abelian case (equation (3.18)), to include the non-Abelian transformations. If both transformations were relevant, the appropriate terms would add in $\mathcal{D}^\mu$. The coupling g_2 is an arbitrary factor which will determine interaction strengths. The W_i^μ must be introduced if the theory is to be invariant under weak isospin transformations. Since they correspond to a particle transforming under space rotations as a vector, they should be realized as spin–one particles, like the photon. Since τ_i is there, equation (4.12) is a 2 × 2 matrix equation.

We want to find how W_i^μ changes under a gauge transformation, since we have no obvious answer analogous to equation (3.5). To find out, we go through a derivation like the one from equation (3.17) through equation (3.20). The basic physics requirement is that

$$D'^\mu \psi' = e^{i\overrightarrow{\epsilon}(x) \cdot \overrightarrow{\tau}/2} \mathcal{D}^\mu \psi , \tag{4.13}$$

since ψ itself transforms that way. Assume that W_i^μ transforms so that

$$W_i'^\mu = W_i^\mu + \delta W_i^\mu, \tag{4.14}$$

and we want to solve for δW_i^μ. As usual, let the transformation be infinitesimal so higher terms in ϵ or in δW such as $\epsilon\,\delta W$ can be neglected, to simplify the calculation. Then the sides of equation (4.13) become

$$
\begin{aligned}
\text{LHS} = D'^\mu \psi' &\simeq \left(\partial^\mu - ig_2\tau_i W_i'^\mu/2\right)\left(1 + i\epsilon_j\tau_j/2\right)\psi \\
&= \left(\partial^\mu - ig_2\tau_i W_i^\mu/2 - ig_2\tau_i\delta W_i^\mu/2\right)\left(1 + i\epsilon_j\tau_j/2\right)\psi \\
&= \partial^\mu\psi - ig_2\tau_i W_i^\mu\psi/2 - ig_2\tau_i\delta W_i^\mu\psi/2 + i\tau_j\partial^\mu(\epsilon_j\psi)/2 \\
&\qquad + g_2\tau_i W_i^\mu\epsilon_j\tau_j\psi/4 \\
&= \left(\partial^\mu - ig_2\tau_i W_i^\mu/2 - ig_2\tau_i\delta W_i^\mu/2 + i\tau_j(\partial^\mu\epsilon_j)/2\right. \\
&\qquad \left. + g_2\tau_i W_i^\mu\epsilon_j\tau_j/4\right)\psi + i\tau_j\epsilon_j(\partial^\mu\psi)/2
\end{aligned}
$$

where a higher order term in $\epsilon\delta W$ has been dropped, and

$$
\begin{aligned}
\text{RHS} &= \left(1 + i\tau_i\epsilon_i/2\right)\left(\partial^\mu - ig_2\tau_j W_j^\mu/2\right)\psi \\
&= \left(\partial^\mu - ig_2\tau_j W_j^\mu/2 + g_2\tau_i\epsilon_i\tau_j W_j^\mu/4\right)\psi + i\tau_i\epsilon_i(\partial^\mu\psi)/2 \; .
\end{aligned}
$$

Comparing, we see the first, second, and fourth terms on the RHS cancel the corresponding terms on the LHS, so

$$
-ig_2\tau_i\delta W_i^\mu/2 + i\tau_i(\partial^\mu\epsilon_i)/2 + g_2\tau_j W_j^\mu\epsilon_i\tau_i/4 = g_2\tau_i\epsilon_i\tau_j W_j^\mu/4
$$

where we have dropped the ψ since every term multiplies it. This is now solvable for δW_i^μ as desired,

$$
\tau_i\delta W_i^\mu = \frac{1}{g_2}(\partial^\mu\epsilon_i)\tau_i + \frac{i}{2}\epsilon_i W_j^\mu\left[\tau_i\tau_j - \tau_j\tau_i\right].
$$

Recognizing the commutator is $2i\epsilon_{ijk}\tau_k$, this becomes

$$
\tau_i\left\{\delta W_i^\mu - \frac{1}{g_2}\partial^\mu\epsilon_i + \epsilon_{ijk}\epsilon_j W_k^\mu\right\} = 0
$$

so we can conclude

$$
\delta W_i^\mu = \frac{1}{g_2}\partial^\mu\epsilon_i - \epsilon_{ijk}\epsilon_j W_k \; . \tag{4.15}
$$

We will not actually use equation (4.15) for further derivations, though it would be used in a more advanced treatment. Our purpose in deriving it was two-fold. First, it gives some practice in working with the formalism and getting used to it. Second, it is non-trivial that a consistent solution exists, and it is such a fundamental matter that it is appropriate for the reader to see the explicit derivation that the non-Abelian theory is fully gauge invariant, *i.e.* there exists a set of transformations for ψ, W_i^μ that allows $\mathcal{D}^\mu\psi$ to transform as ψ itself.

[The form of equation (4.15) is easy to understand, so for interested readers we note its origin. The first term is exactly what is expected by analogy with equation (3.10). The second term will be familiar from classical mechanics as an example of how a vector transforms under rotations. To practice more with the formalism, consider how W transforms under an isospin rotation,

$$\overrightarrow{W}' = e^{i\vec{\epsilon}\cdot\vec{T}}\overrightarrow{W} \tag{4.16}$$

where $\overrightarrow{T}$ is the appropriate representation of the rotation matrices for spin–one,

$$(T_i)_{jk} = -i\epsilon_{ijk} \ , \tag{4.17}$$

and the space-time index μ is not written for simplicity. Then for an infinitesimal transformation,

$$W_i' \simeq (1 + i\epsilon_k T_k)_{ij}\, W_j = W_i + \delta W_i$$

so

$$\begin{aligned}
\delta W_i &= i\epsilon_k(-i\epsilon_{kij})W_j \\
&= \epsilon_{ijk}\epsilon_k W_j
\end{aligned}$$

which is precisely the second term of equation (4.15).]

In equation (4.12) the covariant derivative is written with the understanding that it will act on the doublet representation of $SU(2)$. That is appropriate for us as we will put left-handed fermions (defined in Chapter 5) in such doublets; we have implicitly noted that by labeling the coupling g_2. Two generalizations are necessary.

First, though still in the internal $SU(2)$ weak isospin space, $\mathcal{D}^\mu$ could act on a state ψ in a different representation. If ψ is a state of weak isospin t with $2t+1$ components, let $\overrightarrow{T}$ be the $(2t+1)\times(2t+1)$ matrix operator representation of the $SU(2)$ generators in that basis. Then

$$\mathcal{D}^\mu = \partial^\mu - ig_2\overrightarrow{T}\cdot\overrightarrow{W_\mu} \ . \tag{4.18}$$

For spin–1/2, $\overrightarrow{T} = \overrightarrow{\tau}/2$. We will interchangeably write $\overrightarrow{T}\cdot\overrightarrow{W}$ or $T_i \cdot W_i$, where i is summed from 1 to 3.

Second, we could consider a different internal space, where interactions are invariant under another set of transformations. For an $SU(n)$ invariance, with group generators $\overrightarrow{F}$ in an $(n^2 - 1)$–dimensional vector, and $[F_i, F_j] = ic_{ijk}F_k$, the appropriate $\mathcal{D}^\mu$ to act on the n–dimensional matter state ψ is

$$\mathcal{D}^\mu = \partial^\mu - ig_n \overrightarrow{F} \cdot \overrightarrow{G^\mu} \tag{4.19}$$

where the G^μ are the $(n^2 - 1)$ gauge bosons that must be introduced to have a gauge invariant theory. We will interchangeably write $\overrightarrow{F} \cdot \overrightarrow{G}$ or $F_a \cdot G_a$, where a is summed from 1 to 8 for $SU(3)$.

Apparently nature also knows about an $SU(3)$ internal space, which is called a "color" space as we have already mentioned, as well as about the internal $SU(2)$ isospin space. The appropriate generators F_i are the λ_i of $SU(3)$, described in Appendix B. The λ_i should be thought of as just the generalizations of the Pauli matrices. In both cases, for $SU(n)$ there are $(n^2 - 1)$ matrices.

By adding several terms to ∂^μ we can guarantee that we obtain a covariant derivative $\mathcal{D}^\mu$ that will let us write Lagrangians (and therefore equations) that are invariant under gauge transformations, simultaneously or separately, in all the internal spaces. The full covariant derivative that we presently are aware of can be written

$$\mathcal{D}^\mu = \partial^\mu - ig_1 \frac{Y}{2} B^\mu - ig_2 \frac{\tau_i}{2} W_i^\mu - ig_3 \frac{\lambda_a}{2} G_a^\mu \ . \tag{4.20}$$

The scalar product in $SU(2)$ runs for $i = 1, 2, 3$ while the scalar product in $SU(3)$ runs for $a = 1, 2, \ldots 8$. The couplings g_1, g_2, g_3 are arbitrary real numbers. For the Abelian $U(1)$ symmetry we have written the field that must be introduced as B^μ rather than as the electromagnetic field A^μ, since we do not know ahead of time that nature's $U(1)$ invariance corresponds precisely to electromagnetism; we will use physics arguments to identify the electromagnetic field in Chapter 7. The $U(1)$ term has been written with a "generator" Y in a form analogous to the other terms; for $U(1)$, Y is just a number, though it can depend on the states on which $\mathcal{D}^\mu$ operates. Y is called the $U(1)$ hypercharge generator.

It is worth emphasizing that once the g_i are fixed for any representation, they are known for all representations; measuring g_2 with muon decay fixes it for quark interactions. Once the coupling of W or g to one fermion is measured, their coupling to all fermions and all gauge bosons is known.

The ∂^μ is a Lorentz four–vector, as are all the terms in equation (4.20). The first two terms are singlets (they multiply the unit matrix) in the $SU(2)$ and $SU(3)$ spaces. The third term is a 2×2 matrix in $SU(2)$ and a singlet in the other spaces, and the fourth term is a 3×3 matrix in $SU(3)$ and a singlet in the other spaces. There is no inconsistency in having different size matrices for different terms since they operate in different spaces.

Equation (4.20) is, in a sense, the main equation of the Standard Model. When used in a Lagrangian it leads to the full theory of the Standard Model. It is the culmination of several decades of creative thinking by a number of physicists, leading to the realization that the phase invariance of quantum theory must exist for transformations in new kinds of internal spaces, and that quarks and leptons apparently carry labels that distinguish among three internal spaces. The phase, or gauge, invariance is guaranteed by the form of $\mathcal{D}^\mu$, as we learned in Section 3.3. It is not yet understood why it is these spaces that characterize particles and not others, nor whether additional ones exist. In each case, as in the discussion of gauge invariance for electromagnetism under equation (3.10), additional spin–one gauge boson fields B^μ, W_i^μ, and G_a^μ must exist (one, three, and eight respectively). All of these have now been observed experimentally, as we will see later.

Problems

4.1:   (a) If $\vec{v}^{\,\mu}$ is a vector with components v_1^μ, v_2^μ and v_3^μ, write

$$\mathcal{D}^\mu = \partial^\mu + \vec{\tau} \cdot \vec{v}^{\,\mu}$$

explicitly in a matrix form.

(b) If G_a^μ is an eight–dimensional vector with components G_1^μ, $G_2^\mu \cdots G_8^\mu$, write $\mathcal{D}^\mu = \partial^\mu + \lambda_a G_a^\mu$ explicitly in matrix form.

4.2:   Suppose there is a neutral particle ϵ which is thought to be a strong isospin singlet. To test this, the ratio of the $\epsilon p p$ and the $\epsilon n n$ interaction strengths are measured. What is the expected value of the ratio?

Suggestions for Further Study

Aitchison and Hey, and Moriyasu, have further discussion of non–Abelian gauge theories at an introductory level. Ryder, and Cheng and Li, have more complete treatments.

Non–Abelian gauge invariance plays an important role that we do not cover in this book. It is responsible for the Standard Model being a renormalizable theory, even though gauge bosons and fermions have mass. Although we will remark on this in context in Chapters 7 and 8, the subject is too technical for our level of treatment. However, the reader should appreciate that the non–Abelian gauge invariance is crucial at a theoretical level as well as determining the phenomenological structure of the theory. The history and the impact of the renormalizability of the theory can be traced with the historical treatments mentioned in Chapter 1.

CHAPTER 5

Dirac Notation
for Spin

The basic forms of matter in the Standard Model, the quarks and the leptons, are spin–1/2 particles. To describe their interactions it is necessary to generalize the solutions of the Schrödinger equation to include the presence of spin. In addition we want to write a theory that behaves properly under Lorentz transformations, so to fully incorporate the requirements of relativity we want to use the solutions of the Dirac equation. Spin is fundamental to the formulation of the Standard Model; it treats fermions whose spin is parallel to their momentum differently from fermions whose spin is antiparallel to their momentum!

What we need is the simple and powerful notation that has developed for writing solutions of the Dirac equation. We will not do any extensive calculations with the spin formalism. Although the Feynman rules are written using the Dirac notation in an essential way, we will see that once we have the Feynman rules and the matrix elements, we can estimate all transition rates without using spin. In effect we will learn how to express certain ideas in a powerful notation that greatly simplifies writing down the theory and extracting its consequences. We will derive or motivate what is needed here, without assuming the reader has a previous knowledge of the subject. At the nonrelativistic level, we assume the reader is acquainted with two-component spin physics and is familiar with the Pauli matrices; a brief review is given in Appendix A.

5.1   The Dirac Equation

Dirac wanted to write a relativistic equation to describe the electron including its spin, and to satisfy various constraints. To guarantee conservation of probability, an equation linear in the time derivative, like the Schrödinger equation, was needed. Then relativistic invariance required linearity in the space derivatives as well. So he wrote the most general such equation

$$i\frac{\partial\psi}{\partial t} = \left[-i\left(\alpha_1\frac{\partial}{\partial x^1} + \alpha_2\frac{\partial}{\partial x^2} + \alpha_3\frac{\partial}{\partial x^3}\right) + \beta m\right]\psi . \tag{5.1}$$

The coefficients α_i and β were determined by physics conditions.

From $E^2 = p^2 + m^2$, using the quantum mechanical operators for E and p gives

$$-\frac{\partial^2\psi}{\partial t^2} = \left(-\nabla^2 + m^2\right)\psi \tag{5.2}$$

as one condition on any state ψ . Squaring equation (5.1) gives an equation similar to equation (5.2), so some conditions on the coefficients arise by comparison. Recalling that the Pauli matrices σ_i that describe spin are non–commuting, there is a possibility that α_i and β may not commute, so their order should not be interchanged. Equation (5.1) gives for the operators

$$i\frac{\partial}{\partial t} = -i\left(\alpha_1\partial x_1 + \alpha_2\partial x_2 + \alpha_3\partial x_3\right) + \beta m \tag{5.3}$$

where we have written $\partial x_i = \frac{\partial}{\partial x^i}$. Applying this to equation (5.1) gives (summing repeated indices as always)

$$\begin{aligned}
\frac{\partial^2\psi}{\partial t^2} &= \left[i\left(\alpha_1\partial x_1 + \alpha_2\partial x_2 + \alpha_3\partial x_3\right) - \beta m\right] \\
&\quad \times \left[-i\left(\alpha_1\partial x_1 + \alpha_2\partial x_2 + \alpha_3\partial x_3\right) + \beta m\right]\psi \\
&= \alpha_i^2\frac{\partial^2\psi}{\partial x^{i2}} + \sum_{j>i}\left(\alpha_i\alpha_j + \alpha_j\alpha_i\right)\frac{\partial^2\psi}{\partial x^i\partial x^j} \\
&\quad + im\left(\alpha_i\beta + \beta\alpha_i\right)\frac{\partial\psi}{\partial x^i} - \beta^2 m^2\psi .
\end{aligned} \tag{5.4}$$

Comparing equations (5.2) and (5.4) requires $\alpha_i^2 = \beta^2 = 1$ and

$$\alpha_i\alpha_j + \alpha_j\alpha_i = 0 \qquad i \neq j \tag{5.5}$$

$$\alpha_i\beta + \beta\alpha_i = 0 . \tag{5.6}$$

These conditions look like the anticommutation relations for the Pauli matrices so we expect them to be satisfied by matrices.

5.2   Massless Fermions

If we are considering massless fermions, a case of interest for the formulation of gauge theories, the solution is easy. Then the term with β is absent, and the condition on α_i is $\alpha_i \alpha_j + \alpha_j \alpha_i = 2\delta_{ij}$. This is just the anticommutation relation for the spin–1/2 Pauli matrices, so we can choose

$$\alpha_i = -\sigma_i \tag{5.7}$$

where the minus sign is a convenient convention. Then the Dirac equation is

$$i\frac{\partial \psi}{\partial t} = \overrightarrow{\sigma} \cdot \overrightarrow{p}\, \psi \, , \tag{5.8}$$

and ψ can be interpreted as a two–component spinor.

5.3   Fermions with Mass $\neq 0$

One might hope that a simple choice like $\beta = 1$ would work, so that a term βm could be added to the massless equation (5.8). It does not, as equation (5.6) shows, for example; β must be a nontrivial matrix, and with 2×2 matrices the Pauli matrices plus the unit matrix are the only possibilities, so there can be no 2×2 solutions. A little trial and error shows that the smallest matrices that can work are 4×4. That there is a solution can be seen by example. A particular choice can be written in terms of the Pauli matrices σ_i

$$\alpha_i = \begin{pmatrix} 0 & \sigma_i \\ \sigma_i & 0 \end{pmatrix}, \ \beta = \begin{pmatrix} 1 & 0 \\ 0 & -1 \end{pmatrix} \tag{5.9}$$

where the notation means, for example,

$$\alpha_1 = \begin{pmatrix} \begin{pmatrix} 0 & 0 \\ 0 & 0 \end{pmatrix} & \begin{pmatrix} 0 & 1 \\ 1 & 0 \end{pmatrix} \\ \begin{pmatrix} 0 & 1 \\ 1 & 0 \end{pmatrix} & \begin{pmatrix} 0 & 0 \\ 0 & 0 \end{pmatrix} \end{pmatrix} \text{ and } \beta = \begin{pmatrix} \begin{pmatrix} 1 & 0 \\ 0 & 1 \end{pmatrix} & \begin{pmatrix} 0 & 0 \\ 0 & 0 \end{pmatrix} \\ \begin{pmatrix} 0 & 0 \\ 0 & 0 \end{pmatrix} & \begin{pmatrix} -1 & 0 \\ 0 & -1 \end{pmatrix} \end{pmatrix} .$$

$$\tag{5.10}$$

The notation is very useful, allowing one to work with 2 × 2 matrices at each level. It is a notation closely tied to the physics as well, since the full relativistic description of a spin–1/2 particle is given in terms of the non-relativistic Pauli matrices.

We have, for example,

$$
\begin{aligned}
\alpha_i \beta + \beta \alpha_i &= \begin{pmatrix} 0 & \sigma_i \\ \sigma_i & 0 \end{pmatrix} \begin{pmatrix} 1 & 0 \\ 0 & -1 \end{pmatrix} + \begin{pmatrix} 1 & 0 \\ 0 & -1 \end{pmatrix} \begin{pmatrix} 0 & \sigma_i \\ \sigma_i & 0 \end{pmatrix} \\
&= \begin{pmatrix} 0 & -\sigma_i \\ \sigma_i & 0 \end{pmatrix} + \begin{pmatrix} 0 & \sigma_i \\ -\sigma_i & 0 \end{pmatrix} \\
&= \begin{pmatrix} 0 & 0 \\ 0 & 0 \end{pmatrix} \\
&= 0
\end{aligned}
\tag{5.11}
$$

$$
\alpha_i \alpha_j + \alpha_j \alpha_i = \begin{pmatrix} \sigma_i \sigma_j & 0 \\ 0 & \sigma_i \sigma_j \end{pmatrix} + \begin{pmatrix} \sigma_j \sigma_i & 0 \\ 0 & \sigma_j \sigma_i \end{pmatrix} = 2\delta_{ij} .
\tag{5.12}
$$

5.4   The γ–matrices

The choice of matrices is not unique. Since we are working with 4 × 4 matrices, with 16 elements, there are 16 independent ones, and we can write the results in terms of any set that satisfies the conditions of equations (5.5) and (5.6). It turns out that a very convenient choice is the set of so-called γ-matrices, defined by

$$
\gamma^i = \beta \alpha_i
\tag{5.13}
$$

and

$$
\gamma^0 = \beta .
\tag{5.14}
$$

They can be written as a 4-vector,

$$
\gamma^\mu = (\gamma^0; \gamma^i) .
\tag{5.15}
$$

The main property we will use is

$$
\gamma^\mu \gamma^\nu + \gamma^\nu \gamma^\mu = 2g^{\mu\nu} .
\tag{5.16}
$$

When $i = j$ this gives $(\gamma^i)^2 = -1$. As always, i and j run from 1 to 3, and μ and ν from 0 to 3.

The Dirac equation is

$$i\frac{\partial \psi}{\partial t} = \left(-i\overrightarrow{\alpha} \cdot \nabla + \beta m\right) \psi \ . \tag{5.17}$$

Multiply by β from the left and use equations (5.13) and (5.14). Then

$$i\beta\frac{\partial \psi}{\partial t} = \left(-i\beta \overrightarrow{\alpha} \cdot \nabla + m\right) \psi \tag{5.18}$$

or

$$i\gamma^0\frac{\partial \psi}{\partial t} + i\overrightarrow{\gamma} \cdot \nabla \psi - m\psi = 0 \tag{5.19}$$

which can be written

$$\left(i\gamma^\mu \partial_\mu - m\right) \psi = 0 \ , \tag{5.20}$$

recalling the components of ∂_μ from Chapter 2. This covariant form is simple and convenient.

Since the γ_μ are 4×4 matrices, we can make a scalar by forming $\psi^\dagger \gamma_\mu \psi$ which is of the form

$$(\cdot \quad \cdot \quad \cdot \quad \cdot) \begin{pmatrix} \cdot & \cdot & \cdot & \cdot \\ \cdot & \cdot & \cdot & \cdot \\ \cdot & \cdot & \cdot & \cdot \\ \cdot & \cdot & \cdot & \cdot \end{pmatrix} \begin{pmatrix} \cdot \\ \cdot \\ \cdot \\ \cdot \end{pmatrix} = (\cdot \quad \cdot \quad \cdot \quad \cdot) \begin{pmatrix} \cdot \\ \cdot \\ \cdot \\ \cdot \end{pmatrix} = \text{number} \ . \tag{5.21}$$

It turns out to be more convenient to define

$$\overline{\psi} = \psi^\dagger \gamma^0 \tag{5.22}$$

for any solution ψ, and form scalars in spin space such as $\overline{\psi}\psi$, $\overline{\psi}\gamma^\mu \psi$, etc.

Another very useful quantity is

$$\gamma^5 = i\gamma^0\gamma^1\gamma^2\gamma^3 \ . \tag{5.23}$$

Although we will not go through the details, the sense in which we can call γ^μ a four–vector is that the quantity $\overline{\psi}\gamma^\mu \psi$, which is a scalar in spin space, transforms as a normal four–vector, just like p^μ or x^μ. This can be established by looking at the Dirac equation in two Lorentz frames; the way $\frac{\partial}{\partial x^\mu}$ transforms is known, so the way γ^μ transforms can be deduced, and as the form of equation (5.20) suggests, the appropriate spin scalar is a Lorentz four–vector.

Similarly, $\overline{\psi}\psi$ is a Lorentz scalar. Slightly more subtle, $\overline{\psi}\gamma^5\psi$ transforms as a Lorentz scalar as well, but it is odd under inversion of the coordinate system (a parity transformation) so it is called a pseudoscalar. And $\overline{\psi}\gamma^5\gamma^\mu\psi$ also transforms like a four–vector but has an extra sign change relative to $\overline{\psi}\gamma^\mu\psi$ under inversion of a coordinate system, so it is called an axial vector. We will use all of these properties later on.

5.5   Currents

To see the form of a Dirac current, we proceed just as with the Schrödinger equation in Chapter 2. The Dirac equation is

$$i\gamma^0\frac{\partial\psi}{\partial t} + i\gamma^k\frac{\partial\psi}{\partial x^k} - m\psi = 0 \tag{5.24}$$

where k is summed for $k = 1, 2, 3$. The Hermitian conjugate is

$$-i\frac{\partial\psi^\dagger}{\partial t}\gamma^0 - i\frac{\partial\psi^\dagger}{\partial x^k}\gamma^{k\dagger} - m\psi^\dagger = 0 \ . \tag{5.25}$$

From equations (5.9), (5.13), and (5.14), some explicit multiplications, and the fact that the Pauli matrices are Hermitian, we have

$$\left.\begin{aligned}
\gamma^{0\dagger} &= \gamma^0 \\
\gamma^{k\dagger} &= \left(\beta\alpha^k\right)^\dagger = \alpha^{k\dagger}\beta \\
&= \alpha^k\beta = -\beta\alpha^k = -\gamma^k \\
(\gamma^0)^2 &= \gamma^0\gamma^0 = 1 \\
\gamma^k\gamma^k &= \beta\alpha^k\beta\alpha^k = -1 \ .
\end{aligned}\right\} \tag{5.26}$$

In the last of these equations, there is no sum over k. Now multiply equation (5.25) from the right by γ^0, recall that $\overline{\psi} = \psi^\dagger\gamma^0$, and use equation (5.16) to interchange γ^k and γ^0. Then equation (5.25) gives

$$i\frac{\partial\overline{\psi}}{\partial t}\gamma^0 + i\frac{\partial\overline{\psi}}{\partial x^k}\gamma^k + m\overline{\psi} = 0 \ , \tag{5.27}$$

or

$$i\left(\partial_\mu\overline{\psi}\right)\gamma^\mu + m\overline{\psi} = 0 \ . \tag{5.28}$$

Next multiply equation (5.24) by $\overline{\psi}$ from the left, in order to have a scalar in spin space, and multiply equation (5.28) by ψ from the right, and add. The $m\overline{\psi}\psi$

terms drop out and we get

$$\overline{\psi}\gamma^{\mu}\partial_{\mu}\psi + \left(\partial_{\mu}\overline{\psi}\right)\gamma_{\mu}\psi = 0 \, , \tag{5.29}$$

which can be written as

$$\partial_{\mu}\left(\overline{\psi}\gamma^{\mu}\psi\right) = 0 \, . \tag{5.30}$$

Thus, as expected, we can define a current

$$j^{\mu} = \overline{\psi}\gamma^{\mu}\psi \tag{5.31}$$

which is conserved, *i.e.*

$$\partial_{\mu}j^{\mu} = 0 \, . \tag{5.32}$$

If we consider electrically charged fermions, we can make an electric current by putting the charge in; for the electron

$$j_{\mu}^{\text{electric}} = -e\overline{\psi}\gamma^{\mu}\psi \, . \tag{5.33}$$

But the current for any of the kinds of charges we consider is of the form $\overline{\psi}\gamma^{\mu}\psi$. And as discussed in Chapter 2, the vertex factor in the Feynman rules is essentially the current with the external wave functions removed.

To understand the current a little better, and practice the notation, we can go through some manipulations. Consider a free particle of momentum p^{μ}. Then ∂_{μ} operating on a plane wave solution of momentum p_{μ}, brings down a factor $-ip_{\mu}$, so the Dirac equation (5.20) becomes

$$\left(\gamma\cdot p - m\right)\psi = 0 \tag{5.34}$$

and the Hermitian conjugate equation (5.28) becomes

$$\overline{\psi}\left(\gamma\cdot p - m\right) = 0 \, . \tag{5.35}$$

Multiply (5.34) on the left by $\overline{\psi}\gamma^{\mu}$ and (5.35) on the right by $\gamma^{\mu}\psi$ and add. Then

$$\overline{\psi}\gamma^{\mu}\gamma\cdot p\psi + \overline{\psi}\gamma\cdot p\gamma^{\mu}\psi = 2m\overline{\psi}\gamma^{\mu}\psi \, . \tag{5.36}$$

Writing $\gamma \cdot p = \gamma^\lambda p_\lambda$ and factoring,

$$\overline{\psi}\left(\gamma^\mu\gamma^\lambda + \gamma^\lambda\gamma^\mu\right)\psi p_\lambda = 2m\overline{\psi}\gamma^\mu\psi \, . \qquad (5.37)$$

Using the anticommutator (equation (5.16)), this gives

$$\overline{\psi}\gamma^\mu\psi = \frac{p^\mu}{m}\overline{\psi}\psi \, . \qquad (5.38)$$

Thus, interpreting $\overline{\psi}\psi$ as a probability density, $\overline{\psi}\gamma^\mu\psi$ looks for a free particle just as a current would be expected to look.

5.6   Free Particle Solutions

As we have seen, a great deal can be deduced without discussing the solutions of the Dirac equation. We also need to use a little of the notation and interpretation of the solutions.

It is simplest to choose a specific representation for γ–matrices. We will redefine the γ's and use

$$\begin{aligned}\gamma^0 &= \begin{pmatrix} 0 & 1 \\ 1 & 0 \end{pmatrix}, \\[2mm] \gamma^i &= \begin{pmatrix} 0 & -\sigma_i \\ \sigma_i & 0 \end{pmatrix},\end{aligned} \qquad (5.39)$$

which gives

$$\gamma^5 = \begin{pmatrix} +1 & 0 \\ 0 & -1 \end{pmatrix} \, . \qquad (5.40)$$

Since the γ–matrices are written in 2×2 form, we assume

$$\psi = \begin{pmatrix} \psi_R \\ \psi_L \end{pmatrix} \qquad (5.41)$$

where ψ_L and ψ_R are a pair of two–component spinors; at the moment the labels L and R just serve to identify the two solutions, but they will have an additional interpretation. Then the Dirac equation is

$$\left(\gamma^\mu p_\mu - m\right)\psi = 0 \,,\tag{5.42}$$

or

$$\begin{pmatrix} -m & p_0 + \overrightarrow{\sigma}\cdot\overrightarrow{p} \\ p_0 - \overrightarrow{\sigma}\cdot\overrightarrow{p} & -m \end{pmatrix}\begin{pmatrix} \psi_R \\ \psi_L \end{pmatrix} = 0 \,.\tag{5.43}$$

Separating these equations gives

$$-m\psi_R + \left(p_0 + \overrightarrow{\sigma}\cdot\overrightarrow{p}\right)\psi_L = 0 \,,$$
$$\left(p_0 - \overrightarrow{\sigma}\cdot\overrightarrow{p}\right)\psi_R - m\psi_L = 0 \,.\tag{5.44}$$

Several things can be noticed here.

(a) These equations can be written as

$$\psi_R = \left(\frac{p_0 + \overrightarrow{\sigma}\cdot\overrightarrow{p}}{m}\right)\psi_L \,,$$
$$\psi_L = \left(\frac{p_0 - \overrightarrow{\sigma}\cdot\overrightarrow{p}}{m}\right)\psi_R \,.\tag{5.45}$$

(b) Solutions exist for positive or negative p_0 and can be interchanged by $\psi_L \leftrightarrow -\psi_R$.

(c) If $m = 0$ the two equations separate. Since $\overrightarrow{\sigma}\cdot\overrightarrow{p}$ clearly measures the component of spin along the direction of motion, ψ_R is the large solution for $\overrightarrow{\sigma}\cdot\overrightarrow{p} > 0$ and p_0 positive, or $\overrightarrow{\sigma}\cdot\overrightarrow{p} < 0$ and $p_0 < 0$ with ψ_L having the opposite correspondence. The quantity $\overrightarrow{\sigma}\cdot\hat{p}$ is called the helicity; for massless or relativistic particles, $\overrightarrow{\sigma}\cdot\overrightarrow{p}/p_0 \simeq \overrightarrow{\sigma}\cdot\hat{p}$. Now the meaning of the labels L and R becomes clear. L represents a left–handed positive energy solution, R a right–handed one. If a particle is massless or relativistic, then for a left–handed state with $p_0 > 0$, $\psi_R \ll \psi_L$.

(d) If $m \neq 0$, the two equations do not separate. In particular, we will see that a mass term in a Lagrangian can be interpreted as an interaction between ψ_L and ψ_R.

It is also conventional to separate the space–time dependence. We put

$$\psi = u e^{i\left(\vec{p}\cdot\vec{x} - p_0 t\right)}.$$ (5.46)

Then u satisfies the same momentum space equations we have been writing, since we were implicitly assuming that we were working with energy eigenstates. In general, we write ψ for a full solution, and u for the solution with the space–time dependence factored off. We will only consider free particles with point–like interactions, so the space–time dependence is always a plane wave and always drops out of the discussion.

Finally, we have to choose a normalization for the solutions. The conventional choice is

$$\overline{u}u = 2m.$$ (5.47)

The most important thing to note is that $\overline{u}u$ has the dimensions of mass, or energy (in the natural units we are using). Whenever we approximate a factor such as $\overline{u}u$ or $\overline{u}\gamma^\mu u$, it must be expressed in terms of the available masses and energies.

5.7   Particles and Antiparticles

The way we want to treat particles and antiparticles is to treat both as normal particles. They come in pairs; whichever is the particle, it comes in nature with an antiparticle whose charges are all opposite (electromagnetic charges and any others). When they are fermions, they are all described by solutions of the Dirac equation. We will generally label all fermion solutions as ψ or u, so vertices will be of the form $\overline{\psi}\gamma^\mu\psi$ or $\overline{u}\gamma^\mu u$ whether the wave functions represent fermions or antifermions. To ensure that we are always writing numbers for transition matrix elements, we use u for initial fermions or final antifermions, and $\overline{u}$ for final fermions or initial antifermions.

By proceeding that way, we gain considerable simplicity, but we lose an appreciation of some of the beauty and necessity of a field theory of particles and antiparticles. For a fully consistent treatment, a second–quantized relativistic field theory is necessary. A treatment, in a spirit similar to ours but maintaining the subtleties, is given in a full chapter by Halzen and Martin. In particular, the negative energy solutions of the free particle equations are successfully reinterpreted as describing positive energy antiparticles in the complete theory. In a Feynman diagram, an incoming (outgoing) particle can be turned into an outgoing (incoming) antiparticle by reversing the sign of its four–momentum and of all its charges (electric, color, flavor, *etc.*); that is consistent with the convention for spinors u and $\overline{u}$.

5.8   Left–handed and Right–handed Fermions

Even when particles are not massless, it is useful to separate the upper and lower parts of the wave function. The possibility of doing that, and the methods to do so, are extremely important in the formulation of the Standard Model. It is conventional here to choose a representation of the γ–matrices where γ^5 is given by (equation (5.40))

$$\gamma^5 = \begin{pmatrix} +1 & 0 \\ 0 & -1 \end{pmatrix}. \tag{5.48}$$

We will use this representation of γ^5 in the rest of the book. We will not need a representation for the other γ matrices.

Consider the operators

$$P_L = \frac{1 - \gamma^5}{2} = \begin{pmatrix} 0 & 0 \\ 0 & 1 \end{pmatrix}, \tag{5.49}$$

$$P_R = \frac{1 + \gamma^5}{2} = \begin{pmatrix} 1 & 0 \\ 0 & 0 \end{pmatrix}. \tag{5.50}$$

They are projection operators, *i.e.*

$$\left. \begin{aligned} P_L^2 &= P_L, \\ P_R^2 &= P_R, \\ P_L + P_R &= 1, \\ P_L P_R &= 0. \end{aligned} \right\} \tag{5.51}$$

For any massive fermion described by a solution u of the Dirac equation, we can define a left–handed projection

$$u_L = P_L u. \tag{5.52}$$

Explicitly,

$$u_L = \begin{pmatrix} 0 & 0 \\ 0 & 1 \end{pmatrix} \begin{pmatrix} U_R \\ U_L \end{pmatrix} = \begin{pmatrix} 0 \\ U_L \end{pmatrix} \tag{5.53}$$

so we can think of u_L as the lower components of ψ. Similarly,

$$u_R = P_R u ,\tag{5.54}$$

and

$$u_R = \begin{pmatrix} 1 & 0 \\ 0 & 0 \end{pmatrix} \begin{pmatrix} U_R \\ U_L \end{pmatrix} = \begin{pmatrix} U_R \\ 0 \end{pmatrix} .\tag{5.55}$$

The helicity of a massive fermion can obviously be changed by a Lorentz transformation, since it is possible to go to the rest frame and rotate, so it is not a quantum number that can be used to label the system. Nevertheless, the Standard Model (and nature) treats left–handed fermions differently from right–handed fermions, and the interplay of left versus right and mass is subtle. To elaborate a little on this point, consider the parity operation. Under parity $\vec{x} \to -\vec{x}$, $t \to t$. Then $\vec{p} \to -\vec{p}$. Since angular momenta transform like $\vec{r} \times \vec{p}$, we expect $\sigma \to \sigma$. Therefore, there is a sign change in equation (5.44), and the two solutions go into each other. Thus, if nature is invariant under the parity operation, we expect both solutions to exist. In fact, there are left–handed neutrinos, but no right–handed neutrinos have been observed.

In the case of electrons, both solutions exist. That is, there are both left–handed and right–handed electrons. But they interact differently: e_L can interact directly with a neutrino, but e_R cannot. This is the subtle means nature uses to violate parity invariance. The separation (equations (5.52) to (5.55)) into f_L and f_R for any fermion f is one of the most important technical points in the structure of the Standard Model.

5.9   Useful Relations

In this section we want to note a few relations that will be of use later. As noted above, a current will have the form $\overline{\psi}\gamma^\mu\psi$. This can be written

$$\begin{aligned}
\overline{\psi}\gamma^\mu\psi &= \overline{\psi}\left(P_L + P_R\right)\gamma^\mu\left(P_L + P_R\right)\psi \\
&= \overline{\psi}P_L\gamma^\mu P_L\psi + \overline{\psi}P_R\gamma^\mu P_L\psi \\
&\quad + \overline{\psi}P_L\gamma^\mu P_R\psi + \overline{\psi}P_R\gamma^\mu P_R\psi .
\end{aligned}\tag{5.56}$$

Using the commutation relation, $P_L\gamma^\mu = \gamma^\mu P_R$ and $P_R\gamma^\mu = \gamma^\mu P_L$, so the first and fourth terms vanish. Noting that

$$\overline{\psi}_L = \left(P_L\psi\right)^\dagger \gamma^0 = \psi^\dagger P_L \gamma^0 = \overline{\psi} P_R \tag{5.57}$$

and

$$\overline{\psi}_R = \overline{\psi} P_L , \tag{5.58}$$

we get an important relation,

$$\overline{\psi}\gamma^\mu \psi = \overline{\psi}_L \gamma^\mu \psi_L + \overline{\psi}_R \gamma^\mu \psi_R. \tag{5.59}$$

This says that the helicity is preserved whenever the interaction is of the form $\overline{\psi}\gamma^\mu\psi$.

On the other hand, we know from Chapter 2 that a mass term in the Lagrangian has the form $m\overline{\psi}\psi$. To express $\overline{\psi}\psi$ in terms of left–handed and right–handed states, we have,

$$\begin{aligned}
\overline{\psi}\psi &= \overline{\psi}\left(P_L{}^2 + P_R{}^2\right)\psi \\
&= \left(\overline{\psi}P_L P_L \psi + \overline{\psi}P_R P_R \psi\right) \\
&= \left(\overline{\psi}_R\psi_L + \overline{\psi}_L\psi_R\right) .
\end{aligned} \tag{5.60}$$

Thus a mass term is equivalent to a helicity flip, and conversely.

The parity–conserving electromagnetic interaction of equation (5.59) has both LL and RR terms, *i.e.* they are equally probable. If for some reason only the $\overline{\psi}_L\gamma^\mu\psi_L$ term could occur, we would have

$$\begin{aligned}
\overline{\psi}_L\gamma^\mu\psi_L &= \frac{1}{4}\overline{\psi}\left(1+\gamma^5\right)\gamma^\mu\left(1-\gamma^5\right)\psi \\
&= \frac{1}{2}\overline{\psi}\gamma^\mu\left(1-\gamma^5\right)\psi
\end{aligned} \tag{5.61}$$

which has two pieces, one transforming as a normal four–vector and one as an axial vector; this is called a $V-A$ interaction, or a left–handed interaction or current. If only $\overline{\psi}_R\gamma^\mu\psi_R$ were to occur, it would be a right–handed current, $\frac{1}{2}\overline{\psi}\gamma^\mu\left(1+\gamma^5\right)\psi$. Because of equation (5.57) and the lines above it, only the bottom component of ψ' in any current $\overline{\psi}'\gamma^\mu\psi_L$ can interact. Even if the top component of ψ' corresponds to a physical state, it will not undergo any interaction with the state represented by ψ_L.

5.10   The Dirac Lagrangian

The Lagrangian of a spin–1/2 fermion is important for us, since it describes the quarks and leptons. As with other free particle Lagrangians, it is written because it gives the right answer for the equations of motion, rather than being derived. The appropriate Lagrangian is

$$\mathcal{L} = \overline{\psi}\left(i\gamma^{\mu}\partial_{\mu} - m\right)\psi \ . \tag{5.62}$$

Let us give several arguments as to why this choice is standard. First, for those who are familiar with the Euler–Lagrange equations or who have looked at their derivation in Appendix F, equation (5.62) immediately yields the Dirac equation because $\mathcal{L}$ does not depend on $\partial^{\mu}\overline{\psi}$ so $\partial\mathcal{L}/\partial\overline{\psi} = 0$. A longer but perhaps conceptually simpler approach is to look at the nonrelativistic limit of equation (5.62). Using the solutions of Section 5.6, it can be seen that the dominant term in that limit for $\overline{\psi}\gamma^{\mu}\psi$ is $\overline{\psi}\gamma^{0}\psi \sim p^{0}$, and $\mathcal{L}$ becomes just the kinetic energy, as it should for a free particle. Still another argument is that two terms of the form $\overline{\psi}\psi$ and $\overline{\psi}\gamma^{\mu}\partial_{\mu}\psi$ are the only parity invariant Lorentz scalars that one can write; their coefficients can be fixed by the nonrelativistic limit.

In any case, it is in equation (5.62) that we will put $\partial_{\mu} \rightarrow \mathcal{D}_{\mu}$ in order to obtain the interaction Lagrangian implied by gauge invariance.

Problems

5.1: Calculate P_{L}^{2}, P_{R}^{2}, $P_{L}P_{R}$, $P_{L} + P_{R}$ explicitly.

5.2: Show for a general solution of the Dirac equation that

$$\overline{u}_{L}\gamma^{\mu}u_{R} = \overline{u}_{R}\gamma^{\mu}u_{L} = 0$$

and

$$\overline{u}_{L}\gamma^{\mu}u_{L} \neq 0.$$

5.3: Practice working with γ matrices. Verify explicitly that

$$\gamma^{\mu}\gamma^{\nu} + \gamma^{\nu}\gamma^{\mu} = 2g^{\mu\nu}$$

for $\nu, \mu = 1$, 2 and for $\mu = 2$, $\nu = 1$.

5.4:   For a solution of the Dirac equation $\psi = \begin{pmatrix} \psi_R \\ \psi_L \end{pmatrix}$ write out $\overline{\psi}\psi$ and $\overline{\psi}\gamma^5\psi$ in terms of two-component spinors and matrices. Write ψ_R in terms of ψ_L as in equation (5.45). Compare the behavior of $\overline{\psi}\psi$ and of $\overline{\psi}\gamma^5\psi$ under parity (note the discussions at the end of Sections 5.8 and 5.4).

5.5:   In the Standard Model, as we will see in Chapter 7, the left–handed and right–handed electrons interact differently. It is easy to imagine them having different masses, since interactions in general lead to some mass. Why can we be confident that the theory will indeed give $m_{e_L} = m_{e_R}$ as observed experimentally?

Suggestions for Further Study

Many books have discussions and summaries of Dirac equation physics. A treatment in the same spirit as ours is given in Chapter 5 of Halzen and Martin. The most complete treatment is probably that of Sakurai (1967). Normally getting signs, phases, and factors of two right is a great problem for the student learning to work with the Dirac equation formalism. For us it is not a problem, since our method of calculating results approximately bypasses essentially all of the difficulties. Thus, while the reader is encouraged to pursue these matters further, a good semi-quantitative grasp of the Standard Model does not require more than is presented in this chapter.

CHAPTER 6

The
Standard Model
Lagrangian

Now that we have learned (1) the notation of the Dirac equation in order to express the spin structure, (2) the requirements of gauge invariance that tell us to begin with a free particle Lagrangian and rewrite it with a covariant derivative, and (3) the idea of internal symmetries, we are finally ready to write down the full Lagrangian. In order to describe the particles and interactions known today, three internal symmetries are needed. We do not know yet why there are three, or whether there will be more, or why these three are the ones they are, but it is a remarkable accomplishment to have discovered them. At the present time all experiments are consistent with the notion that the three symmetries are necessary and sufficient to describe the interactions of the known particles. It is easiest to describe how these symmetries act in the language of group theory, so any reader who needs to review that way of describing invariances should turn to Appendix B before proceeding.

All particles appear to have a $U(1)$ invariance. It is like the $U(1)$ invariance or phase invariance described in Sections 2.3, 3.2, and 3.3. That invariance was related to the electromagnetic interaction. However, since the invariance is an internal property of particles, we have no reason to immediately identify it with electromagnetism. We simply have a $U(1)$ or phase invariance whose connection with electromagnetism will be deduced later from physical arguments. The gauge boson required by the invariance of the theory under the $U(1)$ transformations will be called B^μ. The index μ is present since B^μ must transform under spatial rotations the same way the ordinary derivative ∂^μ does, thus guaranteeing the associated particle has spin one. We will reserve A^μ for the name of the photon field. The connection of A^μ and B^μ will be determined in Section 7.3.

All particles appear to have a second internal invariance, under a set of transformations that form an $SU(2)$ group, called the electroweak $SU(2)$ invariance. These then lead to a non-Abelian gauge (phase) invariance, analogous to the strong isospin invariance of Section 4.1. The associated (gauge) bosons necessary to maintain the invariance of the theory are called W_i^μ. The index μ again is required to have space-time transformations that are the same as the ordinary derivative, so the W bosons have spin one. There is one boson for each of the three generators of $SU(2)$ transformations so $i = 1$, 2, or 3. By analogy with Section 4.1, these are called "weak isospin" transformations. Whenever we use indices i, j, k, and l, they will mean weak isospin transformations and each of them can take on values 1, 2, or 3. Just as for the pions of Chapter 4, the physical W particles will have definite electromagnetic charges

$$
\begin{aligned}
W^+ &= \left(-W^1 + iW^2\right)/\sqrt{2} \\
W^- &= \left(-W^1 - iW^2\right)/\sqrt{2} \\
W^0 &= W^3.
\end{aligned}
\tag{6.1}
$$

[Note that we must now distinguish $U(1)$ charge, $SU(2)$ charge, electromagnetic charge, and, soon, $SU(3)$ charge. Nature has given every particle a number of "charge" labels that, taken together, fully describe its (non-space–time) properties.]

All particles appear to have a third internal invariance, under a set of transformations that form an $SU(3)$ group, giving an additional, independent non-Abelian invariance. The associated gauge bosons are labeled G_a^μ, where now $a = 1, 2, \ldots, 8$ since there is one spin–one boson for each of the eight generators of $SU(3)$. The bosons are called gluons, and the theory of particle interactions via gluon exchange is called Quantum Chromodynamics (QCD). The internal charge each particle carries that determines how it interacts with the gluons is called a color charge or QCD charge, and the associated force, the color force. These charges and forces have nothing to do with everyday color, of course, but the name is used because some of the properties of color interactions are analogous to some of the properties of everyday color—particularly that ordinary particles like protons or neutrons are made of three particles (quarks) each of which carries a different color charge, but the charges combine in such a way that the proton and neutron have no color charge (they are color neutral). This is reminiscent of white light being made from the three primary colors.

We will meet more instances where an everyday word (such as color) is used with an entirely different—and entirely precise—meaning in particle physics. As the domains of nature that we study are further removed from our everyday experience, more and more phenomena are encountered that have no familiar counterparts. We have to give them names in order to talk about them. To help make these distant phenomena as human as possible, they are often given familiar names suggestive of the property in question (color, charm, flavor, asymptotic freedom, *etc.*).

6.1   Labeling the Quark and Lepton States

The full Lagrangian, according to our approach, should arise by taking the free particle Lagrangian and replacing the ordinary derivative by the covariant derivative. It will have a part, $\mathcal{L}_{\text{gauge}}$, for the kinetic energies of the gauge fields. The main new physics arises from the terms generated when the covariant derivative is inserted in the quark and lepton kinetic energies; we call this fermion part of the Lagrangian $\mathcal{L}_{\text{ferm}}$.

In order to write the Lagrangian in a compact, easily readable, way, we need to define some notation. Since the notation has to contain information about how each particle transforms under the three internal symmetries, in addition to its space–time properties, a lot of notation is required.

The way the particles behave under the electroweak $SU(2)$ transformation is familiar because spin also transforms under a (different) $SU(2)$ group. Particles with spin-zero are singlets, particles with spin-1/2 are put in doublets $\left(\begin{smallmatrix}\uparrow\\\downarrow\end{smallmatrix}\right)$, particles with spin-one form triplets with $J_z = 1, 0, -1$. For the electroweak $SU(2)$ we have already seen an example of a triplet, the W^i; the states with electric charge $1, 0, -1$ are equivalent to the spin states with $J_z = 1, 0, -1$. Sometimes it is useful to let the three states of W form a vector in the electroweak space, $\overrightarrow{W}$. We have already studied the strong isospin example of an internal $SU(2)$ space for the neutron and proton and pions.

How the observed states transform in the electroweak $SU(2)$ space is an experimental question so far. Someday there may be a theory which tells us, but at present it must be determined by measurement in each case. All known quarks and leptons are observed to be either electroweak $SU(2)$ singlets or parts of electroweak doublets. Later on we will see how to distinguish them experimentally.

The way particles are assigned to electroweak $SU(2)$ states is subtle. Consider the electron state, described by a spinor ψ_e. Separate the left–handed and right–handed components as we have learned to do, defining

$$e^-_R = P_R \psi_{e^-}$$
$$e^-_L = P_L \psi_{e^-} \tag{6.2}$$

where P_R, P_L are the projection operators defined in Chapter 5. Similar separations of left–handed and right–handed spin states are made for every fermion. The electric charge is given as a superscript.

The remarkable thing is that the left–handed and right–handed states transform differently under the electroweak $SU(2)$. Right–handed electrons are electroweak singlets, while left–handed electrons are in electroweak doublets; their partners are left–handed neutrinos. Thus

$$e^-_R = SU(2) \text{ singlet}, \tag{6.3}$$

and define

$$L = \begin{pmatrix} \nu_e \\ e^- \end{pmatrix}_L \tag{6.4}$$

as an $SU(2)$ doublet. When L points "up" in electroweak $SU(2)$ space, it represents ν_{eL}, while when L points "down" in electroweak $SU(2)$ space it represents e^-_L. Rotations in electroweak $SU(2)$ space turn $\nu_{eL} \leftrightarrow e^-_L$, just as rotations in spin space turn spin–up into spin–down, or rotations in the strong isospin space turn neutrons into protons. In spin space the angular momentum raising and lowering operators, which transform as the components of a vector, connect the spin–up and spin–down states. Similarly, in strong isospin space the pions connect neutrons and protons, and in electroweak space we will find that the W bosons connect the members of an electroweak doublet, ν_{eL} and e^-_L. Since e^-_R is a singlet it is not connected to any other state by electroweak transitions, just as a state of spin zero has only one spin state. The notation is standard but a little confusing. The subscript L means left–handed, while the L on the left-hand side of equation (6.4) stands for lepton (doublet).

When we need an index to label the components of L we will use p, q, or r, where each index can take on values 1 or 2, so L_p can represent $L_1 = \nu_{eL}$ or $L_2 = e^-_L$.

The up and down quarks behave in an analogous way. Define

$$Q_{L\alpha} = \begin{pmatrix} u_\alpha \\ d_\alpha \end{pmatrix}_L. \qquad (6.5)$$

We put the left–handed quarks in electroweak $SU(2)$ doublets. The right–handed ones are again singlets,

$$d_{R\alpha}, \ u_{R\alpha}. \qquad (6.6)$$

An additional index, α, is needed to describe how the quarks transform in the color $SU(3)$ space. Just as the basic $SU(2)$ representation is the doublet with two components, the basic $SU(3)$ state is the triplet with three components, so we use the indices α, β, and γ which are equal to 1, 2 or 3 when we need to label the color components. Sometimes for ease of discussion we will refer to color indicies, α, β, and γ, as r, g or b. If a particular color is "↑" in some direction, the combination analogous to ↑↓ (properly symmetrized) is colorless (just as a spin singlet can be constructed) so we speak of color (*e.g.* r) and anticolor ($\bar{r}$) and color singlets ($r\bar{r} + g\bar{g} + b\bar{b}$). All leptons are color singlets so we have not written a color index for them. All quarks are color triplets.

The gluons are the objects which generate the transitions from one quark color to another. The space–time properties of a gluon are like those of a photon, but gluons also carry the color–charge and thus they can change it. Electrically charged particles can change their momentum by emitting or absorbing a photon, but they cannot change their electric charge that way. Colored particles (quarks or gluons) can change both their momentum and their color charge by emitting or absorbing a gluon.

Since gluons can connect any of the color charges r, g, or b to any other, there would appear to be nine gluons required. However, the combination $r\bar{r} + g\bar{g} + b\bar{b}$ is invariant under rotations in the color space (*i.e.* it is "colorless") so in fact there are eight independent color-charge states for gluons; normally it is said that there are eight gluons.

We will close this section with a few remarks. Note that we did not mention a right–handed neutrino when we discussed the leptons, but we did include right–handed quarks u_R and d_R. Present experimental evidence is consistent with that distinction; as we will see later, it may be related to the apparent situation that the neutrinos have masses consistent with zero, while other fermions seem to be massive. If right–handed neutrinos exist, either they are very heavy or they do not interact enough to be produced and detected so far.

Note also that the left–handed and right–handed fermions were put in different $SU(2)$ multiplets (doublet versus singlet). That is of course a violation of parity, since clearly the theory is not invariant under reversal of the component of spin in the direction of motion. As we will see, the known parity violation of weak interactions emerges from this input. Thus the Standard Model theory beautifully describes the parity violation seen in nature, but does not explain it in a fundamental sense. The gain is significant, because the electroweak theory can incorporate parity violation in a natural way, but a deeper understanding is still desirable.

Finally, we have only considered one family of fermions (ν_e, e, u, d). Remarkably, there is good evidence that the theory simply replicates itself for the two other (known) families (ν_μ, μ, c, s) and (ν_τ, τ, t, b). All the elaboration of the structure of the theory that we will carry out in the next section seems to apply to each family; the only distinctions are that the measured masses are to be used in doing kinematics and phase space. It is also remarkable that the known universe is made entirely of the first family. All the heavier particles of the other two families are created at accelerators or in cosmic ray collisions but have short lifetimes ($\stackrel{<}{\sim} 10^{-6}$ sec), decaying quickly into the first family particles. [The ν_μ and ν_τ may be massless or stable, in which case they do not decay. They interact sufficiently weakly that they have no significant effects except perhaps cosmologically.] No reason has yet been found for the existence of three families of particles whose quantum numbers (except for mass) and interactions are identical. We will develop the theory in terms of the first family, returning later on to some of the tests of the theory for the second and third families.

6.2   The Quark and Lepton Lagrangian

With the techniques we have developed, we can quickly incorporate all of the information of the last section into the Lagrangian. For fermions we expect to form the Lagrangian by taking the normal Dirac kinetic energy Lagrangian, and replacing ∂_μ by the covariant derivative $\mathcal{D}_\mu$,

$$\overline{\psi}\gamma^\mu \partial_\mu \psi \rightarrow \overline{\psi}\gamma^\mu \mathcal{D}_\mu \psi , \qquad (6.7)$$

for each fermion. We can write a sum over fermions as described in the last section,

$$f = L,\ e_R,\ Q_L,\ u_R,\ d_R . \qquad (6.8)$$

We expect $\mathcal{D}_\mu$ to have a term for each local gauge symmetry of the theory, so

$$\mathcal{D}_\mu = \partial_\mu - ig_1 \frac{Y}{2} B_\mu - ig_2 \frac{\tau^i}{2} W_\mu^i - ig_3 \frac{\lambda^a}{2} G_\mu^a \,. \tag{6.9}$$

The second term represents the $U(1)$ symmetry. B_μ is the spin–one field needed to maintain gauge invariance, as in Sections 3.2 and 3.3. Y is the generator of $U(1)$ transformations, a constant, but perhaps different for different fermions—we will see below how to determine its value. Since the *form* of the term needed in $\mathcal{D}_\mu$ is determined by gauge invariance, but not its strength, a "coupling strength" g_1 is introduced; its value has to be measured experimentally, as we will discuss below.

Analogous remarks describe the $SU(2)$ and $SU(3)$ terms. They have the spin–one fields needed to maintain the gauge invariance, the three W_μ^i for $SU(2)$ and the eight G_μ^a for $SU(3)$, one for each generator of transformations. They enter in the form $\tau^i W^i$ and $\lambda^a G^a$. Each term has a coupling strength, g_2 or g_3, which has to be measured. As usual, repeated indices are summed, so $\tau^i W_\mu^i = \tau^1 W_\mu^1 + \tau^2 W_\mu^2 + \tau^3 W_\mu^3$, and a is summed from 1 to 8.

In order to be able to write the Lagrangian in a compact form, we introduce one convention: whenever the terms in $\mathcal{D}_\mu$ act on a fermion state of different matrix form, they give zero, by definition. Thus $\tau^i W^i$ is a 2×2 matrix in $SU(2)$ space so it gives zero acting on e_R, u_R, d_R. Similarly, $\lambda^a G^a$ is a 3×3 matrix in color space, so it gives zero acting on the leptons (L, e_R) but is meaningful acting on the quarks. Finally then, keeping equations (6.8) and (6.9) in mind, we can write the full Lagrangian of the fermions,

$$\mathcal{L}_{\text{ferm}} = \sum_{\substack{f = L,\, e_R, \\ Q_L,\, u_R,\, d_R}} \overline{f} i\gamma^\mu \mathcal{D}_\mu f \,. \tag{6.10}$$

Starting from this, whose form we have motivated in the previous chapters, we will see in the following how to calculate quark and lepton interactions. All presently known experimental information on quark and lepton interactions is consistent with the predictions from $\mathcal{L}_{\text{ferm}}$.

6.3   Gauging the Global Symmetries

It may be helpful to look from a slightly different perspective at the $SU(2)$ and $U(1)$ symmetries, a perspective rather close to the historical one and to the ways models are approached. Consider the Dirac kinetic energy Lagrangian for the first generation, with $\nu = \nu_e$,

$$\mathcal{L} = i\bar{e}_R\gamma^\mu\partial_\mu e_R + i\bar{e}_L\gamma^\mu\partial_\mu e_L + i\bar{\nu}_L\gamma^\mu\partial_\mu\nu_L\,. \tag{6.11}$$

What internal symmetries are present? Since the internal transformations should commute with space–time ones, we could put e_L and ν_L in a doublet L as in equation (6.4). The right–handed e_R must be separate so it is a singlet of this symmetry. Then $\mathcal{L}$ is invariant under the global transformations

$$\begin{aligned} L &\to e^{i\tau\cdot\theta/2}L \\ e_R &\to e_R \end{aligned} \tag{6.12}$$

which are $SU(2)$ transformations. There is another symmetry. We can change the fields by a constant phase without changing $\mathcal{L}$, so

$$\begin{aligned} L &\to e^{i\beta}L \\ e_R &\to e^{i\beta'}e_R\,. \end{aligned} \tag{6.13}$$

We cannot change e_L and ν separately without breaking the $SU(2)$ symmetry. This is a $U(1)$ global symmetry.

Next we can "gauge" the theory. That is, we make these local symmetries by introducing potentials W_i^μ for $SU(2)$ and B^μ for $U(1)$, and replace ∂^μ by the covariant derivative $\mathcal{D}^\mu$. Thus, we obtain equations (6.9) and (6.10) by a more deductive approach. Some attempts to extend the Standard Model proceed along these lines, by adding particles and symmetries or relating particles by symmetries, and then gauging the symmetries.

Problems

6.1: If $\mathcal{D}^\mu$ acts on right–handed leptons, show that its form is

$$\mathcal{D}^\mu = \partial^\mu - ig_1 B^\mu.$$

What is it when $\mathcal{D}^\mu$ acts on left–handed leptons?

CHAPTER 7

The
Electroweak Theory
and
Quantum
Chromodynamics

The Lagrangian of equation (6.10) contains a great deal of information. To see the connection to experiment, we will study it piece by piece. The ∂_μ is always present; to simplify the equations we will not write it. We can do all of the $SU(2)$ calculations just with the leptons. Since the color labels of the quarks do not operate in the $U(1)$ or $SU(2)$ space, quarks will behave the same way as leptons for $U(1)$ and $SU(2)$ interactions. At the end of the chapter, we will explicitly write the quark terms.

7.1   The $U(1)$ terms

Specializing to the $U(1)$ terms for the first family leptons, we have

$$-\mathcal{L}_{\text{ferm}}(U(1),\text{leptons}) = \overline{L}i\gamma^\mu \left(ig_1\frac{Y_L}{2}B_\mu \right) L + \overline{e}_R i\gamma^\mu \left(ig_1\frac{Y_R}{2}B_\mu \right) e_R. \quad (7.1)$$

Since Y could have separate values for different fermions, separate labels Y_L, Y_R have been introduced. In $SU(2)$ space L is a doublet, while $g_1 Y B_\mu$ is just a number, so

$$\overline{L}\gamma^\mu L = \overline{\nu}_L\gamma^\mu \nu_L + \overline{e}_L\gamma^\mu e_L. \quad (7.2)$$

Then

$$\mathcal{L}_{\text{ferm}}(U(1),\text{leptons}) = \frac{g_1}{2}\left[Y_L\left(\overline{\nu}_L\gamma^\mu \nu_L + \overline{e}_L\gamma^\mu e_L \right) + Y_R\overline{e}_R\gamma^\mu e_R \right] B_\mu. \quad (7.3)$$

81

Before we interpret this, we need the $SU(2)$ part as well since it will contain terms involving the same particles.

7.2   The $SU(2)$ Terms

Since $\tau^i W^i$ is a 2×2 matrix, the only non–zero lepton term is

$$
\begin{aligned}
-\mathcal{L}_{\text{ferm}}(SU(2), \text{ leptons}) &= \overline{L} i \gamma^\mu \left[i g_2 \frac{\tau^i}{2} W^i_\mu \right] L \\
&= -\frac{g_2}{2} \begin{pmatrix} \overline{\nu}_L & \overline{e}_L \end{pmatrix} \gamma^\mu \begin{pmatrix} W^3_\mu & W^1_\mu - i W^2_\mu \\ W^1_\mu + i W^2_\mu & -W^3_\mu \end{pmatrix} \begin{pmatrix} \nu_L \\ e_L \end{pmatrix} \\
&= -\frac{g_2}{2} \begin{pmatrix} \overline{\nu}_L & \overline{e}_L \end{pmatrix} \gamma^\mu \begin{pmatrix} W^0_\mu & -\sqrt{2} W^+_\mu \\ -\sqrt{2} W^-_\mu & -W^0_\mu \end{pmatrix} \begin{pmatrix} \nu_L \\ e_L \end{pmatrix} \\
&= -\frac{g_2}{2} \begin{pmatrix} \overline{\nu}_L & \overline{e}_L \end{pmatrix} \gamma^\mu \begin{pmatrix} W^0_\mu \nu_L - \sqrt{2} W^+_\mu e_L \\ -\sqrt{2} W^-_\mu \nu_L - W^0_\mu e_L \end{pmatrix} \\
&= -\frac{g_2}{2} \left[\overline{\nu}_L \gamma^\mu \nu_L W^0_\mu - \sqrt{2} \overline{\nu}_L \gamma^\mu e_L W^+_\mu - \sqrt{2} \overline{e}_L \gamma^\mu \nu_L W^-_\mu - \overline{e}_L \gamma^\mu e_L W^0_\mu \right].
\end{aligned}
$$

$$\text{(7.4)}$$

In converting the W^i_μ into charge states W^+_μ, W^0_μ, W^-_μ, we have used the same $SU(2)$ conventions as for the pions in Chapter 4. The step from $\tau^i W^i$ to the 2×2 matrix is the same as the analogous one in Section 4.1. The seven terms in (7.3) and (7.4) are the full lepton content of the fermion Lagrangian. As far as is known, equations (7.3) and (7.4) describe all interactions of leptons (except for the Higgs physics of Chapter 8 and Chapter 21).

7.3   Connection to Experimental Facts;
The Neutral Current

Since we already know something about how electrons and neutrinos interact, the first test of the approach we are pursuing is to be sure that equations (7.3) and (7.4) are consistent with experiment. We saw in Chapter 5 that the electromagnetic interaction of particles of charge Q was

$$
\mathcal{L}_{\text{EM}} = Q A_\mu \left[\overline{e}_L \gamma^\mu e_L + \overline{e}_R \gamma^\mu e_R \right] . \tag{7.5}
$$

In equations (7.3) and (7.4), there are terms of this form, so we have to determine whether they combine properly. However, before doing that we see that there are similar terms involving $\bar{\nu}_L \, \nu_L$ and we know that the (neutral) neutrino does not have an electromagnetic interaction. The relevant terms are

$$\left(-\frac{g_1}{2} Y_L B_\mu - \frac{g_2}{2} W_\mu^0 \right) \bar{\nu}_L \, \gamma^\mu \nu_L \, . \qquad (7.6)$$

So in order to avoid putting $g_1 = g_2 = 0$, which would mean the whole approach is worthless, we can argue that we should assume the electromagnetic field A_μ is a combination of B_μ and of W_μ^0 , and that it is orthogonal to the combination in (7.6). Thus we try defining

$$A_\mu \propto g_2 B_\mu - g_1 Y_L W_\mu^0 \, . \qquad (7.7)$$

If B_μ and W_μ^0 are orthogonal, normalized fields, then the coefficient of $\bar{\nu}_L \gamma^\mu \nu_L$, which we call Z_μ ,

$$Z_\mu \propto g_1 Y_L B_\mu + g_2 W_\mu^0 \, , \qquad (7.8)$$

is indeed orthogonal to A_μ , so the neutrino can have no electromagnetic interaction. We can normalize A_μ and Z_μ ,

$$A_\mu = \frac{g_2 B_\mu - g_1 Y_L W_\mu^0}{\sqrt{g_2^2 + g_1^2 Y_L^2}} \, , \qquad (7.9)$$

$$Z_\mu = \frac{g_1 Y_L B_\mu + g_2 W_\mu^0}{\sqrt{g_2^2 + g_1^2 Y_L^2}} \, , \qquad (7.10)$$

so if W_μ^i and B_μ were normalized to unity, so are A_μ and Z_μ .

Now that it has been possible to combine the neutrino terms in a consistent way, let us return to the electron. From equations (7.3) and (7.4), we have for the electrons the terms,

$$\bar{e}_L \gamma^\mu e_L \left[-\frac{g_1}{2} Y_L B_\mu + \frac{g_2}{2} W_\mu^0 \right] + \bar{e}_R \gamma^\mu e_R \left[-\frac{g_1}{2} Y_R B_\mu \right] . \qquad (7.11)$$

Next solve equations (7.9) and (7.10) for B_μ, W_μ^0 ,

$$B_\mu = \frac{g_2 A_\mu + g_1 Y_L Z_\mu}{\sqrt{g_2^2 + g_1^2 Y_L^2}} \ , \tag{7.12}$$

$$W_\mu^0 = \frac{-g_1 Y_L A_\mu + g_2 Z_\mu}{\sqrt{g_2^2 + g_1^2 Y_L^2}} \ , \tag{7.13}$$

and substitute these in (7.11), since we want to express the electron interaction in terms of the electromagnetic field A_μ (and necessarily, therefore, the new field Z_μ as well).

Then (7.11) becomes

$$
\begin{aligned}
- A_\mu &\left\{ \bar{e}_L \gamma^\mu e_L \left[\frac{g_1 g_2 Y_L}{\sqrt{g_2^2 + g_1^2 Y_L^2}} \right] + \bar{e}_R \gamma^\mu e_R \left[\frac{g_1 g_2 Y_R}{2\sqrt{g_2^2 + g_1^2 Y_L^2}} \right] \right\} \\
- Z_\mu &\left\{ \bar{e}_L \gamma^\mu e_L \left[\frac{g_1^2 Y_L^2 - g_2^2}{2\sqrt{g_2^2 + g_1^2 Y_L^2}} \right] + \bar{e}_R \gamma^\mu e_R \left[\frac{g_1^2 Y_R Y_L}{2\sqrt{g_2^2 + g_1^2 Y_L^2}} \right] \right\} \ .
\end{aligned}
$$

$$\tag{7.14}$$

If this has any chance to be meaningful, the A_μ term must be the usual electromagnetic current as in equation (7.5). The Z_μ terms can be an additional interaction, which must then be checked against experiment. For the A_μ terms, comparing (7.5) and (7.14) says we must have ($Q = -e$, e being a positive number)

$$-e = \frac{g_1 g_2 Y_L}{\sqrt{g_2^2 + g_1^2 Y_L^2}} \tag{7.15}$$

and

$$-e = \frac{g_1 g_2 Y_R}{2\sqrt{g_2^2 + g_1^2 Y_L^2}} \ . \tag{7.16}$$

These fix

$$Y_R = 2Y_L$$

$$Y_L = -e\frac{\sqrt{g_2^2 + g_1^2 Y_L^2}}{g_1 g_2} \ , \tag{7.17}$$

and we also notice that only the combination $g_1 Y_L$ occurs, so we can choose (for convenience) $Y_L = -1$, since a redefinition of g_1 can always absorb any change in Y_L. Putting $Y_L = -1$ gives

$$e = \frac{g_1 g_2}{\sqrt{g_2^2 + g_1^2}} \ . \tag{7.18}$$

So we see that the theory we have been writing can indeed be interpreted to contain the usual electromagnetic interaction (7.5) for electrons and neutrinos (none), plus an additional so-called "neutral current" interaction with Z_μ for both electrons and neutrinos. Before we study the neutral current interactions, we use the suggestive form of (7.18) to define

$$\sin \theta_w = \frac{g_1}{\sqrt{g_2^2 + g_1^2}}$$

$$\cos \theta_w = \frac{g_2}{\sqrt{g_2^2 + g_1^2}} \ . \tag{7.19}$$

Solving,

$$g_2 = \frac{e}{\sin \theta_w} \ ,$$

$$g_1 = \frac{e}{\cos \theta_w} \ , \tag{7.20}$$

so now g_1 and g_2 have been written in terms of the known e ($e^2/4\pi \approx 1/137$ in our natural units), and an angle θ_w, called the electroweak mixing angle, which has to be measured or calculated some other way. All the results we have written are valid for any value of θ_w. Later we will discuss several ways to measure θ_w; its value is approximately given by $\sin^2 \theta_w \approx 0.23$, and is known to about 5%.

Next let us examine the new couplings we found, for both the neutrino and the electron, to the new field Z_μ. From (7.3) and (7.4), using (7.10) and the expression just above for Y_L and the couplings, the neutrino term gives

$$-\frac{\sqrt{g_2^2+g_1^2}}{2}Z_\mu\bar\nu_L\gamma^\mu\nu_L = -\frac{g_2}{2\cos\theta_w}Z_\mu\bar\nu_L\gamma^\mu\nu_L \qquad (7.21)$$

where (7.19) is used for the last step. Note the useful identity:

$$\sqrt{g_2^2+g_1^2} = \left[\frac{e^2}{\cos^2\theta_w}+\frac{e^2}{\sin^2\theta_w}\right]^{1/2}$$

$$= \left[\frac{e^2}{\cos^2\theta_w\sin^2\theta_w}\right]^{1/2}$$

$$= \frac{e}{\cos\theta_w\sin\theta_w}\;. \qquad (7.22)$$

Thus whenever there is a ν_L–Z vertex we can associate a strength factor $g_2/2\cos\theta_w$. We can think of this as the electroweak "charge" of the left–handed neutrino (since no right–handed neutrinos are known, often one leaves off "left–handed" here).

Now consider the interaction of electrons with Z. There is no reason to expect e_L and e_R to have the same interaction since they have been treated differently in the way the theory was constructed. Equation (7.14) gives

$$-Z_\mu\left\{\bar e_L\gamma^\mu e_L\left[\frac{g_1^2-g_2^2}{2\sqrt{g_2^2+g_1^2}}\right]+\bar e_R\gamma^\mu e_R\left[\frac{-g_1^2}{\sqrt{g_2^2+g_1^2}}\right]\right\}\;.$$

For the first square bracket, using (7.19) and various identities,

$$\frac{g_1^2-g_2^2}{2\sqrt{g_2^2+g_1^2}} = \frac{e^2}{2\sqrt{g_2^2+g_1^2}}\left(\frac{1}{\cos^2\theta_w}-\frac{1}{\sin^2\theta_w}\right)$$

$$= \frac{e}{\cos\theta_w\sin\theta_w}\left(-\frac{1}{2}+\sin^2\theta_w\right), \qquad (7.23)$$

while for the second,

$$\frac{-g_1^2}{\sqrt{g_2^2 + g_1^2}} = -\frac{e^2}{\cos^2 \theta_w} \frac{\cos \theta_w \sin \theta_w}{e}$$

$$= \frac{e}{\cos \theta_w \sin \theta_w} \left(-\sin^2 \theta_w \right). \tag{7.24}$$

These last two results have been written in a convenient form to help one notice that they can both be written as

$$\frac{e}{\sin \theta_w \cos \theta_w} \left(T_3^f - Q_f \sin^2 \theta_w \right). \tag{7.25}$$

In (7.25), T_3^f is the eigenvalue of T_3 (the diagonal $SU(2)$ generator analogous to J_z) for any fermion f. Formally, $T_i = \tau_i/2$ for a left–handed doublet, and we have switched to T to emphasize that particle labels are attached. If f is a singlet (e_R, u_R, d_R, etc.) then $T_3^f = 0$. If f is the upper member of a doublet (ν_L, u_L, etc.) then $T_3^f = +1/2$, while if f is the lower member of a doublet (e_L, d_L, etc.), $T_3^f = -1/2$. Q_f is the electric charge of the fermion ($Q_e = -1$, $Q_\nu = 0$, $Q_u = +2/3$, $Q_d = -1/3$) in units of e. Thus (7.25) gives the "electroweak charge" of any fermion, $i.e.$ the strength of the coupling to the Z. From (7.20) and (7.21), we can check that (7.25) applies to the neutrino. We have not explicitly done the derivation for the quarks, but the form of this section would have been exactly the same if we had replaced ν_L by u_L, e_L by d_L, e_R by d_R. Adding u_R (which does not have an analogue in ν_R) would add a term that had the same form as the d_R term. So we do not need to carry out the derivation for quarks explicitly; the result is already present in (7.25).

Let us stop and see what has been accomplished. It has been possible to interpret the electroweak theory so that it contains the ordinary electromagnetic interaction, plus an additional photon–like particle Z (called the Z–boson), that interacts with any fermion having electric charge (the Q_f term in (7.25)) or weak isospin (the T_3^f term in (7.25)) different from zero. The strength of (7.25) is not small—in fact, since $1/\sin \theta_w \cos \theta_w$ is larger than unity, the Z interaction is stronger than the photon interaction! If this theory makes any sense, why were the Z and the associated interactions not discovered long ago?

The new interactions are called "neutral current" interactions, since they are analogous to the charged current weak interactions (we will encounter these in our language in the next section). There must indeed be a new boson, Z, like the photon, if all this can be correct. If it were massless like the photon, it would have been discovered if it existed, so the only way to avoid a contradiction is to assume the Z is massive. Since the mass of Z makes it harder to produce directly (more energy is required), and since the mass occurs in the denominator whenever Z is exchanged (see Section 2.7) in a neutral current interaction, the size of the expected neutral current effects decreases as the mass of Z increases.

The role of mass in the theory will be considered in detail below. We will see that it is possible to give mass to the Z in a consistent way, though how to do that was such a difficult problem that it held up development of the theory for a number of years. Indeed, the mass can be predicted, and the Z was discovered in 1983 at precisely the expected mass. The neutral current effects due to interactions of neutrinos, electrons, quarks, and muons were found in the early 1970's, and provided one of the earliest direct confirmations of the approach we are describing.

7.4   Connection to Experimental Facts; the Charged Current

There are still terms in the lepton Lagrangian that have to be interpreted. The $U(1)$ part of the Lagrangian gave only terms diagonal in the fermions ($i.e.$ $\nu_L \to \nu_L$, $e_L \to e_L$, $e_R \to e_R$), which we have called neutral current transitions, and we have considered these and all the diagonal terms in the $SU(2)$ part of $\mathcal{L}_{ferm}$. From the $SU(2)$ part of equation (7.4), there is also an off–diagonal part, leading to transitions $\nu_L \leftrightarrow e_L$,

$$\mathcal{L}_{\text{ferm}} = \frac{g_2}{\sqrt{2}} \left(\bar{\nu}_L \gamma^\mu e_L W_\mu^+ + \bar{e}_L \gamma^\mu \nu_L W_\mu^- \right). \qquad (7.26)$$

The two terms are Hermitian conjugates of each other so $\mathcal{L}$ is Hermitian as expected.

Note that only e_L is involved; e_L can make a transition into a (left–handed) neutrino by absorbing a W^+ or emitting a W^-, but e_R does not interact with charged W's at all. This is of course the traditional parity violation of the weak interactions. To see that formally, recall from Chapter 5 that

$$\bar{\nu}_L \gamma^\mu e_L = \frac{1}{2} \bar{\nu} \gamma^\mu (1 - \gamma_5) e . \qquad (7.27)$$

So one can see explicitly that the interaction is a coherent sum of vector (γ^μ) and axial–vector ($\gamma^\mu \gamma_5$) currents. An interaction with a current of the form of (7.27) is usually called a $V - A$ charged current interaction. We will study a number of charged current transitions later on; the best known one is neutron "beta–decay". At the hadronic level there is a current of the form of equation (7.27) for $n \to p$ as well (with small corrections for hadronic structure), so a transition as shown in Figure 7.1a can occur. Today all such transitions are formulated in terms of the quark structure of hadrons, so neutron beta–decay is thought of as a transition at the quark level. A d–quark in the neutron turns into a u–quark by emitting $e^- \overline{\nu}$, as in Figure 7.1b. The neutron, dominantly with quark structure udd, automatically becomes a proton (uud) when one of the d–quarks turns into a u–quark.

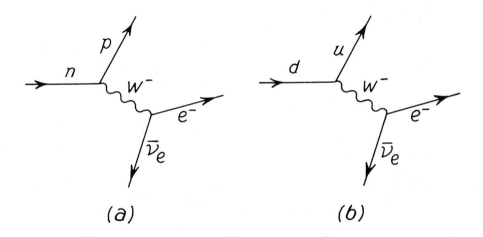

Figure 7.1
W-boson interactions with hadrons, leptons and quarks.

Just as we expected in the previous section to find a Z^0 and the associated neutral current transitions, here we expect to find $W^\pm$ and the associated charged current transitions. In the charged current case the transitions were known to occur (the traditional weak interactions), but with a strength much smaller than that which would naïvely be expected from equation (7.26), of

$$\frac{(g_2/\sqrt{2})^2}{4\pi} = \frac{(e^2/4\pi)}{2\sin^2\theta_w} \simeq \frac{2}{137} \ .$$

Rather than being somewhat larger than electromagnetic interactions, the known "weak" interactions were several orders of magnitude smaller. And the charged $W^\pm$

bosons had not been directly observed. Just as for the Z^0, the solution was to assume the $W^{\pm}$ to be massive, which reduced the size of the transition rates because of the mass in the propagator, and kept $W^{\pm}$ from direct detection because they were too massive to be produced at any available machine. Their discovery in 1983 confirmed that this was indeed the correct view.

Except for dealing with the question of mass we have now seen the content of the leptonic part of the basic Lagrangian. So far the expected terms (from the Lagrangians of Section 2.8) of the form $m\overline{\psi}\psi$ for the fermions, or $m^2 V^{\mu}V_{\mu}$ for the spin–one gauge bosons, have not appeared, so all the fermions and gauge bosons are massless, in contradiction to experiment. After we have made the quark content of the Lagrangian explicit, we will return to the mass problem.

7.5   The Quark Terms

Now we want to consider the remaining electroweak terms of the full Lagrangian, equations (6.9) and (6.10). Let q be a particular quark state ($e.g.$ $Q_{\alpha L}$ or $u_{\alpha R}$ or $d_{\alpha R}$). As is familiar from quantum theory, the quark wave function is a product of factors,

$$q = \begin{pmatrix} \text{space} \\ \text{factor} \end{pmatrix} \times \begin{pmatrix} \text{spin} \\ \text{factor} \end{pmatrix} \times \begin{pmatrix} U(1) \\ \text{factor} \end{pmatrix} \times \begin{pmatrix} SU(2) \\ \text{factor} \end{pmatrix} \times \begin{pmatrix} \text{color} \\ \text{factor} \end{pmatrix}. \quad (7.28)$$

Each factor has some labels, coordinates, or indices. The orthonormality of the wave function holds separately for each factor.

When we consider a term in $\mathcal{L}$ such as $g_2 \gamma^{\mu} \tau^i W_{\mu}^i$, which contains spin and $SU(2)$ pieces, the orthonormality of the wave function automatically gives a factor of unity for all the other pieces, such as the space and the color parts of the wave function. The $SU(2)$ and the spin pieces have non–trivial content, which in this case looks exactly the same for the quarks as for the leptons, since the $SU(2)$ and spin structure of quarks and leptons are the same. Consequently, all the conclusions of the previous sections hold without modification for the quarks. They couple to the same gauge bosons as do the leptons, $W^{\pm}$ and Z^0 and γ. There is a normal electromagnetic coupling to γ, a charged current coupling to $W^{\pm}$ generating transitions $u_L \leftrightarrow d_L$ (but no charged current transitions for u_R or d_R), and neutral current transitions with a universal strength $e(T_3^f - Q_f \sin^2 \theta_w)/(\sin \theta_w \cos \theta_w)$ for each left–handed or right–handed quark, $f = u_L$, d_L, u_R, d_R, with $Q = 2/3$ for u_L, u_R and $Q = -1/3$ for d_L, d_R, and $T_3 = 1/2$ for u_L, $T_3 = -1/2$ for d_L, $T_3 = 0$ for u_R or d_R.

7.6   The Quark QCD Lagrangian

Since the leptons carry no color, the $\lambda^a G^a$ terms in the Lagrangian, equation (6.10), which are a 3×3 matrix in color space since the λ's are 3×3 matrices, give zero contribution to the leptons. For the quarks there is a contribution; with color indices α and β equal to 1, 2, or 3 the terms have the form, for a particular quark q,

$$\frac{g_3}{2} \overline{q}_\alpha \gamma^\mu \lambda^a_{\alpha\beta} G^a_\mu q_\beta \ . \tag{7.29}$$

In the electroweak case we wrote out the equivalent term involving $\tau^i W^i_\mu$, and the W^i were related to states of electromagnetic charge because the interaction with the photon (electromagnetic field) was involved. Here the G^a are eight gluons that are electrically neutral; *i.e.* they have no interactions with the electromagnetic field. They interact with the quarks in a way which is somewhat like a photon interaction, but with one major difference: since the generators λ^a are not all diagonal, the interaction with a gluon can change the color charge of a quark. Since the color charge is harder to directly observe because quarks and gluons are confined, we will not study the structure of equation (7.29) as explicitly as we did for the equivalent electroweak term. A number of ways to observe the effects of color will be discussed in later chapters.

The main point to learn from the presence of the term (7.29) in $\mathcal{L}$ is that the transition from one quark color state to another with the emission of a gluon g is always possible. Since these states all carry momentum, the gluon can carry away or add momentum. Since particles can always be turned into antiparticles by reversing their directions (and momentum and charges), the one vertex describes all the transitions

$$q + g \rightarrow q', \quad q \rightarrow q' + g, \quad g \rightarrow q\overline{q} \ ,$$

$$\overline{q} \rightarrow \overline{q}' + g, \quad \overline{q} + g \rightarrow \overline{q}', \tag{7.30}$$

all with a strength $g_3/2$. We will discuss later how to measure g_3.

7.7   The Second and Third Families

So far we have described the Standard Model without mentioning the muon, the tau, the strange quark, *etc.*. Remarkably, nature seems to have reproduced the same theory for the three families of particles. As far as is known today, all experiments on all known particles are consistent with the replacements:

$$
\left.\begin{array}{c}
\begin{pmatrix} \nu_e \\ e \end{pmatrix} \\[4pt]
\begin{pmatrix} u \\ d \end{pmatrix}
\end{array}\right\} \longrightarrow \quad \begin{pmatrix} \nu_\mu \\ \mu \end{pmatrix} \begin{pmatrix} c \\ s \end{pmatrix} \ \text{or} \ \begin{pmatrix} \nu_\tau \\ \tau \end{pmatrix} \begin{pmatrix} t \\ b \end{pmatrix} . \tag{7.31}
$$

It is unknown whether more families exist, or whether additional quarks or leptons exist that do not fit the pattern of families. The ν_τ and t have not been directly observed, but their existence can be rigorously inferred if the theory is valid; we will go through the reasoning in Chapter 25 as an application of Standard Model techniques. The same set of gauge bosons interacts with all the families of fermions, *i.e.* there is only one set of γ, $W^\pm$, Z^0, g. That e, μ, and τ all have identical interactions is called lepton universality; there is also a universality of the interactions for up-type quarks, and for down-type quarks.

7.8   The Fermion–Gauge Boson Lagrangian

To simplify later use, we bring together in one place the Lagrangian for the interaction of quarks and leptons with photons, $W^\pm$, Z^0, and gluons. All the behavior of quarks and leptons arises from the terms summarized here, given their masses and the charges we have discussed. The relevant part of the $SU(3) \times SU(2) \times U(1)$ Lagrangian is, for the first family,

$$
\begin{aligned}
\mathcal{L} = & \sum_{f=\nu_e,e,u,d} eQ_f \left(\overline{f}\gamma^\mu f \right) A^\mu \\
& + \frac{g_2}{\cos\theta_w} \sum_{f=\nu_e,e,u,d} \left[\overline{f}_L \gamma^\mu f_L \left(T_f^3 - Q_f \sin^2\theta_w \right) \right. \\
& \qquad\qquad \left. + \overline{f}_R \gamma^\mu f_R \left(-Q_f \sin^2\theta_w \right) \right] Z_\mu \\
& + \frac{g_2}{\sqrt{2}} \left[\left(\overline{u}_L \gamma^\mu d_L + \overline{\nu}_{eL} \gamma^\mu e_L \right) W_\mu^+ + \text{h.c.} \right] \\
& + \frac{g_3}{2} \sum_{q=u,d} \overline{q}_\alpha \gamma^\mu \lambda_{\alpha\beta}^a q_\beta G_\mu^a .
\end{aligned} \tag{7.32}
$$

For the second and third families the substitutions $\left(\nu_e, e, u, d \right) \rightarrow \left(\nu_\mu, \mu, c, s \right)$

or (ν_τ, τ, t, b) give the appropriate results. Some numerical relations are

$$\frac{G_F}{\sqrt{2}} = \frac{g_2^2}{8 M_W^2} \tag{7.33}$$

$$g_2 = \frac{e}{\sin \theta_w} \tag{7.34}$$

$$g_1 = \frac{e}{\cos \theta_w} \tag{7.35}$$

$$\alpha = \frac{e^2}{4\pi} \simeq \frac{1}{137} \tag{7.36}$$

$$\alpha_1 = \frac{g_1^2}{4\pi} \simeq \frac{1}{100} \tag{7.37}$$

$$\alpha_2 = \frac{g_2^2}{4\pi} \simeq \frac{1}{30} \tag{7.38}$$

$$\alpha_3 = \frac{g_3^2}{4\pi} \simeq 0.3 \rightarrow 0.1 \ . \tag{7.39}$$

The values of the couplings α, α_1, α_2, α_3 depend on the momentum transfer of the interaction (see Chapter 20); the values of α, α_1, and α_2 are for interactions in the few GeV range or below, but they vary slowly so it is only necessary to consider the variation when precise results are needed. For the strong coupling α_3, the range given is from about 1 GeV to about 100 GeV, with a slow decrease above that range.

7.9   Masses?

We have mentioned that no mass terms were present in the Lagrangians we have written so far. Although all the nice things we have said about the Standard Model and its consistency with experiment are so, there is a major problem: all the fermions and gauge bosons are massless, while experiment shows otherwise.

Why not just add in mass terms explicitly? That will not work, since the associated terms break $SU(2)$ or gauge invariances. For fermions, the mass term should be $m\overline{\psi}\psi$, but (recall Chapter 5)

$$\begin{aligned}
m\overline{\psi}\psi &= m\overline{\psi}(P_L + P_R)\psi \\
&= m\overline{\psi}P_L P_L \psi + m\overline{\psi}P_R P_R \psi \\
&= m\left(\overline{\psi}_R \psi_L + \overline{\psi}_L \psi_R\right) .
\end{aligned} \tag{7.40}$$

However, the left–handed fermions are put into $SU(2)$ doublets and the right–handed ones into $SU(2)$ singlets, so $\overline{\psi}_R \psi_L$ and $\overline{\psi}_L \psi_R$ are not $SU(2)$ singlets and would not give an $SU(2)$ invariant Lagrangian.

Similarly, the expected mass terms for the gauge bosons,

$$\frac{1}{2}m_B^2 B^\mu B_\mu \qquad\qquad (7.41)$$

plus similar terms for the others, are clearly not invariant under gauge transformations $B^\mu \rightarrow B^{\mu\,\prime} = B^\mu - \partial^\mu \chi/g$. The only direct way to preserve the gauge invariance and $SU(2)$ invariance of the Lagrangian is to set $m = 0$ for all quarks, leptons, and gauge bosons. If masses are just put in by hand, the resulting quantum field theory turns out to have infinities for physical quantities.

There is a way to solve this problem, called the Higgs mechanism. The resulting Lagrangian does have $SU(2)$ and gauge invariance broken, but in a very subtle way that preserves the good effects of these invariances. We will now turn to describing the Higgs mechanism. As we will see, it solves the problem technically, but our physical understanding of the Higgs mechanism and associated phenomena is far from satisfactory. The task of improving this situation should be considered the central problem in particle physics today, not only because it is necessary to complete the theory we call the Standard Model, but because any approach to physics beyond the Standard Model will also involve masses and their origin, and consequently the Higgs mechanism.

Problems

7.1: Fill in the table (where appropriate):

Particle	Spin (0, L, R, 1)	Mass	Color (singlet, triplet, octet)	Weak Isospin (0, 1/2, 1)	Electric Charge
ν_{eL}					
e_L^-					
e_R^-					
u_L					
d_L					
u_R					
d_R					
$W^\pm$					
Z^0					
γ					
g					
W^0					
B^0					

7.2: Write the $SU(3) \times SU(2) \times U(1)$ interaction Lagrangian for the fermion family μ_L^-, μ_R^-, ν_μ, c_L, c_R, s_L, s_R. Include color indices for the quarks. Express couplings in terms of e, θ_w, and g_3. Give the vertices for interactions of fermions with gauge bosons.

7.3:   What factor goes at a vertex for $d_L + W^+ \to u_L$? For $d_L + Z \to d_L$? For $u_L + Z \to u_L$? For $u_L + \gamma \to u_L$? For $u_{Lr} + g_{\bar{r}b} \to u_{Lb}$ (where r and b are color labels)?

7.4:   Show that

$$\sum_{L,R} \bar{e}_f \gamma^\mu e_f \left(T_3^f - Q^f \sin^2 \theta_w \right)$$

$$= -\frac{1}{4} \bar{e} \gamma^\mu (a + b\gamma^5) e \ .$$

What are a and b in terms of $\sin^2 \theta_W$? Compare with the equivalent result for a charged current.

7.5:   Explain why detecting the process $\bar{\nu}_\mu + e^- \to \bar{\nu}_\mu + e^-$ would automatically demonstrate the existence of neutral currents, while detecting $\bar{\nu}_e + e^- \to \bar{\nu}_e + e^-$ would not.

Suggestions for Further Study

The books of Quigg, and Halzen and Martin, cover the same material at a graduate level. Commins and Bucksbaum gives detailed applications of the electroweak theory to a variety of processes.

CHAPTER 8

Masses
and the
Higgs Mechanism

In this chapter we will work through the mechanism developed to construct a meaningful gauge theory which can include the possibility that particles have mass. Technically the situation is well under control. Not only can a consistent theory be written, but the masses of the $W^{\pm}$ and Z bosons were calculated (in terms of measured parameters) and these bosons were found with precisely the predicted masses. That was a remarkable achievement, particularly when one keeps in mind that these are fundamental, point–like particles with masses similar to that of a strontium nucleus. The actual theoretical procedure to make W and Z massive is not hard to understand, and the calculations only involve a little algebra. Fermion masses can be included in the theory by a variation of the same mechanism.

On the other hand, the reader should be aware that to appreciate fully the importance of this development an understanding of the infinities of the theory and its renormalization is necessary. Unfortunately, achieving such an understanding is beyond the scope of this book. Further, the situation is not, in some senses, a satisfactory one since there is no deep physical understanding of how and why all mass should arise from such a mechanism.

In essence the assumption is made that the universe is filled with a spin–zero field, called a Higgs field, that is a doublet in the $SU(2)$ space and carries non–zero $U(1)$ hypercharge, but is a singlet in color space. This is meant in much the same sense that space is filled with electromagnetic fields whose sources are electrically charged particles, but in the Higgs case we do not ask (at this level of discussion) about the sources of the Higgs field. The gauge bosons and fermions can interact with this field, and in its presence they no longer appear to have zero mass. A crucial ingredient is that states with one or more Higgs fields are not orthogonal to the ground state (*i.e.* the vacuum) even though these states carry non–zero $SU(2)$ and $U(1)$ quantum numbers. That means the $SU(2)$ and $U(1)$ quantum numbers of the vacuum are non–zero, so the $SU(2)$ and $U(1)$ symmetries are effectively broken. When a symmetry is broken this way, *i.e.* the symmetry is valid for the Lagrangian but not for the ground state of the system, it is said to be a spontaneously broken symmetry.

We will proceed by going through several examples of spontaneously broken symmetries, each one a little more complicated, to the level where the actual case for the Standard Model can be treated. Related discussions are given in several places that will be mentioned at the end of the chapter; often examples from other areas of physics are used to clarify the ideas and may be helpful to people who are already familiar with spontaneous symmetry breaking in other fields.

8.1   Spontaneous Symmetry Breaking

First we examine a very simple case to see the basic physics. Consider a Lagrangian

$$\mathcal{L} = T - V = \frac{1}{2}\partial_\mu\phi\partial^\mu\phi - \left(\frac{1}{2}\mu^2\phi^2 + \frac{1}{4}\lambda\phi^4\right). \tag{8.1}$$

The parameters μ and λ should initially be thought of as simply parameters in the potential. We can require $\lambda > 0$ in order that the potential be bounded below as $\phi \rightarrow \infty$, from general quantum mechanical principles. Note that the theory has a symmetry: it is invariant under $\phi \rightarrow -\phi$. To find the spectrum it is necessary to find the minimum of the potential, which will be the classical ground state of the system. Then one expands the fields around their value at the minimum and determines the excitations. This is the normal procedure for handling perturbations with which the reader is familiar from quantum theory. In field theory, it is conventional to call the ground state the vacuum, and the excitations are particles. Their mass is determined by the form of the Lagrangian near the classical minimum, by comparison with the Lagrangian of Chapter 2. The ϕ^4 term represents an interaction, of strength λ. This Lagrangian is certainly not general, but it is more general than it might appear, since a careful analysis shows that higher powers of ϕ would lead to infinities in physical quantities and must therefore be excluded.

Suppose now that $\mu^2 > 0$. Then obviously the vacuum corresponds to $\phi = 0$, which minimizes the potential. In that case μ^2 can be interpreted as $(\text{mass})^2$ by comparison with Chapter 2, so we have interpreted the theory.

However, there is no physical reason for requiring $\mu^2 > 0$. If $\mu^2 < 0$, we find the minimum of the potential by setting

$$\frac{\partial V}{\partial \phi} = 0 \tag{8.2}$$

which gives

$$\phi\left(\mu^2 + \lambda\phi^2\right) = 0 . \tag{8.3}$$

The lowest energy of the system occurs when both the kinetic energy and the potential energy are minimized. The kinetic energy is minimized by taking $\phi(x) = $ constant. The choice $\phi = 0$ is not a minimum since with μ^2 negative we can get a lower value of the potential. The situation is shown as a graph of potential energy versus ϕ in Figure 8.1.

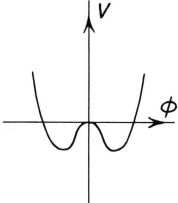

Figure 8.1
Potential energy
versus ϕ.

The choices

$$\phi = \pm\sqrt{\frac{-\mu^2}{\lambda}} \equiv v \tag{8.4}$$

are non–zero values of ϕ which are the minimum of the potential. Since ϕ takes on the value v in the ground state, v is called the vacuum expectation value of ϕ. The field ϕ is called a Higgs field.

To determine the particle spectrum, we must study the theory in the region
of the minimum, so we put

$$\phi(x) = v + \eta(x) \tag{8.5}$$

so that we are expanding around $\eta = 0$. Note we could equally well have chosen $\phi = -v + \eta(x)$, but the physics conclusions will be independent of this choice
since the theory was symmetric under $\phi \to -\phi$. Substituting (8.5) into $\mathcal{L}$ gives

$$
\begin{aligned}
\mathcal{L} =& \frac{1}{2}\left(\partial_\mu \eta \partial^\mu \eta\right) - \left\{\frac{1}{2}\mu^2 \left[v^2 + 2\eta v + \eta^2\right]\right. \\
&+ \frac{1}{4}\lambda \left[v^4 + 4v^3\eta + 6v^2\eta^2 + 4v\eta^3 + \eta^4\right]\Big\} \\
=& \frac{1}{2}\left(\partial_\mu \eta \partial^\mu \eta\right) - \left\{\frac{v^2}{2}\left(\mu^2 + \frac{1}{2}\lambda v^2\right) + \eta v\left(\mu^2 + \lambda v^2\right)\right. \\
&+ \frac{\eta^2}{2}\left(\mu^2 + 3\lambda v^2\right) + \lambda v\eta^3 + \frac{1}{4}\lambda \eta^4\Big\}.
\end{aligned} \tag{8.6}
$$

Using (8.4), the term linear in η vanishes (as it must near the minimum), and $\mathcal{L}$
simplifies to

$$\mathcal{L} = \frac{1}{2}\left(\partial_\mu \eta \partial^\mu \eta\right) - \left(\lambda v^2\eta^2 + \lambda v\eta^3 + \frac{1}{4}\lambda \eta^4\right) + \text{constant}. \tag{8.7}$$

Now the term with η^2 has the correct sign so it can be interpreted as a mass
term. This Lagrangian represents the description of a particle with mass

$$m_\eta^2 = 2\lambda v^2 = -2\mu^2, \tag{8.8}$$

and with two interactions, a cubic one of strength λv and a quartic one of strength
$\lambda/4$. Note that both of these depend on λ, which is a free parameter as far as we
know, and are therefore interactions of unknown strength. The constant can be
ignored since the zero level of the potential can be redefined.

The two descriptions of the theory in terms of ϕ or η must be equivalent if the
problem is solved exactly. If we want a perturbative description it is essential
to perturb around the minimum to have a convergent description. The scalar
particle described by the theory with $\mu^2 < 0$ is a real scalar, with a mass obtained
by its self–interaction with other scalars, because at the minimum of the potential
there is a non–vanishing vacuum expectation value v.

There is no trace of the reflection symmetry $\phi \rightarrow -\phi$ in equation (8.7). A memory of it is preserved in the η^3 interaction term but not in an obvious way. Because the symmetry was broken, in a sense, when a specific vacuum was chosen ($\phi = +v$ rather than $\phi = -v$) the vacuum does not have the symmetry of the original Lagrangian, so the solutions do not.

Many examples exist of similiar situations for physical systems. Here we will go on to repeat the analysis for increasingly more complicated symmetries until we see what happens when the symmetry of $\mathcal{L}$ is the Standard Model $SU(2) \times U(1)$ invariance and when we have the combined Lagrangian of gauge bosons, fermions and Higgs fields. At each stage surprising new features emerge.

8.2   Complex Scalar Field—a Global Symmetry

Suppose that ϕ is a complex scalar, $\phi = (\phi_1 + i\phi_2)/\sqrt{2}$, and

$$\mathcal{L} = (\partial_\mu \phi)^* (\partial^\mu \phi) - \mu^2 \phi^* \phi - \lambda (\phi^* \phi)^2 . \tag{8.9}$$

This is invariant under a global gauge transformation,

$$\phi \rightarrow \phi' = e^{i\chi} \phi , \tag{8.10}$$

so the symmetry of $\mathcal{L}$ is now a global $U(1)$ symmetry rather than a reflection as in the previous section. Written in terms of the (real) components,

$$\mathcal{L} = \frac{1}{2}(\partial_\mu \phi_1)^2 + \frac{1}{2}(\partial_\mu \phi_2)^2 - \frac{1}{2}\mu^2 (\phi_1^2 + \phi_2^2) - \frac{\lambda}{4}(\phi_1^2 + \phi_2^2)^2 . \tag{8.11}$$

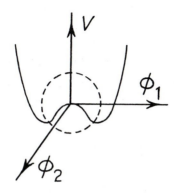

Figure 8.2
Potential energy
as a functions of ϕ_1 and ϕ_2.

In the ϕ_1, ϕ_2 plane (Figure 8.2), the potential energy is clearly a minimum at the origin if $\mu^2 > 0$, and for $\mu^2 < 0$ the minimum is along a circle of radius

$$\phi_1^2 + \phi_2^2 = \frac{-\mu^2}{\lambda} = v^2. \tag{8.12}$$

As before, to analyze the case with $\mu^2 < 0$ we have to expand around $\phi_1^2 + \phi_2^2 = v^2$. We could choose any point on the circle, but to proceed we have to choose *some* point, which will break the symmetry for the solutions. We pick, arbitrarily, the point $\phi_1 = v$, $\phi_2 = 0$, and write, with η and ρ real,

$$\phi = \frac{(v + \eta(x) + i\rho(x))}{\sqrt{2}}. \tag{8.13}$$

Substituting this in equation (8.11), we again find a Lagrangian that can be interpreted in terms of particles and their interactions:

$$\mathcal{L} = \frac{1}{2}\left(\partial_\mu \rho\right)^2 + \frac{1}{2}\left(\partial_\mu \eta\right)^2 + \mu^2\eta^2$$
$$- \lambda v \left(\eta\rho^2 + \eta^3\right) - \frac{\lambda}{2}\eta^2\rho^2 - \frac{\lambda}{4}\eta^4 - \frac{\lambda}{4}\rho^4 \tag{8.14}$$
$$+ \text{ constant }.$$

The first terms are normal kinetic energy terms. The term $+\mu^2\eta^2$ tells us the η field corresponds to a particle of $m_\eta^2 = 2|\mu^2|$. Remarkably, the term in ρ^2 has vanished, implying the ρ field particle has zero mass! It is called a Goldstone boson. There is a general theorem, that whenever a continuous global symmetry (here the $U(1)$ invariance under $\phi \rightarrow \phi' = e^{i\chi}\phi$) is spontaneously broken, the spectrum will contain a massless, spin–zero boson. As expected, since we chose a particular direction, the invariance is no longer present in equation (8.14).

Technically it is clear here how the massless boson arises. The potential is a minimum along a circle. Excitations in the radial direction require pushing up the potential away from the minimum and a mass is associated with the curvature of the potential. Along the circle the potential is flat, so there is no resistance to motion around the circle, which is the meaning of the massless excitation. The Goldstone phenomenon is widespread in physics; we have encountered a simple example. The $U(1)$ symmetry is broken because we had to choose a particular point on the circle to expand around. The presence and particular form of the interaction terms in equation (8.14) provide a memory of the original symmetry, but not in any obvious way.

8.3   The Abelian Higgs Mechanism

This is the third stage of the analysis. In the last section it was surprising to see a massless boson emerge. Here an even more surprising effect occurs. In the next section we will carry out the analysis for the actual situation of the Standard Model, and see masses emerge for the gauge bosons $W^{\pm}$ and Z^0. Finally after that we will consider fermion masses, which may also arise by interactions with a Higgs field but by a quite different mechanism.

Previously we considered a global gauge invariance. Now let us make it a local one, $i.e.$ let us make the Lagrangian invariant under local gauge transformations. We know from our earlier discussions that invariance under a local gauge transformation requires the introduction of a massless vector field A_μ , and we know that we should write $\mathcal{L}$ in terms of the covariant derivative,

$$\partial_\mu \to \mathcal{D}_\mu = \partial_\mu - ig A_\mu \ .$$ (8.15)

The gauge field transforms as

$$A_\mu \to A_\mu{}' = A_\mu - \frac{1}{g}\partial_\mu \chi(x)$$ (8.16)

and ϕ will be invariant under

$$\phi(x) \to \phi'(x) = e^{i\chi(x)} \phi(x) \ .$$ (8.17)

The Lagrangian is then

$$\mathcal{L} = (\mathcal{D}_\mu \phi)^* (\mathcal{D}^\mu \phi) - \mu^2 \phi^* \phi - \lambda (\phi^* \phi)^2 - \frac{1}{4} F_{\mu\nu} F^{\mu\nu}.$$ (8.18)

For $\mu^2 > 0$ this describes the interaction of a charged scalar particle (with $g \equiv e$) of mass μ with the electromagnetic field A_μ , for example. Note there is no mass term for A_μ . We have written the $F_{\mu\nu} F^{\mu\nu}$ kinetic energy terms for the vector field and will carry them along, but they do not enter the analysis. Here we want to choose $\mu^2 < 0$ as in the previous sections. Note that this Lagrangian contains four independent fields or degrees of freedom, the two real scalars ϕ_1 and ϕ_2 , and the two transverse polarization states of the massless vector boson (as expected if A_μ described the photon). We could proceed as before. The algebra gets increasingly complicated, however, so it is worthwhile to use what we have learned in order to simplify the analysis. Equation (8.17) tells us that the theory will be invariant under a gauge transformation of $\phi(x)$. In general ϕ could be written in the form

$$\phi(x) = \eta(x)e^{-i\rho(x)} \tag{8.19}$$

where η, ρ are real, so we can choose to write $\phi(x)$ in the form

$$\phi(x) = \frac{(v + h(x))}{\sqrt{2}} \tag{8.20}$$

with h real, having used a transformation such as $\phi \rightarrow e^{i\chi(x)}\phi$, knowing that if necessary we could find a χ to accomplish that. Note that we could not have done this in the previous section, since we did not have a Lagrangian there that was invariant under a local gauge transformation, only global ones. Now we substitute this in $\mathcal{L}$. Since the original choice of the field A_μ was not fixed by physics we do not bother to rename the A_μ obtained from equation (8.16). So

$$
\begin{aligned}
\mathcal{L} =& \frac{1}{2}\left[(\partial^\mu + igA^\mu)(v + h)\right]\left[(\partial_\mu - igA_\mu)(v + h)\right] \\
& - \frac{\mu^2}{2}(v + h)^2 - \frac{\lambda}{4}(v + h)^4 - \frac{1}{4}F_{\mu\nu}F^{\mu\nu} \\
=& \frac{1}{2}(\partial_\mu h)(\partial^\mu h) + \frac{1}{2}g^2 v^2 A_\mu A^\mu - \lambda v^2 h^2 - \lambda v h^3 \\
& - \frac{\lambda}{4}h^4 + g^2 vhA^\mu A_\mu + \frac{1}{2}g^2 h^2 A_\mu A^\mu - \frac{1}{4}F_{\mu\nu}F^{\mu\nu}.
\end{aligned} \tag{8.21}
$$

Here every term can be interpreted. The surprising result is that there is now a mass term for the gauge boson! But since we started with a gauge invariant theory and made only algebraic transformations, we expect the resulting theory to be gauge invariant as well. The gauge boson mass is the square root of the coefficient of $A_\mu A^\mu/2$,

$$M_A = gv, \tag{8.22}$$

and this is non–zero only when the gauge symmetry is spontaneously broken by the Higgs field acquiring a vacuum expectation value. So the theory is only gauge invariant in a restricted sense. The Lagrangian is gauge invariant but the vacuum is not, because we had to choose a particular direction in the ϕ_1, ϕ_2 space for the potential minimum.

The spectrum is now a single real Higgs boson h, that has mass $2\lambda v^2$, various self interactions and cubic and quartic interactions with the gauge field A_μ, plus a massive vector boson A_μ. Since the massive boson has three spin states (corresponding to $J_z = 1$, 0, or -1 in its rest frame) the number of independent fields is still four, so that is consistent.

What has happened here is that the Goldstone boson of the previous section has become the longitudinal polarization state of the gauge boson. [That can be seen a little more explicitly if the calculation of this section is carried out without the simplifying step of equation (8.20), but using equation (8.13) instead. Then the mass appears for the gauge–vector boson, and a term $A_\mu \partial^\mu \rho$, which apparently allows A_μ to turn into ρ as it propagates. When such cross terms appear one can go to eigenstates by a diagonalization, which can be accomplished here by a gauge transformation, and which eliminates ρ from the Lagrangian. This phenomenon is sometimes referred to as the gauge boson having "eaten" the Goldstone boson.]

The mechanism we have just studied is called the "Higgs mechanism." Technically it is well understood, but at a physical level its meaning is not yet fully grasped in particle physics. In some sense the longitudinal polarization state of the gauge boson, which must exist if it is to be massive in a Lorentz invariant theory where it is possible to go to its rest frame, is the Goldstone boson that would have existed if the theory were not a gauge theory. There is also a neutral spin–zero boson left over that apparently should exist as a physical particle; it is called the Higgs boson. Note that the gauge boson mass is fixed if g^2 and v are known, but the mass of the Higgs boson h depends on the unknown parameter λ. In the next section we are finally ready to add the last bit of complexity needed to fully incorporate the Higgs mechanism into the Standard Model.

8.4   The Higgs Mechanism in the Standard Model

In the Standard Model case one further degree of complexity is needed. The Higgs field, which otherwise carried no quantum number apart from energy and momentum, now is assigned to an $SU(2)$ doublet. Choose

$$\phi = \begin{pmatrix} \phi^+ \\ \phi^0 \end{pmatrix} \tag{8.23}$$

where ϕ^+ and ϕ^0 are each complex fields,

$$\phi^+ = \frac{\phi_1 + i\phi_2}{\sqrt{2}} , \tag{8.24}$$

$$\phi^0 = \frac{\phi_3 + i\phi_4}{\sqrt{2}} . \tag{8.25}$$

In an $SU(2)$ space the two Higgs fields are related by a rotation, like spin–up to spin–down, or the left–handed ν_e to the left–handed electron. The Lagrangian has the same form,

$$\mathcal{L}_\phi = (\partial_\mu \phi)^\dagger (\partial^\mu \phi) - \mu^2 \phi^\dagger \phi - \lambda \left(\phi^\dagger \phi \right)^2 \tag{8.26}$$

but now ϕ is a column, $\phi^\dagger$ a row, so

$$\phi^\dagger \phi = \left(\phi^{+*} \phi^{0*} \right) \begin{pmatrix} \phi^+ \\ \phi^0 \end{pmatrix} = \phi^{+*} \phi^+ + \phi^{0*} \phi^0 \tag{8.27}$$

or, in terms of the real component field,

$$\phi^\dagger \phi = \frac{(\phi_1^2 + \phi_2^2 + \phi_3^2 + \phi_4^2)}{2} . \tag{8.28}$$

As before we study the potential

$$V(\phi) = \mu^2 \phi^\dagger \phi + \lambda \left(\phi^\dagger \phi \right)^2 . \tag{8.29}$$

$V(\phi)$ is invariant under the local gauge transformation

$$\phi(x) \rightarrow \phi'(x) = e^{i \overrightarrow{\alpha}(x) \cdot \overrightarrow{\tau} /2} \phi(x) \tag{8.30}$$

where τ_i are the Pauli matrices and α_i are parameters. Proceeding as before, $V(\phi)$ has a minimum for $\mu^2 < 0$ at

$$\phi^\dagger \phi = \frac{-\mu^2}{2\lambda} = \frac{v^2}{2} . \tag{8.31}$$

From equation (8.28), we see that there are many ways to have equation (8.31) satisfied.

Again, we must choose a direction, this time in $SU(2)$ space, and expand around the minimum. The appropriate choice is what we then call the vacuum, ϕ_0,

$$\phi_0 = \frac{1}{\sqrt{2}} \begin{pmatrix} 0 \\ v \end{pmatrix} , \tag{8.32}$$

i.e. $\phi_3 = v$, $\phi_1 = \phi_2 = \phi_4 = 0$.

As before, we can study the spectrum by expanding around the vacuum, so we choose

$$\phi(x) = \frac{1}{\sqrt{2}} \begin{pmatrix} 0 \\ v + H(x) \end{pmatrix} \tag{8.33}$$

and we will look for the equations satisfied by H. We are guaranteed that we can make this simple choice because for an arbitrary $\phi(x)$ we could make a gauge transformation, $\phi \rightarrow \phi' = \exp(i\overrightarrow{\tau} \cdot \overrightarrow{\theta}(x)/v)\,\phi$, and rotate $\phi(x)$ into the form (8.32). This amounts to "gauging away" three fields, which is consistent with what we learned about the Goldstone theorem: the original symmetry (an $O(4)$ symmetry) was, from equation (8.28),

$$\phi_1^2 + \phi_2^2 + \phi_3^2 + \phi_4^2 = \text{invariant} . \tag{8.34}$$

By choosing a direction we have three broken global symmetries, so three massless bosons, and three fields gauged away. Below we will see these three are just what are needed to become the longitudinal parts of $W^\pm$ and Z^0.

Before we write the covariant derivative and complete the calculation, let us examine a bit further what is happening. The electric charge, the weak isospin eigenvalue T_3, and the $U(1)$ hypercharge Y_H (for the Higgs field) are related by

$$Q = T_3 + Y_H/2 . \tag{8.35}$$

The electric charge assignment of equation (8.23) corresponds to putting $Y_H = 1$. The choice that only the neutral component ϕ^0 gets a vacuum expectation value is very important, since whatever quantum numbers ϕ carries can vanish into the vacuum. If ϕ^+ had a vacuum expectation value then electric charge would not be conserved, contrary to observation.

If the vacuum ϕ_0 is invariant under some subgroups of the original $SU(2) \times U(1)$, any gauge bosons associated with that subgroup will still be massless. Since the Higgs field $\phi(x)$ is a doublet, but only one component gets a vacuum expectation value, clearly the $SU(2)$ symmetry is broken. Since the hypercharge $Y_H \neq 0$, clearly the $U(1)$ symmetry is broken. However, if we operate with the electric charge operator Q on ϕ_0,

$$Q\phi_0 = (T_3 + Y/2)\,\phi_0 = 0 , \tag{8.36}$$

so ϕ_0 (*i.e.* the vacuum) is invariant under a transformation

$$\phi_0 \rightarrow \phi_0' = e^{i\alpha(x)Q}\,\phi_0 = \phi_0 . \tag{8.37}$$

This is also a $U(1)$ transformation, so the vacuum is invariant under a particular $U'(1)$ whose generators are a particular linear combination of the generators of the original $SU(2)$ and $U(1)$. Of course, this $U'(1)$ is the $U(1)$ of electromagnetism, and the gauge boson that remains massless is the photon. The presence of a massless gauge boson was a necessary consequence of electric charge conservation, which forces us to choose a neutral vacuum state.

Finally, let us carry out the algebra to see the Higgs mechanism operate. For the full Lagrangian to be invariant under the transformation in equation (8.30), we know that we replace ∂_μ by the covariant derivative $\mathcal{D}_\mu$ where

$$\mathcal{D}_\mu = \partial_\mu - ig_1 \frac{Y}{2} B_\mu - ig_2 \frac{\overrightarrow{\tau}}{2} \cdot \overrightarrow{W_\mu} \tag{8.38}$$

and B_μ and $\overrightarrow{W_\mu}$ transform as in Chapter 7. Then when ϕ gets a vacuum expectation value, proceeding as in the earlier sections, the Lagrangian contains extra terms

$$\phi^\dagger \left(ig_1 \frac{Y}{2} B_\mu + ig_2 \frac{\overrightarrow{\tau}}{2} \cdot \overrightarrow{W_\mu} \right)^\dagger \left(ig_1 \frac{Y}{2} B^\mu + ig_2 \frac{\overrightarrow{\tau}}{2} \cdot \overrightarrow{W^\mu} \right) \phi . \tag{8.39}$$

Putting $Y = 1$, writing the 2×2 matrices explicitly as in Chapter 7, and putting $\phi = \frac{1}{\sqrt{2}} \binom{0}{v}$, gives for the contribution to $\mathcal{L}$,

$$\frac{1}{8} \left| \begin{pmatrix} g_1 B_\mu + g_2 W_\mu^3 & g_2 \left(W_\mu^1 - i W_\mu^2 \right) \\ g_2 \left(W_\mu^1 + i W_\mu^2 \right) & g_1 B_\mu - g_2 W_\mu^3 \end{pmatrix} \begin{pmatrix} 0 \\ v \end{pmatrix} \right|^2$$
$$= \frac{1}{8} v^2 g_2^2 \left(\left(W_\mu^1 \right)^2 + \left(W_\mu^2 \right)^2 \right) + \frac{1}{8} v^2 \left(g_1 B_\mu - g_2 W_\mu^3 \right)^2 . \tag{8.40}$$

The first term can be rewritten as

$$\left(\frac{1}{2} v g_2 \right)^2 W_\mu^+ W^{-\mu} , \tag{8.41}$$

carefully keeping track of $1/\sqrt{2}$ factors when converting from equation (6.1). For a charged boson the expected mass term in a Lagrangian would be $m^2 W^+ W^-$, so we can conclude that the charged W has indeed acquired a mass $M_W = v g_2/2$!

The second term in equation (8.40) is not diagonal so we have to define new eigenvalues to find the particles with definite mass (B and W^3 are the neutral states with diagonal weak isospin and weak interactions). In fact, we already have the answer in hand, because the combination of B and W^3 appearing in equation (8.40) is just the combination we have called Z_μ (see equation (7.10) and the choice of $Y_L = -1$ described below equation (7.17)). We expect mass terms for Z_μ and for the photon A_μ; for a neutral field there is a 1/2 relative to the charged ones, so mass terms $(M_Z^2 Z_\mu Z^\mu)/2 + (M_\gamma^2 A_\mu A^\mu)/2$ should appear. From equation (8.40) and the normalization of Z in equation (7.10), we can conclude that

$$M_Z = \frac{1}{2}v\sqrt{g_1^2 + g_2^2} \tag{8.42}$$

and

$$M_\gamma = 0$$

since no $A_\mu A^\mu$ term appears. From the discussion above we expected the massless photon. Using the identities of Chapter 7, we can also write

$$M_W/M_Z = \cos\theta_w \ . \tag{8.43}$$

Since B and W^3 mix, the neutral state is not degenerate in mass with the charged ones, unless $\theta_w \to 0$. Once θ_w is measured, the result in equation (8.43) is a prediction of the Standard Model, one which has been found to be consistent with experiments.

A useful quantity to consider is $\rho = M_W/M_Z \cos\theta_w$. Equation (8.43) shows that the Standard Model predicts $\rho = 1$. In fact, it can be shown that even if additional doublets of Higgs fields are present, it is guaranteed that $\rho = 1$. Any deviation from $\rho = 1$ would be an important signal of new physics.

8.5   Fermion Masses

Now that we have available the Higgs field in an $SU(2)$ doublet, it is possible to write an $SU(2)$–invariant interaction of fermions with the Higgs field. To the previous Lagrangian we can add an interaction term for the leptons,

$$\mathcal{L}_{\text{int}} = g_e\left(\overline{L}\phi e^-_R + \phi^\dagger \overline{e^-_R}L\right) . \tag{8.44}$$

Since $L = \begin{pmatrix} \nu \\ e^- \end{pmatrix}_L$ and $\phi = \begin{pmatrix} \phi^+ \\ \phi^0 \end{pmatrix}$, $\overline{L}\phi = \overline{\nu_L}\phi^+ + \overline{e^-_L}\phi^0$ is an $SU(2)$ invariant. Multiplying by the singlet e^-_R does not change the $SU(2)$ invariance. The second term is the Hermitian conjugate of the first. The coupling g_e is arbitrary; neither the presence of such terms nor g_e is determined by a gauge principle.

Following the previous sections, we can calculate the experimental consequences of adding this term by replacing

$$\phi \rightarrow \begin{pmatrix} 0 \\ \frac{v+H}{\sqrt{2}} \end{pmatrix} \qquad (8.45)$$

where v is the Higgs vacuum expectation value, and H is the neutral, physical
Higgs particle. Substituting this into equation (8.44) gives

$$\mathcal{L}_{\text{int}} = \frac{g_e v}{\sqrt{2}} \left(\overline{e^-_L} e^-_R + \overline{e^-_R} e^-_L \right) + \frac{g_e}{\sqrt{2}} \left(\overline{e^-_L} e^-_R + \overline{e^-_R} e^-_L \right) H \ . \qquad (8.46)$$

The first term in equation (8.46) has exactly the form expected for a fermion
mass (recall Chapter 5), so we can write for the electron mass,

$$m_e = g_e v / \sqrt{2} \ . \qquad (8.47)$$

Thus the theory can now *accommodate* a non–zero electron mass. Since g_e is
arbitrary, the value of the electron mass has not been calculated. Rather, we can
invert equation (8.47), so

$$g_e = \sqrt{2} m_e / v \ . \qquad (8.48)$$

The second term in equation (8.46) says that there is an electron–Higgs vertex
in the theory, of strength $g_e/\sqrt{2} = m_e/v$, as shown in Figure 8.3.

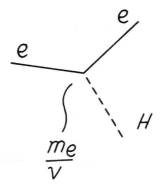

Figure 8.3
Electron–Higgs vertex.

This determines the probability for an electron or positron to radiate a Higgs boson, or for a Higgs boson to decay into e^+e^-. In Chapter 21, we will calculate with this vertex and analogous ones for other fermions when we study how to produce and detect Higgs bosons. Rewriting $\mathcal{L}_{\text{int}}$ to eliminate g_e, we have (using $\overline{e_L^-}e_R^- + \overline{e_R^-}e_L^- = \overline{e}e$)

$$\mathcal{L}_{\text{int}} = m_e \overline{e}e + \frac{m_e}{v}\overline{e}eH \ . \tag{8.49}$$

Some readers may have noticed that no mass term occurred for neutrinos, leaving them with $m_\nu = 0$. Formally, that is because by assumption the theory contains no right–handed neutrino state ν_R, so a term analogous to equation (8.44) cannot be written that will lead to a mass term $\overline{\nu}_R \nu_L$. Note this implies neutrinos do not interact with H. If there were a ν_R, it would be hard to observe; since it would have $T_3 = 0$ and $Q = 0$, it would not couple to $W^\pm$ or Z^0 or γ . We will discuss neutrino masses and how to test for whether they are zero in Chapter 29.

For quarks there is another subtlety (one that would also have occurred for ν_R if it had existed). It is well known in ordinary spin theory (*i.e.* $SU(2)$) that if $\psi = \binom{a}{b}$ is an $SU(2)$ doublet, then so is

$$\psi_c = -i\tau_2 \psi^* = \begin{pmatrix} -b^* \\ a^* \end{pmatrix} . \tag{8.50}$$

Then we can also write terms in $\mathcal{L}_{\text{int}}$ using

$$\phi_c = \begin{pmatrix} -\phi^{0*} \\ \phi^- \end{pmatrix}$$

which becomes, after invoking the Higgs mechanism,

$$\phi_c \rightarrow \begin{pmatrix} -\frac{v+H}{2} \\ 0 \end{pmatrix} . \tag{8.51}$$

Since ϕ has hypercharge $Y = +1$, ϕ_c has $Y = -1$, and still satisfies for each state, $Q = T_3 + Y/2$.

Then for quarks,

$$\mathcal{L}_{\text{int}} = g_d \overline{Q}_L \phi d_R + g_u \overline{Q}_L \phi_c u_R + \text{Herm. conjugate} \ . \tag{8.52}$$

Substituting from equations (8.45) and (8.50), and for Q_L , gives after simple algebra

$$\mathcal{L}_{\text{int}} = m_d \overline{d}d + m_u \overline{u}u + \frac{m_d}{v}\overline{d}dH + \frac{m_u}{v}\overline{u}uH \qquad (8.53)$$

where g_d and g_u have been eliminated in favor of the masses, following the same steps that led to equation (8.48). Again, the quark masses can be included in the description, but since g_d and g_u of equation (8.52) were arbitrary parameters, not related to each other or to g_e, the masses have to be measured. The last two terms of equation (8.53) describe the interaction of d and u quarks with H^0.

The entire procedure of this section can be repeated for the second and third families, giving further pieces of $\mathcal{L}_{\text{int}}$ which come from taking equation (8.49) with $e \rightarrow \mu, \tau$ and equation (8.52) with $u \rightarrow c, t$ and $d \rightarrow s, b$. Since H^0 interacts with a strength proportional to m_f, it couples most strongly to the heaviest fermions.

Now we have concluded our description of the Standard Model. The rest of the text explores the implications, tests, and predictions of the theory.

8.6   Comment on Vacuum Energy

The Higgs mechanism contributes to an important problem when cosmological considerations are introduced. We found the vacuum expectation value $v = \sqrt{-\mu^2/\lambda}$. We can evaluate the Higgs potential, equation (8.29), at the minimum, putting $\phi = v$ there, which gives $V(\phi = v) = V_0 = -\lambda v^4/2$. Since $v = 2M_W/g_2 \approx 246$ GeV, $V_0 \approx 2 \times 10^9 \lambda \text{GeV}^4$. This is apparently the contribution of spontaneous symmetry breaking to the vacuum energy density of the universe.

But, from astrophysics, we have learned that the density of luminous matter in the universe is about one proton per cubic meter on the average, and that the total density of matter is less than about 100 times this number. Thus empirically the total energy density is less than about 10^{-4} GeV/cm^3.

To compare this number with V_0, we need to guess a value to substitute in V_0 for λ (the arbitrary Higgs self-coupling in the Higgs potential), since λ is not known and is not determined by a gauge principle as far as we know. We also have to convert the units. For the latter, 1 GeV$^3 \approx 1.3 \times 10^{41}$cm^{-3}. While λ is not known, if it were eventually to be determined by some fundamental argument such as a gauge principle, presumably $\lambda \gtrsim 1/10$; we will see that the result is not very sensitive to this choice. Combining these, we find $V_0 \approx 2 \times 10^{49}$ GeV/cm^3, larger than the experimental value by a factor of order 10^{54}!

Technically this is not a contradiction, because we can always add a constant to the potential in theories without gravity and cancel V_0, but to do so involves tuning the constant to a part in 10^{54}, which is hardly satisfactory. This is (essentially) what is referred to as the problem of the cosmological constant. If gravity is included, terms of the kind we are considering will contribute to the energy–momentum tensor and, through Einstein's equations, dramatically affect the geometry of space–time. This is another clue that in spite of the remarkable descriptive power of the Standard Model, it is a theory that is incomplete at the fundamental level.

Problems

8.1:   Consider an $SU(2)$ invariant theory. Choose the Higgs field to be an $SU(2)$ triplet of real scalar fields, like W_1, W_2, and W_3 (or like the pion). That choice might seem to be a good way to try to give mass to the gauge bosons, with three W fields and three Higgs fields.

(a) Write the Lagrangian for the Higgs sector.

(b) Assume $\mu^2 < 0$ and $\lambda > 0$, show that the Lagrangian describes a massive scalar of mass $\sqrt{-2\mu^2}$ plus two massless Goldstone bosons.

(c) Write the Higgs Lagrangian in terms of the covariant derivative, $\partial_\mu - ig\,\overrightarrow{T} \cdot \overrightarrow{W_\mu}$. Study the terms that give mass to the W's. Remember, the triplet representation of T_k can be chosen as $-i\epsilon_{ijk}$. Show that only W_1 and W_2 get mass by this procedure.

(d) Comment on the relation of the results of (b) and (c). Comment on the connection to the choice of a Higgs doublet of complex fields of Section 8.4 where all the W's got mass.

8.2:   In working out the Higgs mechanism for the Standard Model, we used the covariant derivative (8.38), and the Higgs field near its vacuum expectation value (8.33), in the Higgs potential (8.26). We put $H(x) = 0$ to find the mass terms for W and Z arising from (8.39). Repeat with $H(x) \neq 0$ and obtain the interactions of H^0 with $W^\pm$ and Z and γ. [Since the mass terms have already been found, the terms that are quadratic in the fields need not be kept.]

8.3:   What is the ratio of couplings for a u–quark to a Z^0 and to an H^0, i.e. $g_{\overline{u}uH^0}/g_{\overline{u}uZ^0}$? Some thought is required to define the ratio.

8.4:   If we had chosen $\phi_1 \neq 0$ instead of $\phi_3 \neq 0$ in Section 8.4, what would be different?

8.5: In Section 2.6, we considered a free complex scalar field. The Lagrangian had a $U(1)$ global symmetry. The consequences could be interpreted in
terms of a pair of charged spin–zero antiparticles. In Section 8.2, we also considered a complex scalar field, and the results of the analysis were interpreted
as two neutral spin–zero particles of different mass. Discuss the relationship of
these two treatments.

Suggestions for Further Study

Most recent books on particle physics have some discussion of spontaneous
symmetry breaking. Aitchison and Hey have a good pedagogical discussion, emphasizing the connections to the physics of superconductivity, ferromagnetism,
and gauge invariance; for readers familiar with such other fields of physics this is
a useful supplement to the direct treatment here. Ryder also devotes a chapter to
a discussion of spontaneous symmetry breaking, with examples from outside particle physics, a proof of the Goldstone theorem, and an extensive list of references
to the original literature both inside and outside of particle physics.

CHAPTER 9

Cross Sections, Decay Widths, and Lifetimes; W and Z Decays

Now that we have developed the fundamental structure of the Standard Model, we want to determine its predictions and relate them to experiment. In general we want to learn how to calculate with the Standard Model.

Any Lagrangian field theory provides the rules to write matrix elements by combining vertices and propagators, as described in Chapter 2. Then observables such as cross sections, decay widths, and lifetimes are determined from the absolute squares of the matrix elements. In keeping with our approach of discussing mainly new physics, we will summarize the results that we will need, occasionally providing heuristic arguments. Reference to complete derivations will be given at the end of the chapter.

Write the S–matrix between final and initial states of momentum P_f and P_i as

$$S_{fi} = \delta_{fi} + (2\pi)^4 \, \delta^4 \left(P_f - P_i\right) \left(-iM_{fi}\right) \prod_{f,i} \frac{1}{\sqrt{2E_{f,i}}} \,, \tag{9.1}$$

which defines the normalization of the matrix element M_{fi}. This gives a transition probability (V is the volume of space)

$$d\Gamma = V \left(2\pi\right)^4 \delta^4 \left(P_f - P_i\right) \overline{|M_{fi}|^2} \prod_{f,i} \frac{1}{2E_{f,i}V} \prod_f \frac{V d^3 p_f}{(2\pi)^3} \,. \tag{9.2}$$

In general we square M_{fi} and sum or average over various degrees of freedom that are not observed, such as spin projections or color projections. We use $\overline{|M_{fi}|^2}$ to denote the appropriate summing and averaging for the calculation at hand. We will need the expression for a two–body cross section, $A + B \to C + D$, which is the transition probability divided by the flux,

$$d\sigma = d\Gamma/\text{flux} . \tag{9.3}$$

In the rest frame of particle B, the flux is v_A/V where v_A is the velocity of A. We can write $v_A = \sqrt{(P_A \cdot P_B)^2 - m_A^2 m_B^2}/E_A m_B$. Using the fact that we are in the rest frame of B, this gives

$$d\sigma = \frac{(2\pi)^4 \, \delta^4(P_C + P_D - P_A - P_B)}{4\sqrt{(P_A \cdot P_B)^2 - m_A^2 m_B^2}} \frac{d^3 P_C}{2 E_C \, (2\pi)^3} \frac{d^3 P_D}{2 E_D \, (2\pi)^3} \overline{|M|^2} . \tag{9.4}$$

All the volume factors have cancelled, as one would expect, and the answer is now in a Lorentz invariant form.

[To see that the factors $d^3 P/2E$ are Lorentz invariant for positive E, note that $\delta(E^2 - \vec{P}^2 - m^2)d^4 P$ is manifestly Lorentz invariant, and can be written

$$\delta(E^2 - \vec{P}^2 - m^2)d^4 P = \delta\left(\left[E - \sqrt{\vec{P}^2 + m^2} \right] \left[E + \sqrt{\vec{P}^2 + m^2} \right] \right) d^3 P dE$$

$$= d^3 P dE \delta\left(E - \sqrt{\vec{P}^2 + m^2} \right) \vartheta(E) / \left[E + \sqrt{\vec{P}^2 + m^2} \right]$$

$$= d^3 P/2E .]$$

For a decay $A \to B + C$, equation (9.2) gives

$$d\Gamma = \frac{1}{(2\pi)^2} \frac{1}{2E_A} \delta^4(P_B + P_C - P_A) \frac{d^3 P_B}{2 E_B} \frac{d^3 P_C}{2 E_C} \overline{|M|^2} . \tag{9.5}$$

Since $\delta^4(P_B + P_C - P_A) = \delta^3(\vec{P_B} + \vec{P_C} - \vec{P_A})\delta(E_B + E_C - E_A)$, the integral over $d^3 P_B$ can be done with the delta function. The energy integral is slightly subtle since E_B is a function of E_C. Remembering that $\delta(f(x)) = \delta(x - x_0)/|f'(x_0)|$ if $f(x_0) = 0$, and that $E_B = [(\vec{P_A} - \vec{P_C})^2 + m_B^2]^{1/2}$, let us evaluate the integral in the rest frame of A. Then the argument of the delta function is $f = E_C + [E_C^2 + m_B^2 - m_C^2]^{1/2} - E_A$, and $df/dE_C = E_A/E_B$. Finally then,

$$d\Gamma = \frac{|\overrightarrow{P_C}|}{E_A^2} \frac{d\Omega_C}{32\pi^2} \overline{|M|^2}. \tag{9.6}$$

The remaining integrations over the angles of C cannot be done until the angular dependence of M is known. Note that the same steps leading from equation (9.5) to equation (9.6) can be applied to the cross section of equation (9.4), giving

$$d\sigma = \frac{|\overrightarrow{P_C}|}{\sqrt{s}\sqrt{(P_A \cdot P_B)^2 - m_A^2 m_B^2}} \frac{d\Omega_C}{64\pi^2} \overline{|M|^2}$$

$$= \frac{|\overrightarrow{P_C}| d\Omega_C}{64\pi^2 s |\overrightarrow{P_A}|} \overline{|M|^2}. \tag{9.7}$$

Equations (9.6) and (9.7) will be used many times in the following. The useful Lorentz scalar variable $s = (P_A + P_B)^2$ has been introduced here (see also Appendix C). The final form of equation (9.7) is evaluated in the center of mass frame of particles A and B. If various degrees of freedom such as spin, color, etc., are observed, then M is the matrix element to produce particles with those degrees of freedom. Any degrees of freedom that are not observed are summed or averaged over in an appropriate way (see below). The way equation (9.6) is used will be illustrated in Section 9.3 below. The width has dimensions of mass and the cross section has dimensions (mass)$^{-2}$. These can be converted to conventional units with the information at the end of Chapter 1.

9.1   The Relation of Lifetime and Resonance Width to the Decay Probability

Consider an unstable particle. The wave function should be of the form $\psi(t) = \psi(0)\exp(-iEt/\hbar)$. If E is real, then $|\psi(t)|^2 = |\psi(0)|^2$ so there is no transition, which is unsatisfactory. We expect

$$|\psi(t)|^2 = |\psi(0)|^2 e^{-t/\tau} \tag{9.8}$$

where τ is the lifetime of the particle described by the state ψ. Then after one lifetime, the probability that the particle has not decayed is $1/e$. Write $E = E_0 - i\Gamma/2$, where for the moment $\Gamma/2$ is simply $-\text{Im } E$, and a result of the form of equation (9.8) is obtained.

To see what Im E means we can Fourier transform,

$$
\begin{aligned}
\widetilde{\psi}(E) &= \frac{1}{\sqrt{2\pi}} \int_{-\infty}^{\infty} dt\, e^{iEt/\hbar} \psi(t) \\
&= \frac{1}{\sqrt{2\pi}} \psi(0) \int_{0}^{\infty} dt\, e^{i(E-E_0)t/\hbar - \Gamma t/2\hbar} \\
&= \frac{i\hbar\psi(0)}{\sqrt{2\pi}} \frac{1}{(E-E_0) + i\Gamma/2}
\end{aligned}
\tag{9.9}
$$

where the integral is from zero to infinity because we assume $\psi = 0$ before $t = 0$. Since $|\widetilde{\psi}(E)|^2$ gives the probability of finding the state ψ with energy E, we have

$$
|\widetilde{\psi}(E)|^2 = \frac{\hbar^2 |\psi(0)|^2}{2\pi} \frac{1}{(E-E_0)^2 + \Gamma^2/4} .
\tag{9.10}
$$

For a decaying state there is not a sharp energy but a range of energies, spread about a central value.

Comparing equation (9.8) with the absolute square of $\psi(t)$, we see that we should also identify the lifetime

$$
\tau = \hbar/\Gamma .
\tag{9.11}
$$

In natural units, of course, $\hbar = 1$.

9.2   Scattering Through a Resonance

Quite often new physics appears as the production of a new particle, followed by its decay. That is like resonant scattering in quantum theory, generalized to allow different initial and final particles. In this section we will consider a process

$$
A + B \to R \to C + D .
\tag{9.12}
$$

For example, A and B could be $e^+ e^-$ or a quark pair, R could be a $W^{\pm}$ or Z^0.

To begin, consider ordinary scattering for spinless particles in quantum theory. Then we will generalize the result to the full set of quantum numbers relevant

to particle physics. The usual partial wave expansion of a scattering amplitude
for spinless particles is

$$f(\theta) = \frac{1}{2i\kappa} \sum_l (2l+1) \left(e^{2i\delta_l} - 1\right) P_l(\cos\theta) \tag{9.13}$$

where κ is the wave number (the magnitude of the center of mass three momentum) and δ_l is the change in phase of the lth partial wave. Squaring this and
integrating over angles gives a total elastic cross section

$$\sigma_{EL} = \frac{\pi}{\kappa^2} \sum_l (2l+1) \left|e^{2i\delta_l} - 1\right|^2 = \frac{4\pi}{\kappa^2} \sum_l (2l+1) \sin^2\delta_l . \tag{9.14}$$

If $\delta_l = \pi/2$ in a single partial wave, the cross section is large and we speak of
a resonance in that partial wave, or a particle of spin l. To exhibit the resonant
behavior more explicitly, we can write

$$e^{2i\delta_l} - 1 = 2ie^{i\delta_l} \sin\delta_l = 2i\sin\delta_l / (\cos\delta_l - i\sin\delta_l) = 2i/(\cot\delta_l - i) . \tag{9.15}$$

Let the total energy of the scattering particles be E, and let $E = E_R$
when $\delta_l = \pi/2$. We can expand $\cot\delta_l$ in a Taylor's series near the resonance,

$$\cot\delta_l \approx \cot\delta_l\left(E_R\right) + \left(E - E_R\right)\left(d\cot\delta_l/dE\right)\big|_{E=E_R} . \tag{9.16}$$

At the resonance $\cot\delta_l = 0$, and $\cot\delta_l$ decreases across the resonance. We define

$$\left(d\cot\delta/dE\right)\big|_{E=E_R} = -2/\Gamma . \tag{9.17}$$

This gives an amplitude

$$f(\theta) \approx \frac{(2l+1)\,P_l(\cos\theta)}{\kappa} \frac{\Gamma/2}{E_R - E - i\Gamma/2} \tag{9.18}$$

and a cross section

$$\sigma_{EL} \simeq \frac{4\pi}{\kappa^2}(2l+1) \frac{\Gamma^2/4}{\left(E - E_R\right)^2 + \Gamma^2/4} . \tag{9.19}$$

It is clear from comparison with the analysis of Section 9.1 why the definition
of equation (9.17) was chosen; Γ measures the rate of change of δ_l near the
resonance. Another way of interpreting Γ is to note that σ_{EL} falls by a factor
of 2 when $E - E_R = \pm\Gamma/2$. The shape described by equation (9.19) is called
a Breit–Wigner resonance; it is shown in Figure 9.1. Γ is the full width at half
maximum (FWHM).

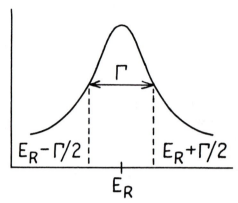

Figure 9.1
Graphical representation of a
Breit–Wigner resonance.

[Note that we can write the Breit–Wigner amplitude in terms of Lorentz scalar variables by multiplying above and below by $E_R + E$. Since $s = E^2$, $s_R = E_R^2 = m_R^2$, and $E_R + E \simeq 2\sqrt{s_R} = 2m_R$ whenever the resonance approximation is valid, this gives

$$\frac{\Gamma/2}{E_R - E - i\Gamma/2} \simeq \frac{m_R\Gamma}{(s_R - s) - im_R\Gamma} \, .]$$

$$(9.20)$$

Now we can generalize to the real situation where the particles may carry color and where the decay can occur into several final states. The key to seeing how to include the color degree of freedom is to understand the factor $2l + 1$ in equation (9.19) for a resonance of spin l. It simply says that after integrating over angles the $2l + 1$ states of different l_z all contribute equally. Similarly, if the final particles have spin, all $2s + 1$ states of each particle can be produced (for photons or gluons that have only transverse polarizations, $2s + 1$ is replaced by 2).

For the initial states it is a little more subtle. Suppose two spin–1/2 particles combine into a spin–zero resonance. The initial particles can combine into a triplet or a singlet state, with three of the four possible combinations being in the triplet. Only the singlet can give the spin–zero resonance. A little thought will convince the reader that the proper way to proceed is to average over initial spins, *i.e.* sum over all spin projections but divide by $(2s_A + 1)(2s_B + 1)$.

A similar argument holds for the color states. The color multiplicity for a quark is three, for a gluon it is eight, and for a color singlet state is unity. Some possible resonances could have any of these multiplicities.

Suppose, then, that particles A and B have spins s_A, s_B and color multiplicities c_A, c_B and scatter through a resonance R with spin s_R and color multiplicity c_R . The cross section is

$$\sigma(A + B \rightarrow R \rightarrow CDE\ldots)$$

$$\approx \frac{\pi}{\kappa^2} \left[\frac{(2s_R + 1)\, c_R}{(2s_A + 1)\,(2s_B + 1)\, c_A c_B} \right] \frac{\Gamma_{AB}^R \Gamma_f^R}{(E_R - E)^2 + \Gamma_R^2/4} . \qquad (9.21)$$

Here Γ_R is still the total width, but Γ_{AB}^R is the partial width for $R \rightarrow AB$, which measures the probability for the transition $AB \rightarrow R$. The partial width for the final state is Γ_f^R; we have allowed for any final process where $R \rightarrow C + D + E + \cdots$. If the total production is desired, then we sum over all ways R can decay, which just gives the total width, so $\Gamma_f^R \rightarrow \Gamma_R$.

When Γ_{AB}^R and Γ_f^R are being computed all spin states should be summed over, as described above.

Often, when we use equation (9.21), we will make one kinematical change. We define the Lorentz scalar variable $s = \left(p_A + p_B\right)^2$. Then in the center of mass system the total energy is $E = E_A + E_B$ and $s = E^2$. Similarly, $s_R = E_R^2 = m_R^2$ (the mass of a single state produced at rest and its total energy are the same). In relativistic notation, $\kappa^2 = \left[s - \left(m_A + m_B\right)^2 \right] \left[s - \left(m_A - m_B\right)^2 \right] / 4s$. Since the cross section is small if $E \neq E_R$, we can replace E by E_R or vice-versa anywhere but in the Breit–Wigner denominators. Then multiplying through by $\left(E_R + E\right)^2 / \left(E_R + E\right)^2 \simeq 4s / \left(E_R + E\right)^2$, we get

$$\sigma(A + B \rightarrow R \rightarrow C + D + E\cdots)$$

$$\approx \frac{4\pi s}{\kappa^2} \left[\frac{(2s_R + 1)\, c_R}{(2s_A + 1)\,(2s_B + 1)\, c_A c_B} \right] \frac{\Gamma_{AB}^R \Gamma_f^R}{\left(s - m_R^2\right)^2 + m_R^2 \Gamma_R^2} . \qquad (9.21')$$

Often equation (9.21') is used instead of equation (9.21) because s is a Lorentz scalar variable and is usually more convenient. Any numerical differences between equation (9.21) and equation (9.21') simply show the inadequacies of a resonance approximation. Note this result implies that resonance cross sections cannot be arbitrarily large. At large s, $\sigma < \left(16\pi\Gamma_{AB}^R\right) / \Gamma_R m_R^2$, ignoring the statistical factors. Thus the resonance mass and width and branching ratios limit the size of σ.

9.3   The W Width

In Chapter 10, we will use equation (9.21') to calculate the production cross sections for $W^{\pm}$ and Z^0. A and B will be quarks, which occur as the constituents of protons, and the $W^{\pm}$ and Z^0 will decay to all states allowed by the Standard Model.

To calculate the $W^{\pm}$ width, we must, according to equation (9.6), calculate the matrix element absolute squared, summed and averaged over spins, and evaluate the angular integral to get Γ. There will be a partial width for each channel into which W can decay. Since we need estimates but we do not need detailed calculations, it will not be difficult to do that. We will find, in a number of examples throughout the book, that calculating decay widths or cross sections that are approximately correct does not require any expertise beyond what we have developed.

We always begin by looking at the Lagrangian, equation (7.32), to see what transitions are allowed. The possibilities for W^+ are (from the lightest family),

$$W^+ \rightarrow e^+\nu_e, \; u\overline{d} \; . \tag{9.22}$$

In addition, from other families we can have the channels obtained by replacing e by μ, τ; ν_e by ν_μ, ν_τ; u by c, t; d by s, b . These give

$$W^+ \rightarrow \mu^+\nu_\mu \; , \; \tau^+\nu_\tau \; , \; c\overline{s}, \; t\overline{b} \; . \tag{9.23}$$

Since all of the final particles are spin-1/2 fermions, and since the Lagrangian tells us that they all have the same vertex factor $(g_2/\sqrt{2})$, we only need to calculate any one of these decays. The others are the same.

The matrix element for $W^+ \rightarrow e^+\nu_e$ is then

$$M = \frac{g_2}{\sqrt{2}}\epsilon_\mu \overline{e}\gamma^\mu P_L \nu \tag{9.24}$$

where ϵ^μ is the polarization wave function of the W, and $\overline{e}$, ν represent the spinor wave functions of the positron and neutrino.

We need to calculate the sum over final spins and average over initial spins of $|M|^2$. We could learn the technical details necessary to do that. Instead, however, we will keep in mind that our goal is not necessarily to become working particle physicists but to achieve a useful, semi–quantitative, understanding of particle physics and its aims and achievements. Consequently, we will calculate with very simple approximations, here and later. Because of the P_L in equation (9.24) that only allows left–handed ν_e, e^+ to interact with W, we will only count one spin state each for e^+, ν_e. The W has three polarization states but we average over them for an initial W, so the sum and average over spin states just gives a factor of unity. We know from equation (5.47) that the product $\bar{e}\nu$ has dimensions of mass, so the matrix element squared has dimensions of $(\text{mass})^{-2}$. Neglecting the positron and neutrino masses, the only quantity that can supply the dimensions is M_W. We do not know whether to use M_W, $M_W/2$, $2M_W$, etc., from a dimensional argument, but our goal is only to get an *estimate* of Γ_W. When doing estimates, we will always use a naïve choice from the mass parameters that are present, and replace spinor factors $\bar{u}u$ by the appropriate mass or energy. Hence we put $\epsilon_\mu \bar{e}\gamma^\mu P_L \nu = M_W$. With the spin factors counted as above giving a factor of 1, we get

$$\overline{|M|^2} \simeq g_2^2 M_W^2 / 2 \ . \tag{9.25}$$

The correct answer is $g_2^2 M_W^2 /3$, so our estimate is off by a factor $3/2$. For all our calculations, we will finally quote the correct value so the formulas are useful for reference purposes. But, for anyone not actively working as a particle physicist, it hardly matters to have such results accurate to better than a factor of two or three, so readers should be entirely confident of their ability to estimate and understand interaction rates.

From equation (9.6) we then have in the W rest frame,

$$d\Gamma \simeq \frac{P_e}{M_W^2} \frac{d\Omega_e}{32\pi^2} \frac{1}{3} g_2^2 M_W^2 \ . \tag{9.26}$$

Since $\overline{|M|^2}$ comes out to be independent of the angles of the final particles for unpolarized W's, $\int d\Omega \to 4\pi$. Neglecting positron and neutrino masses implies that the electron's energy is $M_W/2$, so $P_e \approx P_{0e} \simeq M_W/2$. Thus

$$\Gamma_W^{e\nu} = \frac{\alpha_2 M_W}{12} \tag{9.27}$$

where $\alpha_2 = g_2^2/4\pi$. From the way the calculation went, it is clear we could have written, without any calculations, that $\Gamma_W^{e\nu} = \delta \alpha_2 M_W$, where δ is a pure number. Typically $\delta \simeq 1$ for the total width of any particle, summed over all final channels (see equation (9.28) below).

There is one subtlety for the quark channels. The W decay could be into any of the quark colors α, *i.e.* $W^+ \rightarrow u_r \overline{d}_r$ or $W^+ \rightarrow u_b \overline{d}_b$, *etc.*, so the quark channels contribute three times as much as $W^+ \rightarrow e^+ \nu_e$. [The technical way to see the factor of three is to note that the W is a color singlet, so if the quarks carry color indices α, β then the matrix element in color space is $\delta_{\alpha\beta}$, and its square summed over final colors is $M^2 = \sum_{\alpha\beta} \delta_{\alpha\beta}\delta_{\alpha\beta} = \sum_\alpha \delta_{\alpha\alpha} = 3$.] Thus the total number of channels is

$$
\begin{array}{cc}
W^+ \rightarrow e^+ \nu_e & 1 \\
u\overline{d} & 3 \\
\mu^+ \nu_\mu & 1 \\
c\overline{s} & 3 \\
\tau^+ \nu_\tau & 1 \\
t\overline{b} & 3 \\
\hline
& 12
\end{array}
$$

so the total width is the sum of all of these, or

$$\Gamma_W^{\text{TOT}} = \alpha_2 M_W . \tag{9.28}$$

This is computed in the approximation that all final fermions are very light compared to M_W. That is true for all but the t-quark; its mass is not known yet, but it cannot be lighter than about 23 GeV or it would have been produced and detected in $e^+ e^-$ collisions. So there is a kinematical correction needed for $1/4$ of the channels; if m_t is too large the $t\overline{b}$ channel could be kinematically excluded. Since we are only interested in approximate results, we will not worry about these details. Finally, then, the total width of the W is given by equation (9.28), the partial width to $l^+ \nu_l$ by equation (9.27), and the partial width to $u\overline{d}$ by $3 \times$ [equation (9.27)]. If we had carried through our approximation method, we could have "calculated" for $W^+ \rightarrow e^+ \nu_e$ and gotten $\alpha_2 M_W / 8$, a good estimate for equation (9.27), or guessed equation (9.28), which would have been exactly correct. Note that the factor of three for color is our first testable prediction involving the color force. The Standard Model predicts that a $W^\pm$ decays three times more often into quarks than into leptons. Both the existence of a property that is different for quarks and leptons, and the numerical value of the number of colors, are involved. The prediction is consistent with data, but it is not yet tested extremely accurately.

Numerically, $\alpha_2 \simeq 1/30$ and $M_W \simeq 82$ GeV (Chapters 10 and 11 describe how these are measured), so

$$\Gamma_W^{TOT} \simeq 2.7 \text{ GeV} \tag{9.29}$$

$$\Gamma_W^{e\nu} \simeq 0.23 \text{ GeV} \tag{9.30}$$

$$\Gamma_W^{u\bar{d}} \simeq 0.68 \text{ GeV} \tag{9.31}$$

giving numbers to two places. Note that $W^{\pm}$ is a rather narrow resonance, with $\Gamma/M = \alpha_2 \simeq 1/30$. The widths for $\mu\nu_\mu$, $\tau\nu_\tau$, $c\bar{s}$ are the same as those above, while that for $t\bar{b}$ is a little smaller because the final masses are larger and the phase space consequently a little smaller.

That the quarks are confined and appear as jets of hadrons (see Chapter 15) does not affect any of the above results, since one sums over all of the ways a quark could hadronize. However, computing the partial width for a particular hadronic final state would require a detailed knowledge of the bound state wave function of the hadron, and is beyond the present state of the art.

9.4   The Z^0 Width

From our basic Lagrangian, we can write the Z width as well. Z^0 has only flavor–diagonal interactions,

$$
\begin{aligned}
Z^0 &\to e^+e^- \\
&\to \nu_e\bar{\nu}_e \\
&\to u\bar{u} \\
&\to d\bar{d}
\end{aligned}
\tag{9.32}
$$

for the first family. As usual the other families occur too. Again, the masses in the final state can be ignored except for $t\bar{t}$, and the quark channels have three times the partial width because they have the three color channels.

Consider e^+e^-. From equation (7.32), the matrix element is

$$M_{e^+e^-} = \frac{g_2\epsilon_\mu}{\cos\theta_w}\left[\overline{e_L^-}\gamma^\mu e_L^-\left(-\frac{1}{2}+\sin^2\theta_w\right)+\overline{e_R^-}\gamma^\mu e_R^-\left(0+\sin^2\theta_w\right)\right]. \tag{9.33}$$

This is similar in form to equation (9.24), with a little more complicated spin structure. The factors in parentheses are $T_3 - Q\sin^2\theta_w$ as expected from equation (7.25). Recalling from Chapter 5 that $\overline{e_L^-}\gamma^\mu e_L^- = \overline{e_L^-}\gamma^\mu P_L e_L^-$, we see that each term is of the same form as for the W decay. Further, since a left–handed particle cannot turn into a right–handed one (*i.e.* flip its spin) if it is massless, the two terms in equation (9.33) will not interfere in the approximation that we neglect the final masses. (So we expect corrections of the order m_f/M_W for each fermion f.)

Consequently, we can calculate $\Gamma_Z^{e^+e^-}$ by repeating the steps from equation (9.24) to (9.28), or simply by taking $\Gamma_W^{e\nu}$ and replacing

$$e^+e^- : \quad \left(\frac{g_2}{\sqrt{2}}\right)^2 \to \left(\frac{g_2}{\cos\theta_w}\right)^2 \left[\left(-\frac{1}{2}+\sin^2\theta_w\right)^2 + \left(0+\sin^2\theta_w\right)^2\right]$$

$$(9.34)$$

$$= \frac{g_2^2}{\cos^2\theta_w}\left[\frac{1}{4}-\sin^2\theta_w + 2\sin^4\theta_w\right],$$

and $M_W \to M_Z$. For the other fermions, the replacements are

$$\overline{\nu}_e\nu_e : \quad \left(\frac{g_2}{\sqrt{2}}\right)^2 \to \frac{g_2^2}{\cos^2\theta_w}\left[\left(\frac{1}{2}-0\right)^2 + (0+0)^2\right] = \frac{g_2^2}{4\cos^2\theta_w}, \quad (9.35)$$

$$\overline{u}u : \quad \left(\frac{g_2}{\sqrt{2}}\right)^2 \to \left(\frac{g_2}{\cos\theta_w}\right)^2 \left[\left(\frac{1}{2}-\frac{2}{3}\sin^2\theta_w\right)^2 + \left(0-\frac{2}{3}\sin^2\theta_w\right)^2\right]$$

$$(9.36)$$

$$= \frac{g_2^2}{\cos^2\theta_w}\left[\frac{1}{4}-\frac{2}{3}\sin^2\theta_w + \frac{8}{9}\sin^4\theta_w\right],$$

$$\overline{d}d : \quad \left(\frac{g_2}{\sqrt{2}}\right)^2 \to \left(\frac{g_2}{\cos\theta_w}\right)^2 \left[\left(-\frac{1}{2}+\frac{1}{3}\sin^2\theta_w\right)^2 + \left(0+\frac{1}{3}\sin^2\theta_w\right)^2\right]$$

$$(9.37)$$

$$= \frac{g_2^2}{\cos^2\theta_w}\left[\frac{1}{4}-\frac{1}{3}\sin^2\theta_w + \frac{2}{9}\sin^4\theta_w\right].$$

We can put $\alpha_2 \to \alpha_2/2\cos^2\theta_w$ for $\overline{\nu}_e\nu_e$, *etc.*, to get the final numbers, with analogous substitutions for the others,

$$\Gamma_Z^{e^+e^-} = \frac{\alpha_2 M_Z}{24\cos^2\theta_w}\left(1 - 4\sin^2\theta_w + 8\sin^4\theta_w\right),$$

$$\Gamma_Z^{\bar{\nu}_e\nu_e} = \frac{\alpha_2 M_Z}{24\cos^2\theta_w},$$

$$\Gamma_Z^{\bar{u}u} = \frac{3\alpha_2 M_Z}{24\cos^2\theta_w}\left(1 - \frac{8}{3}\sin^2\theta_w + \frac{32}{9}\sin^4\theta_w\right), \qquad (9.38)$$

$$\Gamma_Z^{\bar{d}d} = \frac{3\alpha_2 M_Z}{24\cos^2\theta_w}\left(1 - \frac{4}{3}\sin^2\theta_w + \frac{8}{9}\sin^4\theta_w\right).$$

A factor of three for each color has been inserted for the quarks, as discussed above.

In Chapter 11 we will discuss in detail how $\sin^2\theta_w$ is measured. The best value at present is $\sin^2\theta_w = 0.23$, with an accuracy of about 5%. Using this in the above partial widths we get:

$$\Gamma_Z^{e^+e^-} = 0.08 \text{ GeV}$$

$$\Gamma_Z^{\bar{e}\nu_e} = 0.16 \text{ GeV}$$

$$\Gamma^{\bar{u}u} = 0.28 \text{ GeV} \qquad (9.39)$$

$$\Gamma^{\bar{d}d} = 0.36 \text{ GeV}.$$

For the first family this gives 0.88 GeV, so for all three families,

$$\Gamma_Z^{\text{TOT}} = 2.64 \text{ GeV}, \qquad (9.40)$$

almost exactly the same as Γ_W^{TOT} (accidentally).

One of the most important things about Γ_Z^{TOT} is that any new particle x that has non-zero weak isospin, and therefore couples to Z^0 according to equation (7.25), will appear in Z^0 decays. For example, if additional families exist their neutrinos may have masses below $M_Z/2$, so the decays $Z \rightarrow \bar{\nu}_x\nu_x$ could occur and increase Γ_Z^{TOT}. Since each neutrino pair contributes 160 MeV to Γ_Z^{TOT}, a measurement to an accuracy better than about 100 MeV is necessary to detect such a contribution. Such measurements are expected to be made for the first time at the linear electron collider called SLC, at SLAC, in late 1987 or early 1988. The present best value for Γ_Z^{TOT} is $\Gamma_Z^{\text{TOT}} < 5.6$ GeV at 90% confidence level, from the UA 2 group at CERN.

From equation (9.11), we note that Z or $W^\pm$ lifetimes are $\tau_W = 1/\Gamma_W =$ (1/2.64) GeV^{-1}. From equations (1.1) or (1.2), 1 GeV^{-1} $= 6.6 \times 10^{-25}$ sec, so $\tau_W = 2.5 \times 10^{-25}$ sec. In such a time the distance a W could travel is $\gamma c \tau$. Since $\gamma = E_W/M_W$ will be less than about 10 for any machine in the next few decades, and $c = 3 \times 10^{10}$ cm/sec, the distance a W typically travels is less than 10^{-14} cm. Detectors will probably be able to observe a separation of production and decay points of somewhat less than 10^{-2} cm, but not much less than that. Thus a W will always decay before it can be detected, and its presence must be deduced from its decay products.

9.5   Branching Ratios

The various ratios $\Gamma_W^{a\bar{b}}/\Gamma_W^{TOT}$ and $\Gamma_Z^{f\bar{f}}/\Gamma_Z^{TOT}$ are called branching ratios; they determine the production rates and the decay signatures that allow W's and Z's to be detected. Combining the various partial widths calculated in the previous two sections, we have

$$\text{BR}(W^\pm \to e^\pm \nu) = \text{BR}(W^\pm \to \mu^\pm \nu) \simeq 1/12$$
$$\text{BR}(W^\pm \to u\bar{d}) = \text{BR}(W^\pm \to c\bar{s}) \simeq 1/4$$
$$\text{BR}(Z^0 \to e^+ e^-) = \text{BR}(Z^0 \to \mu^+ \mu^-) \simeq 0.030$$
$$\text{BR}(Z^0 \to u\bar{u}) = \text{BR}(Z^0 \to c\bar{c}) \simeq 0.106 \qquad (9.41)$$
$$\text{BR}(Z^0 \to d\bar{d}) = \text{BR}(Z^0 \to s\bar{s}) \simeq 0.136$$
$$\text{BR}(Z^0 \to \nu_e \bar{\nu}_e) = \text{BR}(Z^0 \to \nu_\mu \bar{\nu}_\mu) = \text{BR}(Z^0 \to \nu_\tau \bar{\nu}_\tau) \simeq 0.061$$

and the $\tau\nu$, $b\bar{b}$, and $t\bar{b}$ modes can be filled in by the reader. Note that the branching ratios depend on the factors of three inserted for quarks, so confirmation of these branching ratios confirms the presence of colors, the number of colors, and the $SU(2) \times U(1)$ couplings.

Problems

9.1:   Suppose a fourth family exists, with a complete set of particles including a fourth neutrino ν_4 of mass similar to the other neutrinos, and gauge interactions identical to those of the other families. What would the branching ratio be for $Z \rightarrow \nu_4 \bar{\nu}_4$? What about the branching ratio for $W^{\pm} \rightarrow L_4^{\pm} \nu_4$ where L_4 is the charged lepton of the family?

9.2:   Suppose we want to measure the cross section for colliding ν_{μ} with right–handed eletrons, e_R, in order to determine $\sin^2 \theta_w$. (i) How does the cross section vary with θ_w? (ii) Estimate the cross section assuming $m^2{}_Z \gg s \gg m_e^2$. Proceed by (a) drawing any diagrams that allow the scattering to occur, using the rules of Chapter 7; (b) put in the vertex factors and propagators; (c) make approximations as described in Section 9.3 to obtain the matrix element; (d) go from the matrix element to the cross section using the results of the beginning of this chapter.

9.3:   From the formulas of Section 9.2, argue that the largest a cross section can be for a physical process $a + b \rightarrow c + d$ is $\sigma \leq 16\pi/(2s_A + 1)(2s_B + 1)s$ if $\sqrt{s}$ is large compared to external masses and the scattering occurs through a resonance.

Suggestions for Further Study

The derivation of a transition probability or a cross section from the scattering amplitude is covered at some level in all quantum theory books. One of the most careful treatments, for readers who wish to look into the subtleties, is in Gottfried; even that treatment does not include wave packet spreading. A fundamental treatment is given by deWit and Smith. For particle physics, where essentially all interactions are point–like, the simplest derivations are satisfactory.

CHAPTER 10

Production
of $W^\pm$ and Z^0

Once the prediction of the $W^\pm$ and Z^0 masses was firm and the theory on which it was based was being confirmed in a variety of ways, it was a matter of high priority to actually find the electroweak gauge bosons directly. But their expected masses were very large, 80–100 GeV. [Since the masses depended on $\sin^2\theta_w$, as in equations (8.41) and (8.43), determining their precise value depended on the quality of the measurements of $\sin^2\theta_w$, which had accuracies only of perhaps 30% at the time. In addition, carefully including corrections to the simplest theory led to shifts to higher masses by almost 3%; shifts of about 3 GeV cannot be ignored when planning a machine.] Thus it was necessary to be able to produce a particle with mass about 100 GeV, which required $\sqrt{s} \gtrsim 100$ GeV, at a sufficient rate to detect enough events to establish there was a signal.

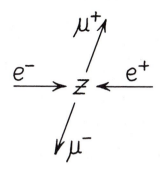

Figure 10.1
e^+e^- collide to make a Z^0,
which decays to $\mu^+\mu^-$.

One could hope to produce Z^0's at a machine where e^+e^- beams collided, as in Figure 10.1. But the e^+e^- colliders planned or working at the time had $\sqrt{s} \lesssim 40$ GeV. Because electrons radiate energy easily, being very light, it is difficult to accelerate them to the necessary energies. Such colliders are only now under construction; one at SLAC, called SLC, using newer ideas to make a linear collider, is expected to produce Z^0's this way in 1987. Another, at CERN, called LEP, is expected to produce Z^0's in 1989. As remarked in the Appendix on kinematics, it is much easier to get to higher energies at a collider than at a fixed target machine; fixed target machines with beam energies over one-hundred times greater than what existed would have been required.

So the only way to produce the gauge bosons was to use the quarks in the proton. This had the additional advantage that both the charged $W^\pm$ and the neutral Z^0 could be produced, as in Figure 10.2, where one possibility is shown for each.

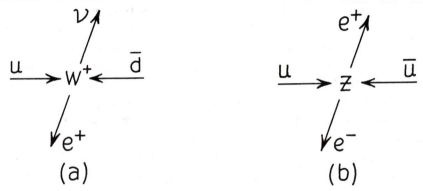

Figure 10.2
Production of W and Z at a hadron collider.

(At e^+e^- colliders only Z^0 can be produced when $\sqrt{s} \simeq M_Z$; to obtain W's at a detectable rate, $\sqrt{s} > 2M_W$ is needed.) The quarks, however, are in hadrons, so one has to accelerate the hadrons and then collide them appropriately. In Chapter 18, we will study more systematically the quark and gluon structure of hadrons; here we take a very simple point of view and work out what is needed to estimate the rate for producing and detecting $W^\pm$ and Z^0. At CERN the decision was made to collide protons and antiprotons, since they could be accelerated in opposite directions in one ring of magnets, rather than requiring two rings to accelerate and collide protons.

10.1   Getting Quarks to Collide

A proton is made of quarks and gluons, and one can approximately write

$$p = \underbrace{uud}_{\substack{\text{valence} \\ \text{quarks}}} + \underbrace{u\bar{u} + d\bar{d} + \cdots}_{\text{sea quarks}} + \underbrace{g + g + \cdots}_{\text{gluons}}. \tag{10.1}$$

The first uud have the right combination of spin, electric charge, and color (singlet) to be a proton; a proton at rest is effectively composed of these as far as its electroweak properties go. In a relativistic theory, creation of pairs is always going on, so protons contain $u\bar{u}$ pairs, $d\bar{d}$ pairs, and so on, each with some probability. The gluons are the bosons exchanged to provide the forces to hold the quarks together.

As one might expect, the momentum of a proton is about equally divided between quarks and gluons. Since there are three quarks, each carries on the average about $1/6^{\text{th}}$ of the proton's momentum—when we need a number, we will use 0.15 for the quark fraction. More quantitatively, if the proton has four momentum P and the quark of interest has four momentum p_i, we define $x_i P = p_i$. For a given kind of quark small values of x_i are most probable since quark–antiquark pairs frequently appear and disappear (such quarks are called "sea quarks"). As x_i gets larger, the probability of finding the appropriate quark with momentum fraction x_i decreases. Around $x_i = 0.15$ there is still a significant probability, and after that it decreases rapidly.

Whenever collisions occur that involve mass scales or momentum transfers large compared to those normally present in a proton, one can define a "structure function", $F_{q/p}(x)$, that gives the probability of finding a quark q carrying momentum fraction x in a proton p. Similar structure functions exist for gluons. Also, similar distributions exist for antiquarks, *etc.*, in an antiproton. The structure functions are measured in various experiments; we will describe them further in Chapter 18. Then the cross section for proton–antiproton collisions to give $u\bar{d} \rightarrow W^+$ is

$$\sigma = \int dx_1 dx_2 \, F_{u/p}(x_1) F_{\bar{d}/\bar{p}}(x_2) \hat{\sigma}(u\bar{d} \rightarrow W^+) \tag{10.2}$$

where $\hat{\sigma}$ is the "constituent cross section" for the process $u + \bar{d} \rightarrow W^+$. $\hat{\sigma}$ depends on the u and $\bar{d}$ momenta, and thus on x_1 and x_2. The full cross section is obtained by integrating over all allowed x_1 and x_2. We calculate $\hat{\sigma}$ in the Standard Model (just below) and convolute it with measured structure functions to obtain the p–$\bar{p}$ cross section.

10.2   The Constituent Cross Section

We can obtain the constituent cross section from equation (9.21′), writing $\hat{s}$ for the square of the $u\bar{d}$ center of mass energy,

$$\sigma\big(u+\bar{d}\to W^+\to f\big) = \frac{4\pi\hat{s}}{\kappa^2}\,\frac{3}{2\times2\times3\times3}\,\frac{\Gamma^W_{u\bar{d}}\,\Gamma^W_f}{\big(\hat{s}-M_W^2\big)^2+M_W^2\Gamma_W^2}\,. \qquad (10.3)$$

Neglecting the u, $\bar{d}$ masses compared to M_W, $\kappa^2=\hat{s}/4$, so the numerical factor simplifies to $4\pi/3$. We only want an estimate so we will work in the narrow width approximation (which is rather good since $\Gamma_W/M_W \ll 1$). Using the approximation

$$\delta(x-z) \simeq \frac{\Gamma}{\pi}\,\frac{\sqrt{x}}{\big(z-x\big)^2+x\Gamma^2} \qquad (10.4)$$

we get

$$\hat{\sigma} \simeq \frac{4\pi^2}{3}\,\frac{\Gamma^W_{u\bar{d}}\,\Gamma^W_f}{\Gamma_W\,M_W}\,\delta\big(\hat{s}-M_W^2\big)\,. \qquad (10.5)$$

The delta function is expected, to conserve energy.

10.3   The W Production Cross Section

To extract the x_1, x_2 dependence of $\hat{\sigma}$, we have to do some kinematics. Let p and $\bar{p}$ be the proton and the antiproton four momenta, and p_u, $\bar{p}_d$ the quark momenta. Then, following Appendix C and recalling $p_u = x_1 p$, $\bar{p}_d = x_2\bar{p}$,

$$s = (p+\bar{p})^2 \qquad (10.6)$$
$$\hat{s} = (p_u+\bar{p}_d)^2 \qquad (10.7)$$

so

$$\hat{s} = (x_1 p + x_2\bar{p})^2 = \big(x_1^2+x_2^2\big)m_p^2 + 2x_1 x_2 p\cdot\bar{p}$$
$$s = 2m_p^2 + 2p\cdot\bar{p}\,,$$

where we used the fact that $p^2 = \bar{p}^2 = m_p^2$. Since $\hat{s}\simeq m_W^2 \gg m_p^2$ and $s>\hat{s}$, we can drop the m_p^2 terms. Then we get

$$\hat{s} \simeq x_1 x_2 s \ . \tag{10.8}$$

Using this in the delta function gives

$$\delta(\hat{s} - M_W^2) = \delta(x_1 x_2 s - M_W^2) = \frac{1}{s} \delta\left(x_1 x_2 - \frac{M_W^2}{s}\right). \tag{10.9}$$

Finally, we can use this and $\hat{\sigma}$ in equation (10.2), giving

$$\sigma_W = \frac{4\pi^2}{3} \frac{\Gamma_W^{\mathrm{TOT}}}{sM_W} \mathrm{BR}(W^\pm \to u\overline{d}) \mathrm{BR}(f) \times$$
$$\int dx_1 dx_2 u(x_1) \overline{d}(x_2) \delta(x_1 x_2 - M_W^2/s) \tag{10.10}$$

where we have written $u(x_1)$ for $F_{u/p}(x_1)$, $\overline{d}(x_2)$ for $F_{\overline{d}/\overline{p}}(x_2)$, and expressed the results in terms of the initial and final branching ratios. From equation (9.41), the initial $\mathrm{BR}(W^+ \to u\overline{d}) = 1/4$. We will discuss the final state branching ratio $\mathrm{BR}(f)$ below when we consider how to detect W. Note that σ_W is proportional to $\Gamma_W^{\mathrm{TOT}}/M_W$.

If we had instead calculated $u\overline{u} \to Z^0$, some obvious changes would be required. In fact, equation (10.10) gives the cross section for production of any spin one, color singlet state from quarks, with obvious changes for the branching ratios, structure functions, and labels. Note that σ is essentially a product of various factors: (i) the numerical factor, that depended on the spins and colors (for particles or beams of different spins or colors, this factor would change), (ii) the factor $\Gamma_W^{\mathrm{TOT}}/M_W$, which will always be present, (iii) the initial and final branching ratios, and (iv) a factor that depends on the probabilities of finding the beam particles in a proton and only depends on the particle being produced through its mass.

To estimate the integral in equation (10.10), we proceed as follows. In practice, we would just look up the structure functions (see Chapter 18) and evaluate the integral. To get a feeling for the result, we can make some analytic approximations. A reasonable form for $u(x)$ is

$$u(x) \simeq \frac{35}{16\sqrt{x}} (1 - x)^3 \ ; \tag{10.11}$$

the normalization is fixed so $\int u(x)dx = 2$ because there are two u–quarks in a proton. Very crudely, the d–quark function, $d(x)$, can be taken to be $1/2 \ u(x)$

since there is one d–quark in a proton. This form says that the probability of a quark carrying a large fraction of the proton momentum decreases as $(1-x)^3$, and increases (in an integrable way) as $1/\sqrt{x}$ at small x. Such a form is not very different from the more complicated measured values.

The integral in equation (10.10) is (defining $\tau = M_W^2/s$)

$$I = \int dx_1 dx_2 \; u(x_1)\overline{d}(x_2)\delta(x_1 x_2 - M_W^2/s)$$

$$= \int_\tau^1 u(x) d\left(\frac{\tau}{x}\right)\frac{dx}{x} \; . \tag{10.12}$$

Substituting from equation (10.11) into equation (10.12) gives an integral that can be done analytically. For $M_W = 82$ GeV and $\sqrt{s} = 630$ GeV at CERN ($\tau = 0.017$),

$$I = \left(\frac{35}{16}\right)^2 \frac{1}{2\sqrt{\tau}} \int_\tau^1 dx \frac{(1-x)^3\,(x-\tau)^3}{x^4} \; .$$

Because of the x^{-4}, the result for I is significantly larger than the separate integrals over u or d; the answer is $I \simeq 16.4$. For the measured structure functions, the answer is somewhat smaller.

Putting all the factors together, we find (using Section 1.5 to convert from GeV^{-2} to cm^2)

$$\sigma_W \simeq \frac{4\pi^2}{3}\frac{(2.6 \text{ GeV})}{(82 \text{ GeV})}\frac{1}{4}\text{BR}(f) \times 16.4 \times \frac{1}{(630)^2}$$

$$\simeq 2 \times 10^{-33}\text{BR}(f) \text{ cm}^2. \tag{10.13}$$

10.4   The W Decay and Total Event Rate

Finally we have to decide what final states are detectable. If a $W^\pm$ is produced and decays $W^\pm \to e^\pm \nu$, and the W is essentially at rest, the electron often appears at large angles to the original beam, opposite a neutrino (which escapes detectors because it interacts so weakly), giving an event with the clear signature of an electron (or positron) opposite large missing momentum. A similar signature occurs for $W^\pm \to \mu^\pm \nu$. If $W^\pm \to \tau^\pm \nu$, the τ decays quickly (we will compute its lifetime in Chapter 19) and is harder to see. If $W^\pm \to q\overline{q}$, the quarks appear as jets and can be detected (see Chapter 15), but it is hard to identify an individual flavor of quark (u, d, c, s, ...). Consequently, any scattering of

$q q$, $q\bar{q}$, $q g$, $g g$, $\bar{q} g$ which gives a similar final state with two quark or gluon jets can mask the $q\bar{q}$ from $W^\pm$. Since these scatterings can occur by gluon and quark exchange and are characterized by the strength of the QCD interaction, they give larger cross sections, and $W^\pm \rightarrow q\bar{q}$ is difficult to detect.

From this discussion we see that (i) since we could produce either W^+ or W^-, the total cross section is twice that of equation (10.13), and (ii) we should hope to detect the $W^\pm \rightarrow e^\pm \nu_e$ and $\mu^\pm \nu_\mu$ modes, which give a final branching ratio of 1/6. Then

$$\sigma_W \simeq 0.6 \times 10^{-33} \text{ cm}^2. \tag{10.14}$$

In Chapter 12, we will discuss characteristics of accelerators a little more; for now we note that the number of events is given by

$$N = \sigma L T \tag{10.15}$$

where the luminosity L is a characteristic of accelerators, T is the running time, and σ the cross section. For $T = 10^7$ sec (a full year is $\pi \times 10^7$ sec) and σ given by equation (10.14), an accelerator has to be designed with $L \geq 10/0.6 \times 10^{-26}$ cm^{-2} sec^{-1} $= 1.6 \times 10^{27}$cm^{-2} sec^{-1} to get 10 events, assuming reasonable detection efficiency. For detection efficiencies less than unity, L has to be increased accordingly. The $p\bar{p}$ collider at CERN was successfully designed to achieve at least this event rate, and was able to produce and detect $W^\pm$ and Z^0. After a few years running at an improving luminosity, they have detected several hundred $W^\pm$ and several dozen Z^0.

10.5 Measurement of Z^0 and $W^\pm$ masses

After a gauge boson is produced and decays, the decay products can be studied to measure the mass and width. Consider, for example, $Z^0 \rightarrow e^+ e^-$, where $e^\pm$ have four momenta $p_\pm$. Then $P_Z = p_+ + p_-$, so

$$\begin{aligned} P_Z^2 = M_Z^2 &= 2m_e^2 + 2p_+ \cdot p_- \\ &\simeq 2E_+ E_- - 2|\vec{p}_+| \, |\vec{p}_-| \cos\theta_\pm \\ &\simeq 2E_+ E_- (1 - \cos\theta_\pm) \end{aligned} \tag{10.16}$$

where $p_\pm = (E_\pm, \vec{p}_\pm)$ and $\theta_\pm$ is the angle between $\vec{p}_+$ and $\vec{p}_-$. When m_e is neglected, $|\vec{p}_\pm| = E_\pm$. By measuring E_+, E_-, and $\theta_\pm$, a value can be calculated for M_Z for each event. When plotted the values should fall on a Breit–Wigner curve with a width Γ_Z centered at M_Z, with some spreading due to experimental resolution.

For $W^\pm$, the situation is a little more complicated. In a collision of hadrons, the u and $\bar{d}$ that make the $W^\pm$ have a hard collision, but the remaining quarks and gluons in the hadron mainly go down the beam direction and are not detected; they carry away energy. When the $W^\pm$ decays to $e^\pm \nu$, the electron direction and energy can be measured well, but the neutrino has to be inferred from momentum conservation. Since the magnitude and direction of the momentum carried off by the soft hadron constituents that went down the beam direction is not known accurately, there is uncertainty in the ν momentum, especially in the longitudinal direction. Two related techniques have been used to circumvent this problem.

For the first, called the Jacobian Peak technique, we assume the W is produced at rest. The event is shown in Figure 10.2a, when a proton and an antiproton collide, a W is produced at rest, and decays to $e^\pm$ and an invisible ν. The electron has a transverse momentum p_T^e . Then $p_T^e = (M_W/2)\sin\theta$ where θ is the angle between the electron direction and the beam direction. The essential point is to note that the p_T^e is a measure of M_W, so we want to plot $d\sigma/dp_T^e$. In terms of the usual $d\sigma/d\cos\theta$, this is given by

$$\frac{d\sigma}{dp_T^e} = \frac{d\sigma}{d\cos\theta}\frac{d\cos\theta}{dp_T^e} . \tag{10.17}$$

The last factor is

$$\frac{d\cos\theta}{dp_T^e} = \frac{1}{dp_T^e/d\cos\theta} = -\frac{2}{M_W}\frac{\sin\theta}{\cos\theta} \tag{10.18}$$

and

$$\cos\theta = \sqrt{1 - (2p_T^e/M_W)^2}$$

$$= \frac{2}{M_W}\sqrt{\left(\frac{M_W}{2}\right)^2 - p_T^{e\,2}} \tag{10.19}$$

so

$$\frac{d\cos\theta}{dp_T^e} = \frac{2p_T^e}{M_W}\frac{1}{\sqrt{\left(\frac{M_W}{2}\right)^2 - p_T^{e\,2}}} . \tag{10.20}$$

Because of this last factor, which arises from the Jacobian (the change of variables), $d\sigma/dp_T^e$ will peak at $p_T^e = M_W/2$ and drop rapidly, as sketched in Figure 10.3. Any momentum of the W, and the width of the W, will spread the curve out, but the effect of such corrections can be estimated and taken into account.

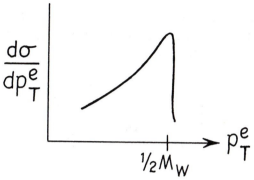

Figure 10.3
Variation of $d\sigma/dp_T^e$
with p_T^e showing a Jacobian peak.

A related and somewhat better technique follows from defining the "transverse mass" of the W,

$$M_T^2 = (E_T^e + E_T^\nu)^2 - (\vec{p}_T^{\,e} + \vec{p}_T^{\,\nu})^2, \tag{10.21}$$

in terms of actual measurable tranverse quantities; for each particle $E_T^{i2} = p_T^{i2} + m_i^2$ and for e, ν we can drop m_i . Then

$$M_T^2 \simeq 2p_T^e p_T^\nu \left(1 - \cos\theta_{\nu e}\right). \tag{10.22}$$

A distribution similar to that of Figure 10.3 occurs when $d\sigma/dM_T$ is plotted versus M_T, with a peak at M_W. Effects of Γ_W can be calculated, and the spreading below M_W is due to longitudinal momentum of the ν. If the W is not produced at rest but with a momentum $\vec{\kappa}$, then $\vec{p}_T^{\,e} \to \vec{p}_T^{\,e} + \vec{\kappa}$ and $\vec{p}_T^{\,\nu} \to \vec{p}_T^{\,\nu} - \vec{\kappa}$ so the correction is of the order κ^2/p_T^e , which is normally quite small at the CERN collider. The transverse mass is a variable with a simple intuitive interpretation and it will certainly be used extensively in future analyses of data.

10.6   The W Spin and Decay Asymmetry

Both the spin of the W, which we expect to be one, and the $SU(2)$ structure of the theory [that $W^{\pm}$ couples only to left–handed quarks and leptons, or right–handed antiquarks and antileptons] are confirmed by a striking aspect of the behavior of the W.

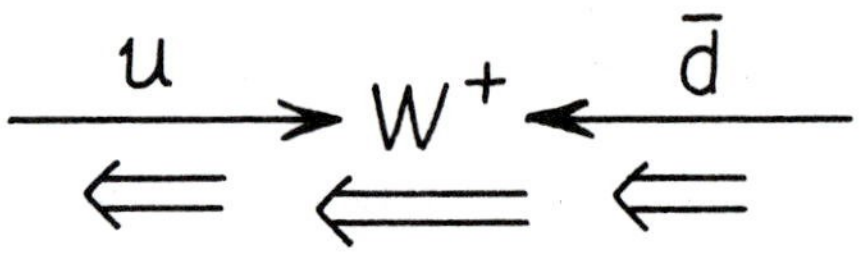

Figure 10.4
Polarized W^+ produced in a $p\overline{p}$ collider.

At the $p\overline{p}$ collider, some of the W's will be produced by a left–handed u in the p and a right–handed $\overline{d}$ in the $\overline{p}$, making a W^+ polarized as shown in Figure 10.4. In the figure, momenta are shown as single lines and spins as double lines. Then W^+ decays to $e^+\nu_e$. Decays can occur in any direction, of course. Consider decays where the $e^+\nu_e$ direction is near the beam direction. Then the e^+, which is an antiparticle, will be right–handed and the ν_e left–handed. So to give a total spin projection adding up to the W spin projection, we must have the configuration of Figure 10.5. Sending the e^+ in the opposite direction will not work!

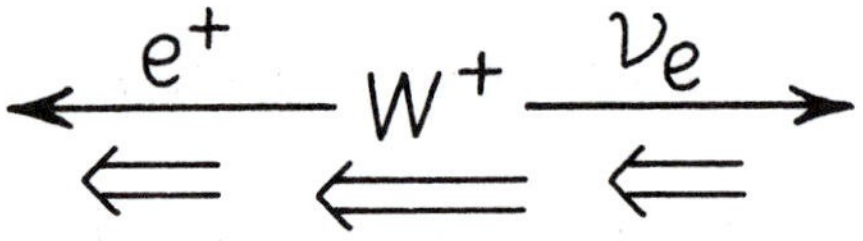

Figure 10.5
Spin configuration of W decay.

Thus we expect the e^+ to go predominantly in the direction opposite to the direction of the original proton—not only a striking prediction but a counterintuitive one, since the flow of electric charge reverses. If we make W^- from $\overline{u}d$, the same conclusion holds. When the prediction is made a quantitative one, taking into account the angle θ that the decay e^+ makes relative to the beam direction, the result is that the positrons should have a distribution $(1 + \cos\theta)^2$ relative to the $\overline{p}$ direction. The data are in good quantitative agreement with that, as shown in Figure 10.6. Both the V–A form of the coupling and that the W has spin one are confirmed by this data, though a rather technical argument is required to make the latter connection.

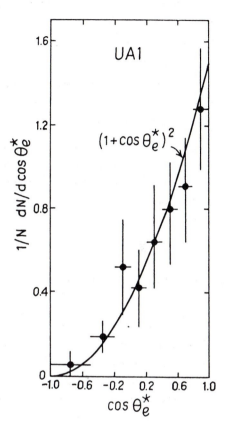

Figure 10.6
Data for W decay asymmetry.

Problems

10.1:   Suppose in a collision two oppositely charged particles emerge into the detector with four momenta measured to be

$$p_1 \simeq (43; 0, 0, 43)$$
$$\text{and } p_2 \simeq (48; 0, 0, -48)$$

with units in GeV. How would you interpret the event?

10.2:   Show that the basic argument of Section 10.6 holds for W^- produced by $\bar{u}d$ and decaying to $e^- \bar{\nu}_e$.

10.3:   Estimate the number of Z^0's produced and detected at the CERN collider. Rather than fully repeating the calculation of Sections 10.1–10.4, just estimate the ratios of Z to $W^\pm$ at each stage of the argument. For the constituent cross section, it would be a good approximation to use the branching ratio numbers of Chapter 9, and one could estimate the effect of the larger M_Z by calculating $dI/d\tau$.

10.4:   Estimate the number of neutral Higgs bosons produced at the CERN collider if $M_H \simeq M_Z$. Rather than repeating the full calculation, just estimate the ratio of H^0 production to Z^0 production by the same mechanism we have considered for $W^\pm$ and Z^0. Can you think of other mechanisms to give comparable or larger numbers of H^0?

10.5:   Draw Feynman diagrams to show that both single Z^0's and single W's can be produced at e^+e^- colliders, but the rate for single Z^0's should be considerably larger. Argue that the rate for producing a W^+W^- pair should be larger than the rate for a single $W^\pm$, at least if $\sqrt{s}$ is not too large compared to M_W.

Suggestions for Further Study

The story of the search for the W and Z is told in several places. Watkins is a member of the UA1 collaboration so his book has a lot of experimental detail, and tells the story from that interesting point of view. The book by Taubes tells the story from an unusual perspective. W and Z production is not discussed in any book in much more detail than we give; Collins and Martin mention a few more complications that have to be considered in a full treatment.

CHAPTER 11

Measurement
of Electroweak
and
QCD Parameters;
the Muon Lifetime

We have now seen the most remarkable prediction of the Standard Model worked out, the existence of the $W^{\pm}$ and Z^0 as fundamental, point-like particles. They are apparently as fundamental and point-like as the photon, even though they are almost one-hundred times as heavy as a proton, having about the same mass as a strontium nucleus. Both the discovery of the W and the Z, and the successful prediction of their properties, are extraordinary accomplishments.

To completely specify the Standard Model and test many of its predictions, we must first determine the parameters of the Standard Model as formulated in Chapters 6 and 7. Many predictions can be calculated from the measured masses of the gauge bosons and the fermions, plus α, $\sin^2 \theta_w$, and the QCD coupling.

In this chapter, we describe the way couplings are measured. In Chapter 23 we will discuss masses and how quark masses are defined in a little more detail.

The electroweak couplings are determined if we measure the W mass, the fine structure constant, and $\sin^2 \theta_w$, for example, or alternatively, we could measure the muon decay rate, $\sin^2 \theta_w$, and M_W. Our purpose here is not to discuss or argue which procedure is best, but to illustrate several ways the parameters are determined, both for the intrinsic interest of the question and to illustrate

143

Standard Model techniques. For measurements of the fine structure constant, $\alpha = e^2/4\pi$, the best techniques are from areas of physics outside of particle physics, and we refer the reader to the Particle Data Tables and references therein. We will close the chapter with comments on the measurement of the QCD coupling.

11.1   Measurement and Significance of $\sin^2 \theta_w$

One can view θ_w in various ways. It was introduced in Chapter 7 simply as a parameter that basically fixed the ratio of $U(1)$ and $SU(2)$ couplings. It enters into any process where there is a virtual or real Z^0 involved, since the Z^0 coupling to fermions is (f = left–handed or right–handed fermion) $\overline{f}\gamma^\mu f \left(T_3^f - Q^f \sin^2 \theta_w \right)$, so the cross sections for $\nu_\mu p \rightarrow \nu_\mu X, \overline{\nu}_\mu p \rightarrow \overline{\nu}_\mu X, \nu_\mu e^- \rightarrow \nu_\mu e^-, \overline{\nu}_\mu e^- \rightarrow \overline{\nu}_\mu e^-$, $\overline{\nu}_e e^- \rightarrow \overline{\nu}_e e^-, e^+ e^- \rightarrow \mu^+ \mu^-, e^+ e^- \rightarrow b\overline{b}, e^- d \rightarrow e^- X$, and many others, all depend on θ_w. In addition, $M_W / M_Z = \cos\theta_w$ as we saw in Chapter 7.

In fact, this situation provides a profound test of the Standard Model, because it predicts that the *same* θ_w occurs in all processes. If θ_w could take an arbitrary and different value in each process, there would be no unification of weak and electromagnetic processes. An additional constraint is that $0 \leq \sin^2 \theta_w \leq 1$. If such a parameterization had been attempted where it did not belong, $\sin^2 \theta_w$ could have any value and the attempt to impose a single value in a variety of processes would lead to nonsense. Figure 11.1 presents the current data for $\sin^2 \theta_w$ evaluated from various reactions, and clearly shows that all the values are consistent. The current best value for $\sin^2 \theta_w$ is

$$\sin^2 \theta_w = 0.229 \pm 0.004 \tag{11.1}$$

as compiled by W. Marciano. Even if averaging data and combining errors is optimistic, presumably $\sin^2 \theta_w$ is now known to an accuracy of 3% or better. It is hoped that precision experiments at $e^+ e^-$ colliders will lead to errors of less than ± 0.001 in a few years.

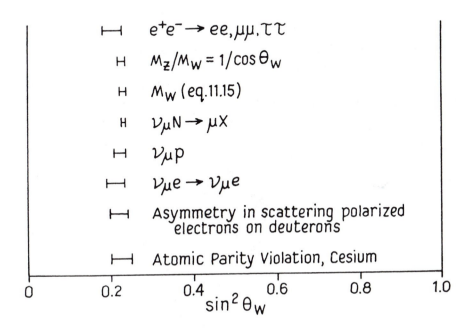

Figure 11.1

*This figure shows the remarkable agreement of independent mea-
surements of $\sin^2 \theta_w$. If the weak and electromagnetic interactions
were not unified, the attempt to describe them in terms of one
parameter could have allowed $\sin^2 \theta_w$ to have any value, including
$\sin^2 \theta_w < 0$. Most of the values here were compiled by W. Marciano
in a report to the 1986 International Conference on High Energy
Physics, Berkeley, July 1986; radiative corrections are included.
The $e^+ e^-$ values are from a compilation of B. Naroska; without
further input $e^+ e^-$ data is not yet able to directly measure $\sin^2 \theta_w$,
so a value has been used for M_Z to arrive at the result shown.*

11.2   Muon Decay

In this section we will go through the calculation of the decay

$$\mu^- \to \nu_\mu\, e^-\, \overline{\nu}_e \tag{11.2}$$

for two reasons. First, it provides a way to accurately measure g_2/M_W. Second, it is a typical fermion decay, a prototype for any decay of the form $f' \to f h' \overline{h}$ where f', f, h', h are quarks or leptons. The technique used will be instructive and the answer will be one we use several times in later chapters. The full calculation requires considerable effort with Dirac equation solutions, *etc.*, so we will do it with our methods. We want a meaningful estimate of the rate without extensive calculations.

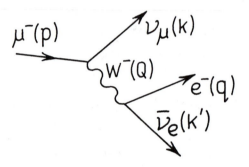

Figure 11.2
Muon decay.

The only diagram for the decay is Figure 11.2, where four–momenta are labled in parentheses. At the vertices there are factors $(g_2/\sqrt{2})\,\overline{u}\gamma^\lambda P_L u$, and (ignoring the spin subtleties) the propagator of the W boson is $1/\left[Q^2 - M_W^2\right]$. Putting all these together the amplitude is approximately

$$M = \frac{g_2^2}{2}\left(\overline{\nu}_\mu \gamma^\lambda P_L \mu\right) \frac{1}{Q^2 - M_W^2} \left(\overline{e}\gamma_\lambda P_L \nu_e\right). \tag{11.3}$$

The momenta will not be larger than the mass of the muon, m_μ, since it provides all the energy for the decay, so Q^2 is very small when compared to M_W^2 and can be dropped to a good approximation. Putting in P_L, we then have

$$M \simeq \frac{g_2^2}{8M_W^2}\left(\overline{\nu}_\mu \gamma^\lambda (1-\gamma_5)\,\mu\right)\left(\overline{e}\gamma_\lambda (1-\gamma_5)\,\nu_e\right). \tag{11.4}$$

This is the basic Standard Model prediction for μ decay. The $\overline{\nu}$, μ, $\overline{e}$, ν represent

appropriate external wave functions for the fermions. In this case, it had been
known for some years before the final formulation of the Standard Model that
equation (11.4) led to a good description of muon decay, and the form of equation (11.4) was a major input to the formulation. Note that the matrix element,
in either equation, has the form of a product of two currents. Both currents
have the structure $\gamma^\lambda - \gamma^\lambda \gamma_5$; the first term transforms like a space–time vector
under Lorentz tranformations, and the second term like an axial vector, so this
is called a $V\!-\!A$ interaction. The vertex $\gamma^\lambda P_L$ is how we will generally write the
$V\!-\!A$ interaction.

The coefficient is defined to be the Fermi coupling,

$$G_F/\sqrt{2} = g_2^2/8M_W^2 \;. \tag{11.5}$$

The $1/\sqrt{2}$ is a convention of historical origin. Note that G_F has dimensions
of $1/(\text{mass})^2$. Measurement of the muon lifetime and mass will determine G_F .

Next we need $\overline{|M|^2}$, *i.e.* the absolute square of the matrix element summed
and averaged over spins. Since M is dimensionless, the answer must be

$$\overline{|M|^2} = CG_F^2 m_\mu^4 \tag{11.6}$$

where C is a constant of order unity, because all other fermion masses are negligible when compared to m_μ . This arises technically because the true answer
is a sum of terms each of the form $p \cdot k\, q \cdot k'$ with all permutations, and every
momentum is proportional to m_μ . Or one can replace each factor $\bar{u}\gamma_\lambda P_L u$ by
m_μ , which has the correct dimensions and must be the mass that enters since
all others are small. Maintaining our policy of counting spin states to estimate
the numerical factor, there are two spin states each for μ and e, one for each
ν, and an initial $1/2$ for the average, so we use $C = (2 \times 2 \times 1 \times 1)/2$. [As in
other places, we will get an approximately correct answer. Since our goal is to
give meaningful estimates, there is no point in trying to refine the estimation
procedure to improve its accuracy. As usual, after showing that the reader can
obtain useful estimates by these procedures, we will quote the correct answer
and use it in the following so the reader is well informed about correct answers.]

From equation (9.2) for the decay width, we then have for muon width,

$$d\Gamma_\mu = \frac{1}{(2\pi)^5} \frac{2G_F^2 m_\mu^4}{2m_\mu} \delta^4(p - q - k - k') \frac{d^3q}{2q_0} \frac{d^3k}{2k_0} \frac{d^3k'}{2k_0'} \;. \tag{11.7}$$

To get the full width, we have to complete the phase space integration. The
integral over $\delta^4 d^3k\, d^3k'/k_0 k_0'$ is the same as the one we did for the cross section

in Chapter 9, and is given by 2π; there it was evaluated in the overall center of mass, while here we are in a different system. However, because the factor $\delta^4 d^3k\, d^3k'/k_0 k_0'$ is Lorentz invariant (as shown in Chapter 9), we can evaluate that piece in the $\vec{k} + \vec{k'}$ rest frame and then use the answer in the muon rest frame. Then we can write $d^3q = q^2 dq\, d\Omega_q$ and note $q dq = q_0 dq_0$. Then

$$d\Gamma_\mu = \frac{1}{(2\pi)^4} \frac{G_F^2 m_\mu^3}{8} q\, dq_0\, d\Omega_q \ . \tag{11.8}$$

Since there is no dependence on the angles of $\vec{q}$, $\int d\Omega_q = 4\pi$. The maximum value q_0 (the electron energy) can take on is $m_\mu/2$, since then both neutrinos go off opposite to the e^- with total energy $m_\mu/2$. In the approximation that $m_e \ll m_\mu$,

$$\int\limits_0^{m_\mu/2} q\, dq_0 = \frac{1}{2}\left(\frac{m_\mu}{2}\right)^2 = \frac{m_\mu^2}{8} \ , \tag{11.9}$$

so

$$\Gamma_\mu \approx \frac{G_F^2 m_\mu^5}{32\,(2\pi)^3} \ , \tag{11.10}$$

or

$$\Gamma_\mu \approx \frac{3}{4}\frac{G_F^2 m_\mu^5}{192\pi^3} \ . \tag{11.11}$$

By dimensions, the answer must be of this form, and even the π^3 is necessarily there from the form of $d\Gamma$.

As the reader may guess, the correct answer is

$$\Gamma_\mu = \frac{G_F^2 m_\mu^5}{192\pi^3} \ . \tag{11.12}$$

This result turns out to be extremely general and useful, because nature has arranged fermion masses in a hierarchy so that every fermion decays into three much lighter fermions to a very good approximation. Thus equation (11.12) gives the decay width per final channel of every fermion we know of: μ, τ, c, b, t (substituting the appropriate mass for m_μ). Note that it grows as the fifth power of the mass. This works until $m_f > M_W$; for larger fermion masses the dominant decay would be to real W's and the formula for the width changes. When improved accuracy is wanted, it is of course possible to do the calculation fully and include all masses, obtaining a small phase space correction; for almost all purposes equation (11.12) is adequate, and we will use it several times in the following chapters.

From the experimental value for the muon lifetime, taken from the Particle Data Tables,

$$\tau_\mu = \frac{1}{\Gamma_\mu} = 2.19703 \pm 0.00004 \times 10^{-6} \text{ sec} \tag{11.13}$$

we get (if all radiative corrections are included)

$$G_F = 1.16637 \pm 0.00002 \times 10^{-5} \text{ GeV}^{-2}. \tag{11.14}$$

In general, we will not keep track of the errors nor the full accuracy. They are included here so the reader has a sense of how well these values are known. Using $g_2 = e/\sin\theta_w$, $e^2/4\pi = 1/137$, and equation (11.5), we can write this as

$$M_W = \frac{37.42}{\sin\theta_w} \text{ GeV} \tag{11.15}$$

which is how the numerical value of M_W was predicted historically. Putting $\sin^2\theta_w = 0.23$ gives $M_W = 78$ GeV. [When a consistent treatment is performed, including higher order diagrams, to relate all these parameters to data and to each other, M_W increases by about 3 GeV, so the expected experimental value is about 81 GeV, with an uncertainty of about 2 GeV from errors in both experimental input and theoretical input.]

So far we have seen how to measure, say, α, M_W, and θ_w. These are equivalent to G_F, g_1, and g_2, by equation (7.20) and the above. Finally we need to discuss α_3. In Chapter 20, we will examine a very important and very fundamental result of the Standard Model—that the coupling strengths are not constants but depend on the momentum transfer in the interaction being discussed. This is true for all the g_i, and for $\alpha = e^2/4\pi$ as well as for the QCD coupling $\alpha_3 = g_3^2/4\pi$, although in atomic physics it is usually not emphasized. Here we will take the variation of the couplings into account approximately, by specifying the energy scale where the couplings are measured. In Chapter 20, we will be more precise, though a completely precise treatment is beyond the scope of our discussion.

11.3   Measurement of α_3

To measure g_3, it is necessary to find and isolate processes where the quark–quark–gluon vertex or the gluon–gluon–gluon vertex enters. In Chapter 15, we will examine more carefully how quarks and gluons are observed—because of the properties of the QCD interaction they begin as objects bound in a hadron and they appear as "jets" of hadrons, mostly pions, rather than as a single electrically charged or neutral object. That leads to some extra subtleties in determining the numerical value of g_3. In addition, because α_3 is a stronger coupling, it is necessary to calculate more corrections to the theory in order to reliably extract a numerical value from data. But the essential aspects of measuring g_3 are not different from those of the other couplings.

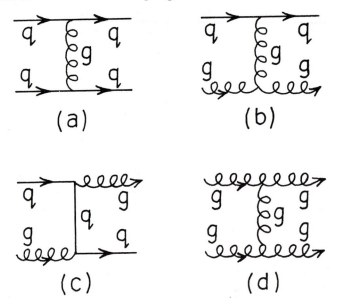

Figure 11.3
Quark and gluon interactions.

A number of different processes can be studied to measure g_3. The simplest is a scattering process, such as $q + q \rightarrow q + q$, as in Figure 11.3a. The initial quarks and gluons are in hadrons, *i.e.* in colliding protons, so all of the diagrams in Figure 11.3, and more, give significant contributions. The final quarks and gluons appear in detectors as jets. All these diagrams have a factor g_3 at each vertex, so each amplitude is proportional to α_3, and the cross section proportional to α_3^2. The value of α_3 is extracted by carefully calculating the expected cross sections, working in regions where the contributions shown are expected to dominate, and comparing with experiment.

Another thing that can happen in collisions is for a quark or gluon to radiate a gluon. Figure 11.4 shows one way to do that, and the reader can imagine attaching the gluon line to any of the q or g lines in Figure 11.3.

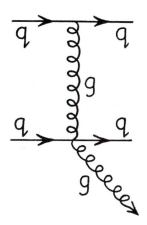

Figure 11.4
Radiation of a third jet in a hadron collision.

The rate for this is smaller by α_3 (times a calculable phase space factor) than the rate for the processes in Figure 11.3, and the processes of Figure 11.4 appear in the detector as three-jets. Thus the ratio of the three jet rate to the two jet rate is a measure of α_3.

In e^+e^- collisions a $q\bar{q}$ pair can be produced as in Figure 11.5a, with a rate proportional to α^2. A gluon can be exchanged between the quarks as in Figure 11.5b, adding a piece to the amplitude proportional to α_3. Since α_3 is not so small, this is an observable correction ($\sim 5\%$), and a measure of α_3. Or a gluon can be radiated, as in Figure 11.5c, giving a three-jet final state, again with a rate proportional to α_3 relative to the two-jet final state.

All of these effects have been observed, and all give similar results,

$$\alpha_3 \simeq 0.15 . \tag{11.16}$$

This is known to an accuracy of 20-25%, and is valid in the region from about 3–40 GeV. For momentum transfers small or large compared to this range, α_3 will be different. We will return to these subtleties in Chapter 20.

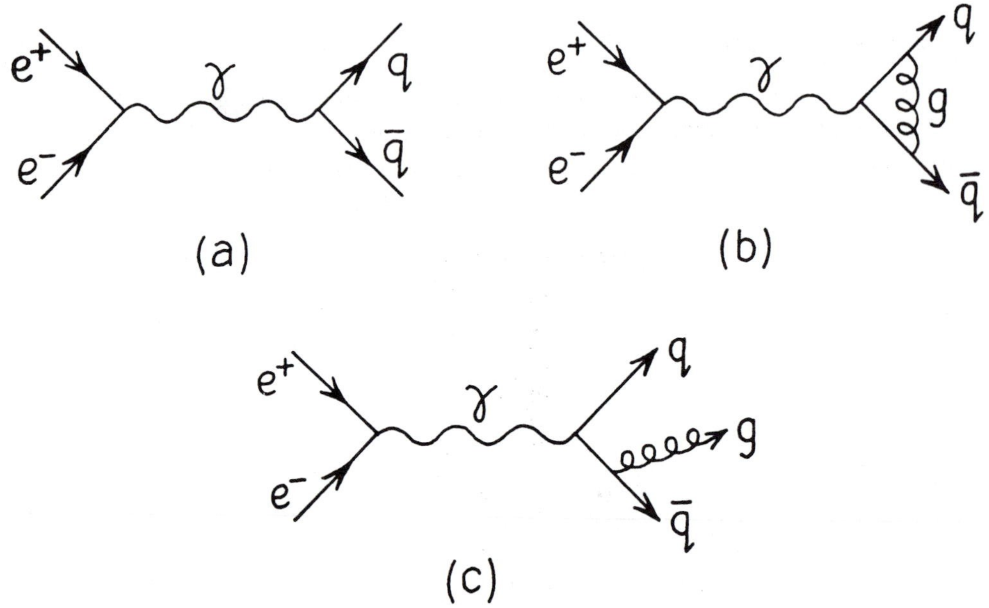

Figure 11.5
Radiation of a third jet in e^+e^- collisions.

11.4   Comments on Standard Model Parameters

For completeness here we include a few comments on the general situation regarding parameters in the Standard Model. Some of the remarks will be fully understood only after additional chapters have been studied.

To use the Standard Model for calculating the results of experiments, we need to input a variety of parameters. Most of them are masses. All the fermion masses have to be input. Probably they should not be thought of as many separate parameters, since once we achieve an understanding of the origin of mass (if ever), most or all of them will be calculable. There are four parameters we will discuss only in Chapter 22—essentially angles which describe a rotation between the quarks in a weak eigenstate basis and the quarks in a mass eigenstate basis. It is hoped that an understanding of fermion masses will also allow these parameters to be calculated. The couplings α_1, α_2, and α_3 have to be measured as we have discussed; eventually perhaps they will be calculable in terms of one coupling if the Standard Model can be embedded in a larger theory. The Higgs boson mass is not calculable in the Standard Model because the physical origin of the $\lambda\phi^4$ term in the Higgs potential is unknown.

For all of these, the situation is very much like the one to which we are historically accustomed. In quantum theory, or quantum electrodynamics, the electron mass and the fine structure constant are measured and input; from there on we can calculate. It is essentially the same in the Standard Model. The progress is that all of the weak, electromagnetic, and strong interactions are now included in what the theory describes. In addition, although the Higgs mechanism is not understood, at least the description of how mass arises is now incorporated into the theory instead of being outside of it; perhaps that indicates that we are closer to gaining insight into the mass problem.

Finally, for completeness, we mention one rather subtle parameter (called the θ-parameter) that is set to zero or a very small value in the Standard Model, in order to be consistent with measurements, but whose value does not appear to be constrained by the structure of the theory. It is the coefficient of a term that could appear in the QCD Lagrangian. If it did, time reversal invariance (or CP invariance—see Chapter 24) would be violated in strong interactions. It is analogous to what would occur in electromagnetism if a term proportional to $\overrightarrow{E} \cdot \overrightarrow{B}$ were added to the usual Lagrangian (equation (2.15)). Since $\overrightarrow{E} \cdot \overrightarrow{B}$ changes sign under parity and CP, these quantities would no longer characterize classical systems and quantum mechanical eigenstates. Rather than forcing this term to be very small by making its coefficient very small, it would be preferable to have a theoretical argument as to why such a term could not be present. So far no one has found a compelling argument, though there are possibilities. This is an important area of research which is outside the scope of our treatment. The lack of understanding of why the θ-parameter is small is one more indication that additional physics will be discovered to indicate how to improve the conceptual foundations of the Standard Model.

Suggestions for Further Study

Many of the properties of $W^{\pm}$ and Z^0 examined in Chapters 9 and 10, and the parameters of the Standard Model, are studied with full precision in the lectures of L. Maiani given at the 1984 TASI school.

CHAPTER 12

Accelerators
—Present
and Future

Before we consider further the tests and predictions of the Standard Model, it is useful to discuss the experimental facilities. Some understanding of the accelerators and detectors is essential to understand how the experiments are done, why some experiments are done and not others, *etc.* We will not discuss in any detail how accelerators work. On the other hand, it seems appropriate to have three goals, (1) to indicate how high energy beams of quarks and gluons and electrons are obtained, (2) to survey the available machines for doing particle physics, and (3) to describe the plans and hopes for accelerators over approximately the next decade. It does not require much of a crystal ball to do the latter, since the large construction times and large costs of new facilities imply that existing decisions almost completely determine what can happen until about 1994, and plans currently under discussion greatly restrict what can happen for even longer.

The accelerators provide the beams and make them collide. Then it is necessary to detect the collision products and interpret what happened. Detectors do that, and the next chapter will be devoted to a description of how they do it.

12.1   Parameters of Accelerators

From the point of view of the physics purpose of an accelerator, four of its properties are the main ones. The first is the type of particle being accelerated. Since the acceleration is done by electromagnetic fields, any long–lived, electrically charged particle can qualify. In practice, for particle physics the main possibilities are electrons (e^-), positrons (e^+), protons (p), and antiprotons ($\overline{p}$). Beams of other particles can be obtained by hitting a target with a proton beam, which causes all varieties of particles to emerge, and carefully selecting particles of the desired type by bending or stopping the others. This has led to experiments with neutrinos, pions, kaons, photons, lambdas, muons, and others.

The second basic property is the final energy to which the particles are to be accelerated. The initial protons or electrons are obtained from ion generators and have energies in the eV to KeV range. They are accelerated as they travel through regions with electric fields. The paths of the particles can be controlled with magnetic fields, and they can be made to travel in circles so they can be accelerated over long total distances in order to achieve higher energies. As we saw in the discussion of W and Z production, energies in the region of hundreds of GeV have been achieved, with the acceleration process giving increases in energy of over 10^9.

Unfortunately, there is a limit to what can be achieved with circular accelerators. For electrons, the machine (called LEP) under construction at CERN in Geneva, Switzerland, can have electron and positron beams up to 100 GeV if the full complement of accelerating devices is eventually installed. But any electrically charged particle that is forced to travel in a circle radiates photons and loses energy, and the cost of supplying that energy rises so rapidly that LEP is expected to be the last circular electron–positron collider.

The energy loss per turn for a particle of energy E, mass m, and charge e is given by

$$\Delta E/2\pi R = 4\pi e^2 \beta^2 \gamma^4/3R \,, \tag{12.1}$$

where R is the radius of curvature, $\beta = v/c \simeq 1$, and $\gamma = E/m$. Protons and antiprotons lose less energy per turn, by a factor $(m_e/m_p)^4 \simeq 10^{-13}$, than electrons or positrons. Consequently it is much easier to accelerate protons to a higher energy. With protons, however, the energy is shared among the constituents. For subtle reasons, the energy per constituent decreases slowly as the collision energy increases, so it becomes improbable that a particular quark or gluon carries more than 5–10% of the proton's energy. Nevertheless, the 10^{-13} wins, and the particle physics community in the United States is planning to build a circular machine to collide two proton beams each of energy 20,000 GeV (the SSC).

The third fundamental property of an accelerator is its luminosity L. For a circular collider, suppose the particles of each beam come in bunches, with k particles per bunch, and n bunches around the ring. Let A be the cross-sectional area of the beams, and f the frequency with which the bunches go around. Then the luminosity is $L = fnk^2/A$, with units $\text{cm}^{-2}\,\text{sec}^{-1}$. The number of events of a particular kind in a time T is

$$N = \sigma L T ,\qquad(12.2)$$

where σ is the cross section for events of the particular kind to occur. The energy and luminosity of a machine can be specified in its design. They are in principle independent. The cross section comes from nature. This definition of L is a natural one for colliding beam machines; for fixed target ones, or for study of secondary beams (which played a larger role during the development of the Standard Model than they do now or are expected to in the future) the properties of the target become involved, but the concept is still useful.

Before the 1980's luminosity seldom was a limiting factor but now it has become a major one. The reason is that cross sections, having dimensions of $1/(\text{mass})^2$, fall as the square of the mass scale which the new machine is designed to probe. Thus as physics probes frontier questions and moves to higher energies to do so, the cross sections decrease. To maintain an event rate that allows a new phenomenon to be observed, the luminosity must increase. Our ability to probe further is limited by both energy and luminosity.

The fourth characteristic, which we examine in the next section, is what fraction of the energy is useful for producing new phenomena.

12.2   Useful Energy

The primary purpose in going to higher energies is to be able to study phenomena at higher mass scales, or by the uncertainty principle, to be able to probe interactions at shorter distances. But different arrangements make very different use of the available energy.

We can consider several categories. In each case, the relevant quantity is the center of mass energy available for creation of new particles, so we calculate $\sqrt{s}$. We are most interested in high energies, so we can usually neglect masses. Let the colliding particles have four momenta p_a and p_b. Then $s = (p_a + p_b)^2$.

a. Fixed Target

The beam momentum is $p_a = (E_a, \vec{p}_a)$ and the target momentum is $p_b = (m_b, 0)$, so

$$s \approx 2 p_a \cdot p_b = 2 E_a m_b \,. \qquad (12.3)$$

Then $\sqrt{s}$ only grows as $\sqrt{E_b}$, so the available energy grows very slowly. Fixed target interactions are needed to study collisions involving particles that cannot be accelerated in colliders, and to get relatively intense sources of particles such as kaons to study for rare decays, but they are not so useful when higher energy is needed.

b. pp Collider

Assuming head–on collisions, we can take $p_a = \left(E, \vec{p} \right)$ and $p_b = \left(E, -\vec{p} \right)$. Then

$$s = 4E^2, \qquad (12.4)$$

so $\sqrt{s}$ grows as E. However, as seen from equation (10.8) or Chapter 18, the constituents only have available energy

$$\sqrt{\hat{s}} \simeq \sqrt{x_1 x_2} \sqrt{s} \,. \qquad (12.5)$$

As the value of x_i increases, the probability of a collision at that x_i decreases, so there is a tradeoff. If the probability is reasonably large at $x_i \simeq 0.1$, then $\sqrt{\hat{s}} \simeq \sqrt{s}/10$ and grows approximately as E. There is a tail of useful events at $\sqrt{\hat{s}} > \sqrt{s}/10$, especially at a machine with larger luminosity.

c. $e^+ e^-$ Collider

Here the calculation gives the same result as equation (12.4), so $\sqrt{s}$ grows as E, and all the energy is useful. Unfortunately, as remarked above, because of the larger energy loss due to synchrotron radiation, at the present time the largest energy for constituent collisions can be obtained in *pp* collisions. To avoid the synchrotron radiation problem considerable effort is currently being focused on developing high energy, high luminosity, linear electron colliders. In 1987, the first such machine, called SLC, will begin to operate at SLAC with 50 GeV/beam. It has an energy gradient of about 25 MeV/meter, so a one–thousand GeV beam would require forty kilometers, which is not very feasible. At circular colliders the beams can be made to collide repeatedly, while at a linear collider only one collision is possible. To be useful a collider with one–thousand GeV beams would have to have a luminosity over one-hundred times larger than the luminosity of SLC, and it is not known how to achieve that. So there is considerable uncertainty as to whether a very high energy e^+e^- collider will be possible.

Since the constituents in a proton carry about 10% of the proton momentum, very crudely the available energy in an e^+e^- collision needs to be about 10% that in a pp collision. In practice, because of the limited luminosity of a pp collider, an e^+e^- collider of somewhat less than 10% the energy of a pp collider can hope to do similar physics.

12.3   Present and Approved Facilities

In this section we list the present facilities available for research, including those actually under construction, with a few remarks about the parts of their physics program that are of greatest interest from the point of view of this book. Although the contents of this section must become out of date as time goes on, given the long construction times and large costs of new facilities, this list is expected to give an essentially complete picture of where new experimental information could come from until about 1994. The list includes the main high energy facilities, but does not aim for a complete list of all facilities.

CERN, in Geneva, Switzerland, has a fixed target proton beam of 400 GeV, called the SPS, and has the $p\bar{p}$ collider where $W^\pm$ and Z^0 were discovered. The collider runs at $\sqrt{s} = 630$ GeV, and has run at $L \simeq 3 \times 10^{29} \mathrm{cm}^{-2} \sec^{-1}$. It is being upgraded in luminosity, by adding an accumulator (ACOL) for $\bar{p}$'s, to $L \gtrsim 10^{30} \mathrm{cm}^{-2} \sec^{-1}$ by the beginning of 1988. A circular e^+e^- collider, LEP, is under construction and should begin to take data in 1989. LEP is designed so that the total e^+e^- energy is just equal to M_Z, so the machine can sit at the Z^0 resonance and have an enhanced event rate, allowing study of any final state into which Z^0 can decay, and testing of a variety of Standard Model predictions.

The DESY accelerator complex, in Hamburg, Germany, has had two e^+e^- colliders running in the past. One, PETRA, ran for $\sqrt{s}$ up to 46 GeV, with L around 10^{31} cm^{-2} sec^{-1}. The other, DORIS, concentrated on the lower energy region where $e^+e^- \rightarrow b\bar{b}$ and associated reactions occurred copiously. Now a higher energy ep collider, called HERA, is under construction there. It will have an electron beam of 30 GeV colliding with a proton beam of 800 GeV, and a luminosity goal of nearly 10^{32} cm^{-2} sec^{-1}.

At KEK, the Japanese national particle physics laboratory, at Tsukuba, near Tokyo, the main facility is an e^+e^- collider called TRISTAN, which will take data in 1987. It has a maximum $\sqrt{s}$ of 60 GeV, and a luminosity of over 10^{31} cm^{-2} sec^{-1} is expected.

In Beijing an e^+e^- collider, BEPC, is under construction. It is designed to have $\sqrt{s}$ up to 5.6 GeV, with high luminosity, in order to study charmed quarks, the ψ/J system, and tau leptons. Collisions are scheduled to occur in 1989.

In the Soviet Union a 3 TeV proton beam is under construction. It will be used for fixed target experiments scheduled to begin in 1992. Collisions with a 500 GeV $\bar{p}$ beam are planned for the future, and construction of a second 3 TeV ring is also a possibility. This facility, called UNK, is located at Serpukhov. At Novosibirsk there is an e^+e^- collider, VEPP-4. It has a maximum energy of 11.6 GeV and is often run at about 5 GeV. It has run with partially polarized beams. A higher energy e^+e^- collider is planned.

Brookhaven National Laboratory, on Long Island, has a 32 GeV fixed target accelerator, the AGS. At the present time it has several unique experiments searching for effects of neutrino masses and rare or forbidden decays of kaons.

At Cornell, in Ithaca, NY, an e^+e^- collider called CESR has a maximum energy of 13 GeV and has generally run to optimize production of b–quarks and associated physics. It has recently been upgraded to $L \sim 5 \times 10^{31}$ cm^{-2} sec^{-1}.

Fermi National Accelerator Laboratory (Fermilab), near Chicago, has the highest energy beam that can be used for fixed target experiments, a 1000 GeV (1 TeV) proton beam. In addition it has constructed a $\bar{p}$ beam that can also be accelerated to 1 TeV, so it has a $p\bar{p}$ collider that can achieve $\sqrt{s} = 2$ TeV, which should begin to do physics in early 1987. The expected luminosity is 10^{30} cm^{-2} sec^{-1}. It is intended that the luminosity should be upgraded to 5×10^{31} by 1992.

At SLAC, in Palo Alto, CA, there are several electron accelerators. The original two kilometer linac has been upgraded to accelerate electrons and positrons to 50 GeV as part of a new facility called SLC. When the electrons and positrons have reached full energy, they are bent around a half circle in opposite directions and made to collide. SLC is designed for collisions at the Z^0 mass (similar to LEP). The design luminosity of SLC is 6×10^{30} GeV. It will turn on in 1987; because it is the first attempt at a linear e^+e^- collider it could be some time until the design luminosity is achieved. A circular e^+e^- collider, PEP, is also running at SLAC, with $\sqrt{s}$ up to 30 GeV. The e^+e^- collider called SPEAR, where a number of important discoveries such as the c–quark and the τ–lepton were made, is still running as a facility to study charmed particles and the ψ/J in detail.

12.4   Future Facilities

There is of course great uncertainty about what will be built in the future. The United States particle physics community has proposed a 40,000 GeV (40 TeV) pp collider called the SSC (Superconducting Super Collider). The cost estimated in the proposal, for the machine, is \$3 billion. The construction time would be 6–7 years at full funding, so taking into account the time needed for the funding process and site selection, the SSC could not operate before 1995.

In Europe there are plans to upgrade LEP to "LEP II" with 100 GeV/beam, and discussions about a multi–TeV pp collider in the LEP tunnel.

A number of laboratories in the United States, in Europe, and in Japan have research programs which study how to make linear e^+e^- colliders with energy gradients over 100 MeV/meter, and high enough luminosity to study new electroweak phenomena at the energies in question. We will see that the cross section for $e^+e^- \rightarrow \mu^+\mu^-$ is $\sigma \sim \alpha^2/s$ as is essentially clear dimensionally. Since σ falls like $1/s$, and most (not all) other point–like cross sections have a similar behavior, the luminosity needed to see events grows correspondingly, so the luminosity for a 1 TeV e^+e^- collider would have to be over 100 times that for SLC or LEP—a remarkable challenge. It is not yet known whether the technology and the cost of such machines will be controllable to the point where one can be constructed.

From the 1950's until recently, particle physics has relied on traditional accelerator technology. Even the planned SSC is largely a traditional device. To further probe fundamental particle interactions in the next century, new techniques will be needed to produce larger acceleration gradients. At the same time, it will be essential to produce much higher luminosities. Fortunately, a major revitalization of the important field of accelerator physics started in the early 1980's, and promising new ideas have begun to emerge.

Problems

12.1:   Suppose an accelerator has been constructed that provides a one TeV proton beam. Now it is necessary to decide how to utilize it. Four possibilities are:

(a) hit a fixed target,

(b) collide it with a 50 GeV electron beam to study ep collisions,

(c) collide it with another proton beam, also accelerated to one TeV, and

(d) collide it with an antiproton beam, made from a source of $\bar{p}$'s ($\bar{p}$'s are harder to obtain than protons, of course).

What is the maximum available energy for production of new particles in each case? What other considerations might be important in making a decision?

12.2:   The cross section for $e^+e^- \to \mu^+\mu^-$ falls as $1/s$ as the energy increases. If future e^+e^- colliders are being considered, say at $\sqrt{s} = 500$ GeV and at $\sqrt{s} = 4000$ GeV, and the cross section is 0.9 nb at $\sqrt{s} = 10$ GeV where 10^4 events are collected in a year, what increases in luminosity are needed at such future machines to measure the cross section, or one of similar size, to about ten percent accuracy?

Suggestions for Further Study

Fraunfelder and Henley give an elementary introduction to accelerators; Perkins and Hughes go into somewhat more detail than we do about how accelerators function.

CHAPTER 13

Experiments
and
Detectors

This chapter provides a brief description of how information is obtained about the particles that emerge from a collision. In order to fully understand how the theory and its predictions are tested, it is necessary to have some understanding of how the experiments function.

A further goal is to identify the main detectors from which results could emerge during the period up to the mid-1990's. At colliders the products of an interaction can emerge in any direction, so detectors must cover essentially all of the 4π solid angle—they are called 4π detectors. The cost of a detector that can identify hadrons, photons, electrons, muons, and also know when something has escaped, has become so large that only a few detectors at each machine can be funded or find manpower for construction and operation and analysis, and they generally will have a rather long lifetime, being up-graded once or twice to do more physics as the machine improves. Consequently the detectors have taken on a life of their own and results will often get quoted as "CDF reports $\cdots$" or "MARK II discovered $\cdots$". We will list the main detectors which have been important and those which will be the most common sources of results over the next decade. The size and complexity of detectors implies that they are very costly (an average cost for detectors currently under construction is about $50 million) and take several years to build, so it is now possible to specify most or all of the detectors which will provide results over the next decade.

13.1   What Emerges From A Collider

The beams available to collide quarks, leptons, and gauge bosons are electrons and positrons $(e^{\pm})$, quarks (q), and gluons (g) where the quarks and gluons are carried in hadrons. When the accelerated hadrons are $p^{\pm}$, the quarks are mainly u– and d–quarks. In addition, quarks and electrons emit photons and $W^{\pm}$ and Z^0 particles at sufficient rates to effectively make photon and W and Z beams under appropriate conditions. We will not distinguish quarks from antiquarks for our purposes here. Various particles will emerge from the collisions. By detecting what emerges, with what momentum, at what angle, and how often, we hope to test the present understanding of particles and their interactions and to find any new effects.

Any particle that lives longer than about 10^{-11} sec will enter the detectors. The possibilities are $e^{\pm}$, $\mu^{\pm}$, γ, ν, and hadrons $(n, p, \pi^{\pm}, K^{\pm}, K^0, \Lambda, \cdots)$. For our purposes, we can lump all the hadrons together, since they all interact rather strongly with matter, while the other long–lived particles do not. Quarks and gluons appear as jets of hadrons; their behavior is explained further in Chapter 15.

Each particle behaves in a characteristic way that allows it to be identified. The hadrons interact with nuclei and lose energy in collisions that produce other hadrons. The photons and electrons lose energy rapidly by scattering and by radiating other photons and electrons; the difference between them is that the electron is electrically charged so it leaves a track in a detector sensitive to charged particles and will curve in a magnetic field. The muons are much heavier than electrons so they lose energy much less easily than electrons do and will go through large amounts of material without slowing down. The neutrinos will not be directly noticed by the detectors, but they carry away momentum which can be observed by applying momentum conservation to the other particles; to measure missing momentum it is necessary to be confident that not only the direction but also the energy of each particle emerging from the original collision has been well measured.

Having observed the final particles, it is possible to work backwards to deduce which unstable particles were produced. The simplest test is to see if any of the pairs of particles could have come from the decay of known unstable ones. If the unstable particle had a four–momentum P and mass M, then $P^2 = M^2$, and for a two-body decay $P = p_1 + p_2$ where p_1 and p_2 are the momenta of the detected pair of particles. Then

$$M^2 = (p_1 + p_2)^2 = m_1^2 + m_1^2 + 2p_1 \cdot p_2$$
$$= m_1^2 + m_2^2 + 2E_1 E_2 - 2\vec{p}_1 \cdot \vec{p}_2 \ . \tag{13.1}$$

If $\vec{p}_1$, $\vec{p}_2$, E_1, and E_2 are measured, then M^2 can be calculated, and if the M obtained corresponds to a known mass for a particle that can decay to the observed ones, we have learned that the original collision really produced the particle of mass M. Continued analysis along such lines allows reconstruction of the original event, and allows an experimenter to determine whether every event that occurred is an expected kind of event or whether something new has happened.

There is one further distinction that is important for understanding detectors at hadron colliders. When two hadrons collide, the total cross section is essentially a geometrical one, determined by their size of about 10^{-13} cm; $\sigma_{TOT} \gtrsim 10^{-26}$ cm^2. Most of the collisions are called "soft." That means that the colliding hadrons might stay together and just scatter elastically, or they might break up but only very slightly, so the outgoing particles are in a group and follow a path not far from the beam direction. Occasionally, however, there is a collision between quarks and/or gluons. Then the scattering can occur at large angles, giving some collision products with large transverse momenta relative to the beam direction; these are called "hard" collisions. The products of soft collisions have transverse momenta of about 0.5 GeV at most. In hard collisions, transverse momenta of up to $\sqrt{s}/2$ can be observed. A typical separation is at about 10 GeV, though the precise value of transverse momentum chosen is rather arbitrary. In a typical soft event at the CERN collider (at $\sqrt{s} = 630$ GeV), all the soft particles (perhaps ~ 100 of them) add up to about 25 GeV of energy in directions transverse to the beam, even though no single particle has more than half a GeV. The detectors must be able to absorb this soft energy without losing their ability to see all the hard tracks. Most collisions are "soft" and do not involve an interaction of quark and gluon constituents. The study of the soft collisions, and the behavior of much of σ_{TOT}, involves non–perturbative aspects of QCD, and long–range aspects of hadronic binding; it is a difficult subject, outside the scope of our treatment. At e^+e^- colliders, of course, all the collisions involve point–like constituents and are hard.

13.2　　Triggering

At high energy hadron colliders, the total cross section for all collisions is $\sigma_{\text{TOT}} \simeq 100$ mb (to within a factor of 2 depending on the energy), while the cross section for an "interesting" electroweak event is $\sigma_{\text{EW}} \sim \alpha_2^2/M_W^2 \lesssim 1$ nb. This gives $\sigma_{\text{EW}}/\sigma_{\text{TOT}} \lesssim 10^{-8}$. The main goal of experimentation at a high energy hadron collider is to find the rare events which test in detail the predictions of the electroweak theory and QCD for perturbative, short–distance collisions, and perhaps eventually to find some even more rare events which tell us how to extend the Standard Model. The problem for detectors is to pick out the occasional event at the level of one in 10^8.

That is a complicated problem, of course, and here we only want to indicate schematically how it might be done. The essential ingredients are the use of some materials with extremely fast response times, and the use of fast electronics. Consider the detector arrangement of Figure 13.1.

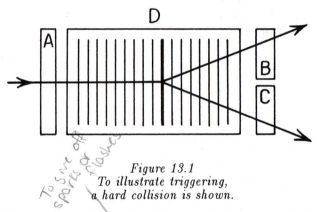

Figure 13.1
To illustrate triggering,
a hard collision is shown.

The units A, B, and C are (say) scintillators which have rather fast responses, on the order of 100 ns or less. They scintillate when a charged particle goes through and excites some of their atoms, which then de-excite by photon emission. The light can be collected by phototubes. The central chamber, D, has plates shown as vertical lines, to which a voltage can be applied. When a charged particle goes through the chamber, it ionizes molecules in a gas which fills the chamber. The electrons move to the plates and provide a record of the track. In a soft collision essentially all of the produced particles go along the path of the beam, shown entering from the left. In almost all soft collisions only one of the detectors, B or C, will have particles going through it. However, in most hard collisions particles will be produced at larger angles. Thus the scintillators are checked for every collision, but only when detectors A, B and C all have a track is a voltage applied to the plates and an event collected for analysis. Since the ions in the chamber D generally remain for microseconds after a collision, there is plenty of time to check the scintillators and decide to pulse the chamber and record the event. With sufficiently clever setups, it is possible to select samples where the events of interest are a large fraction of the total sample, and still have a high efficiency for collecting the events of interest.

A new and subtle aspect arises whenever triggering is used, because some completely different and unexpected kind of event could be occurring but may not be collected by the trigger. In our example, the particle going through B or C might be a new kind of particle which interacts too weakly to cause scintillation. Biases about what is interesting are built into the trigger. If the biases are not well enough informed, major discoveries could be missed. At future machines, it will be increasingly hard to find something one is not looking for, and ability and intuition will be required to pick out new signals.

13.3   Elements of Large Detectors

A large detector is a composite of a number of systems, each designed to accomplish certain tasks. Here we briefly describe the most common ones. Specific detectors will emphasize some, or add particular capabilities.

(a) Magnet. Most detectors are partially or entirely in a large magnet, so that the tracks of charged particles are curved and their momenta can be measured.

(b) Tracking. Most detectors have some chambers near the region where interactions occur to detect the tracks of charged particles emerging from the event.

(c) Calorimetry. Calorimeters absorb the energy carried by particles and measure it. Since all particles passing through matter lose energy, by putting appropriate kinds and amounts of matter in place, it is possible to determine accurately how much energy was transferred in a collision. Some materials are very sensitive to electomagnetic energy (so they deal with $e^{\pm}$ and γ), and others to hadronic energy. Calorimeters are natural detectors for jets of hadrons, which is how quarks and gluons appear, and for $e^{\pm}$ and γ, so they will play a major role in high energy detectors. They are also very important in deciding whether energy might have been carried off by a new kind of particle.

(d) Vertex detectors. The particles b, c, and τ all have lifetimes from about 2.5 to 10×10^{-13} sec. They will then travel a distance $d = \gamma c \tau$ where the Lorentz factor $\gamma = E/m$, E being the energy they are produced with. Then $d \gtrsim 75\gamma$ microns. Detectors for colliders now exist that have resolutions on the order of 50 microns, so they will often be able to determine that a secondary vertex occurred separated from the primary interaction vertex. This has two very useful consequences. First it is possible to know when b, c, and τ are produced at the primary vertex, as well as e, μ, γ and light q or g. Second, it gives a sample of c, b, and τ that can be directly studied. Such a sample may be very important if it is necessary to directly examine the heavier families in order to understand them better.

(e) Muon chambers. Muons do not lose energy very easily, so they will go through large amounts of iron without stopping. To measure their momenta precisely takes a very large detector element which can surround the parts of the detector discussed above.

(f) In addition, extensive use of fast electronics and computing capability is generally required.

13.4   Major Detectors

Since much of the collider-era physics comes from detector groups, it is useful to list briefly the major detectors of the past few years and the next few years. Since detector costs are large and detector construction times are long, our list will be largely correct until about 1995, regardless of the decisions made in the meantime.

a. CERN

$\bar{p}p$ collider ($S\bar{p}pS$)

There are two major detectors that have taken data at the CERN collider, called UA1 and UA2. Both discovered $W^{\pm}$ and Z^0 events in 1983, and both are undergoing up-grades for running in 1988 at higher luminosity.

LEP

Four major detectors are under constuction for LEP (named ALEPH, DELPHI, OPAL, and L3). All are 4π detectors and all but L3 will have strong magnetic fields with central tracking. L3 has very high precision muon chambers inside a large but weaker magnet and very good $e^{\pm}$ and γ calorimetry. All have high resolution vertex detectors.

b. DESY

DORIS

Studies of $b\bar{b}$ production and the Υ region are expected to continue for several years, using the ARGUS detector.

HERA

Two major detectors are under construction to operate initially at HERA. They are called ZEUS and H1.

c. Fermilab

Tevatron Collider

One 4π detector, CDF, should be in place by early 1987 when the Tevatron begins to run for physics purposes. CDF has a large solenoidal magnet for tracking, electromagnetic and hadronic calorimeters, and muon detection. A second detector, D0, has no central magnetic field. It is designed to have very good calorimetry, measure hadron and electron energy very well, detect muons particularly well, and detect any missing energy.

d. SLAC

SPEAR

The MARK III detector will continue to study c, τ and charmonium physics. It will be the only detector taking data in this important region until the beginning of operation of the Beijing e^+e^- collider.

PEP

PEP will continue running after a luminosity up-grade, with the TPC detector in place. TPC will have a vertex detector, and is able to identify separate hadrons (π, K, and p) as well as $e^\pm$ and $\mu^\pm$.

SLC

The MARK II detector, earlier versions of which have made many discoveries at SPEAR and PEP, has been up-graded to take initial data at SLC, where there is only one interaction region. A new 4π detector, SLD, is scheduled to replace MARK II in 1989, with very good 4π coverage for all particles and very high resolution vertex detection.

e. Cornell

CESR

There are two major detectors at CESR, the CUSB and the CLEO collaborations. Both are being up-graded to take advantage of the increased luminosity at CESR.

f. KEK

TRISTAN

There will be three detectors at TRISTAN. TOPAZ and VENUS are general 4π detectors. AMY is smaller and emphasizes precise detection of leptons and photons.

13.5   How the System Works

A few words about how the system functions may be instructive. There are about 1500 Ph.D. particle physicists in the United States, and a somewhat larger number in Europe. The world total is somewhat over 4000. About 2/3 are experimenters. These scientists can make proposals to a laboratory to use its accelerator and facilities to carry out some experiment. The proposals are reviewed and may or may not be approved by a committee of peers, typically two–thirds of whom are from outside the laboratory. The considerations are fundamentally scientific but may also involve such questions as whether sufficient manpower is available and whether the cost is appropriate for the potential information that can be gained. The committees are advisory to the laboratory directors, but normally their advice is taken. A similar procedure occurs when a large group proposes a large detector at a new accelerator, particularly if there are competing proposals.

In the United States, the funding for the laboratories comes from the Department of Energy except for CESR at Cornell which is funded by the National Science Foundation, while both the Department of Energy and the National Science Foundation fund particle physics research at universities. The budget for CERN is international; there are about 13 member countries, each contributing more or less in proportion to its GNP. The laboratories have a limited amount of funding for research activities. When an experiment is approved the experimenters must apply for and justify the funding to construct the detectors and to analyze the data.

13.6   Research and Development

Because of the need to work at high luminosities, the triggering problem described in Section 13.2, and the problem of collecting and analyzing large amounts of data, many major developments in detector technology will be needed to do particle physics at accelerators. The important events may be rare relative to the number of Standard Model events, with cross sections small compared to similar Standard Model cross sections. A major and continuing emphasis on research into new detector technologies will be essential to properly utilize planned and future machines.

Problems

13.1:   Suppose a new kind of particle (called a Majorana neutrino) N is produced at the Tevatron collider at FNAL. The unique feature it has that permits it to be identifed is that it is its own charge conjugate, $N = \overline{N}$. Thus once N (or $\overline{N}$) is produced, half the time it decays as N and the other half of the time as $\overline{N}$. Its interactions are given completely by the vertices below:

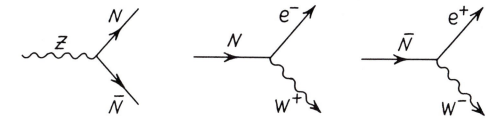

If one N or a pair of $N + \overline{N}$ are produced, show that the signature of something new is the occurrence of "like–sign dilepton" events, e.g. $e^+ e^+$ + jets or missing energy or additional leptons. List some Standard Model processes that will occur and study whether any of them are likely to give like–sign dileptons. What major properties of detectors would be essential to find such events?

13.2:   Suppose Z^0's are produced in the process $q + \overline{q} \rightarrow Z^0 + g$. What fraction of Z^0's would not appear in the detector? How could such events be recognized anyhow?

Suggestions for Further Study

The same sources as for Chapter 12 will be useful.

CHAPTER 14

Low Energy
and
Non–Accelerator
Experiments

The previous two chapters emphasized high energy colliders and their detectors. In order that the reader not get the impression that the only frontier is the high energy one, here we briefly mention other directions from which major discoveries might come. It turns out that our basic subject, the Standard Model and its tests, is indeed more naturally the domain of high energy, high luminosity machines. That is because the natural scale of the Standard Model is on the order of M_W or M_Z, *i.e.* on the order of 100 GeV. Testing the high energy predictions of the Standard Model requires TeV energies in the quark, lepton or gluon collisions.

Nevertheless, our purpose is also to prepare the reader to understand future developments in particle physics, both by providing an explanation of the Standard Model as the foundation on which anything new will stand, and by knowing where the Standard Model is conceptually incomplete. Several kinds of experiments are in progress or planned which could extend the Standard Model in new directions. Some use low energy secondary beams at accelerators, and others are truly non–accelerator experiments. They include: (1) searches for neutrino mass effects from neutrino oscillations, from detection of solar neutrinos, or from nuclear β decays (see Chapter 29); (2) searches for neutrinoless double β decay, which is sensitive to neutrino masses, right–handed currents, and any new light particles which might couple to neutrinos; (3) searches for rare or forbidden decays of kaons or muons; (4) searches for dark matter (mentioned in Chapter 28); (5) searches for nucleon decay (see Chapter 27); (6) searches for monopoles. Additional topics could be included.

The rare decays of quarks and leptons are an area where some very important tests of the Standard Model are being carried out. From the vertices of Chapter 7, plus the rotation between weak and mass eigenstates for the quarks described in Chapter 22, all the quarks and leptons except the lightest ones can decay. But a variety of decays are forbidden. The Standard Model separately conserves electron, muon, and tau number, for example (*i.e.*, the total number of $\tau^- + \nu_\tau - \tau^+ - \overline{\nu}_\tau$, particle minus antiparticles, does not change in a process, *etc.*). Thus many decays do not occur in the Standard Model. Examples are

$$\mu \not\to e\gamma$$
$$\tau \not\to \mu\gamma$$
$$\tau \not\to \mu\mu\mu$$
$$\mu \not\to eee$$
$$K^0 \not\to \mu e$$
$$B^0 \not\to \mu e$$
$$B^0 \not\to \tau\mu$$
$$K^+ \not\to \pi^+ \mu e$$

and so on. It is a prediction of the Standard Model that these decays do not occur, and so far they are not observed; branching ratio limits range from 10^{-4} to 10^{-11}. Yet many ideas for understanding the origin of mass, or related questions, lead to theories where some of these decays are induced at low levels. Finding them, or showing they are not present, may be one of the most important ways to gain insight into the physics that lies beyond the Standard Model.

The detectors for non-accelerator experiments often have to solve new kinds of problems and to be based on innovative devices and techniques. Progress in these areas may depend crucially on detector development.

CHAPTER 15

Quarks, Confinement, Light Mesons, Baryons, Jets, and Glueballs

Electrons and protons have electric charge. Opposite charges attract, and they bind into hydrogen atoms held together by the electromagnetic force. The hydrogen atom has a ground state, and an infinite tower of radial excited states for each angular momentum, and an infinite number of orbital angular momenta. The energy levels are modified by $\vec{L} \cdot \vec{S}$ forces, all of which are spin–dependent electromagnetic forces. In particle language, the bound state potential is generated from an infinite set of photon exchange diagrams. No finite set of perturbation theory diagrams will give a bound state, so actual calculation of the binding effects is a complicated problem; the fully relativistic bound state problem is not yet solved even in quantum electrodynamics.

Since quarks have electric charge, they will form "atoms" as well. But quarks feel another force, the QCD or color force mediated by gluons. Since the gluon force is considerably stronger than the electromagnetic force, its properties will determine which states are bound, and what spectroscopic patterns are expected. From a space–time point of view, a single gluon is like a single photon, but because of the self–interaction of gluons the multigluon diagrams can lead to quite different properties, and indeed they do.

From the point of view of the spectroscopic rules, the electromagnetic force is very simple. Atoms only form from opposite charges. That a very different result will hold for QCD is suggested by observing that the electromagnetic force is characterized by a $U(1)$ symmetry, while the QCD interaction is characterized by an $SU(3)$ symmetry. The difference leads to the existence of baryons!

Since the QCD force is very complicated, and perturbation theory arguments based on a few diagrams are not expected to give the dominant effects for bound state questions, we will describe the situation by writing a few rules, showing that the observed states obey the rules, and motivating the rules from QCD–based arguments. Considerable work is going on to solve QCD by non–perturbative methods, particularly lattice calculations; all of the results being obtained there are consistent with the perspective we are taking here.

15.1   Confinement of color and color singlet hadrons

The essential point to make is that it is believed that QCD has the property that the potential energy of two colored particles increases (linearly) with the distance between them.

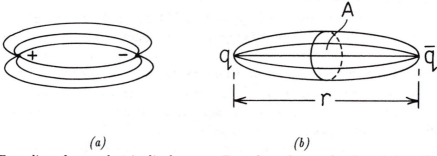

(a)   (b)

Force lines for an electric dipole.   *Force lines for a color force interaction.*

Figure 15.1

Figure 15.1a shows lines of force for a dipole electric field connecting two electric charges. Figure 15.1b shows the lines of force expected for the gluon field between a quark and an anti–quark, at a distance r. The cross-sectional area A is expected to remain constant as r increases, because of the gluon self–coupling. The number of field lines depends only on the total (color) charge so it does not change. Therefore the field energy grows as the volume of the flux tube, *i.e.* as r.

This has two major consequences. (1) Since it would take an infinite supply of energy to separate off a q or $\bar{q}$, they are "confined" to be forever within hadrons. (2) The energy supplied in a collision goes into producing hadrons.

Figure 15.2
Production of a meson jet.

Confinement implies that colored particles (quarks and gluons) appear as jets. To understand this, suppose in a collision a lot of energy is transferred to a quark in a hadron. It moves off relative to the other quark(s) it was bound with. As the distance between (say) a q and $\bar{q}$ pair grows, the energy in the color field grows, until enough energy is present to create a pair of physical mesons. Then it is energetically favorable to fragment into mesons, as in Figure 15.2. This process repeats until the original energy is dissipated, producing a number of hadrons. Experimentally, a 10 GeV quark fragments into about seven hadrons, while a 100 GeV quark gives 15 or so hadrons. Since pions are the lightest hadrons, they form the majority of hadrons in a jet.

Since a jet is an energetic quark or gluon, both of which carry color charge, while only color singlet hadrons will appear in the detectors, the color must some-how be compensated by soft hadrons. This is an example of a non–perturbative problem that is presently too complex for the theory to deal with rigorously. One testable observation concerns the separation of production and hadroniza-tion. The production of energetic quark or gluon jets should occur in perturbative electroweak or QCD processes. After production, they will hadronize indepen-dently, though the presence of other colored quarks or gluons could have some effect on the hadronization. Thus a given kind of jet should behave essentially the same way no matter how it originates. Testing that result requires a good technique for separating quark jets from gluon jets, which is not yet a fully solved problem, but all data is consistent with jets fragmenting essentially inde-pendently of their production.

The above arguments for the confinement of color and for jets are suggested by several approaches, such as attempts to study QCD on a lattice, or analogies with electromagnetic fields. The interested reader can consult the sources listed at the end of the chapter for further details. No rigorous proof is yet known, but that is not surprising since the problem is highly non–linear and non–perturbative. Note that what is confined is color, not electric charge. No known principle even suggests that particles of fractional electric charge should not occur in nature, but as we presently understand it, QCD implies that only color–singlet particles should appear. Only combinations of quarks and gluons (and any colored objects discovered in the future) that are color–singlets can be separated to distances greater than about a fermi, and appear in detectors.

Although the arguments for confinement are not quite rigorous, they are widely accepted and believed, and the mechanisms that operate can be understood. At the most basic level, like colors repel, just as like electric charges repel. For attraction it is a little more subtle: unlike colors will attract in a quantum state which is antisymmetric under interchange of color labels, but they will repel in a symmetric state. The situation is familiar to anyone who has experience with strong isospin and nuclei; a proton and a neutron (like two different color states) will attract in the antisymmetric state and make a deuteron if the spin and orbital states are appropriate, while they repel in the symmetric states that would give dineutrons or diprotons.

When a color and its anticolor are involved, the force is attractive and mesons form. When colors are being combined, we can start with pairs. There are three pairs for colors r, g, and b: rg, rb, or gb. A third quark will be repelled by each pair unless it is of the third color, so only triplets with all colors in antisymmetric combinations will be bound. That is exactly the situation for baryons.

A fourth quark brought up to a baryon will be repelled by one quark and attracted by the others. There will be less attraction than in the baryon state. Detailed calculations are necessary to see how forces balance and what states might exist, *i.e.* to do "quark chemistry." In general, though, both a meson and a baryon have no net color, so the color forces can be saturated locally, just as for a neutral atom. Residual forces exist between atoms but they are much smaller than the forces which form atoms. Similarly, small forces (nuclear forces) exist between hadrons, but the main color forces form the hadrons.

The description of the forces is along the following lines. For electrostatics, we think of lines of force that begin on positive charges and end on negative ones. The number of lines of force is determined by the strength of the electric field, which is proportional to the strength of the charges and decreases like $1/r^2$ a distance r away from a charge. The force between two charges is determined by the component of the field in the direction joining the charges. In the color case, we can also think of lines of color force. The number of lines is determined by the strength of the color charges. Because the gluons also carry the color charge, the force between the charges does not decrease with distance, so the density of lines stays constant as the distance between color charges increases. Since the force is independent of distance, the work done to separate color charges increases with the distance. Once enough energy is put into the system to produce $q\bar{q}$ pairs as pions, it is energetically favorable to do so, and the quarks are effectively confined.

Although we cannot calculate the force, we can make a crude estimate of its strength. If the potential effectively contains a term Kr that approximately describes the long range behavior, there will be a constant force K. In order to give hadrons whose size is about one fermi, and whose mass is about one GeV, K must be (within a factor of a few) about one GeV/fermi, which is 10^{15} GeV/meter. To convert that to everday units, note that one Newton is one $kg \cdot m/s^2$, which is about 2.2×10^{-6} GeV^2 in natural units. Also, 1 lb = 4.45 Newtons. Then 10^{15} GeV/m $\simeq 0.2$ GeV$^2 \simeq 10^5$ Newtons $\simeq 2 \times 10^4$ lb $\simeq 10$ tons is approximately the strength of the color force.

The Coulomb force is $-e^2/r^2$ between electron and proton. The QCD force due to gluon exchange, we have just argued, is constant with distance. We might have naïvely written a force proportional to α_s/r^2 since gluon exchange is like photon exchange. The constant force cannot come out of a perturbative calculation; it must arise from highly non–linear contributions of many gluons. Various approaches have been followed to study non–perturbative phenomena in QCD. The most productive one so far is called lattice gauge theory, where the goal is to make progress in finding solutions by working on a lattice with a minimum distance scale, so the theory is cut off in momentum transfer, and by introducing a variety of techniques such as statistical mechanical methods for handling complicated systems. Such treatments are beyond the scope of this book.

15.2   Color–Singlet Hadrons

To determine what we expect the spectrum of QCD to be, we need to see how to make color–singlet states from quarks and gluons. The quarks are $SU(3)$ triplet states and the gluons are color octet states.

To understand the results more easily, first consider the situation with ordinary $SU(2)$, *i.e.* angular momentum. If we have some particles of spin–1/2 (analogous to quarks) and some of spin–one (analogous to gluons), how can we construct spin singlets? The spin–singlet is a state invariant under rotations, just as a color singlet is a state invariant under rotations in color space. If we have two particles of spin–1/2, we know we can combine them to make a spin–one state with $S_z = 1$, 0, or -1 plus a spin–zero state. By symmetry, the spin–zero state is the antisymmetric combination $[|\uparrow\downarrow\rangle - |\downarrow\uparrow\rangle]/\sqrt{2}$, while the $S_z = 0$ component of the spin–one state must be the symmetric combination since it goes with the symmetric states with $S_z = \pm 1$, $|\uparrow\uparrow\rangle$ and $|\downarrow\downarrow\rangle$. Here we use a place notation, where particle one is in the first position and particle two is in the second.

Now turn to color. We only need repeat the above arguments for $SU(3)$. Here the fundamental state has three components instead of two, which we think of as the three quark colors, and label r, g, b. A new feature does occur. Now there are two ways to make singlets from quark states alone (instead of one way for spin–1/2 states). One is the equivalent way to the $SU(2)$ singlet, $\left(r\bar{r} + g\bar{g} + b\bar{b}\right)/\sqrt{3}$, where each quark is labeled by its color. This is obviously invariant under rotations in color space, $r \leftrightarrow b \leftrightarrow g$; it is like the singlet spin combination. This color singlet requires us to form states of quark–antiquark, using the property that the antiquark has color charge opposite to that of the quark. These are of course the meson states, which we will enumerate below.

We can form another color singlet from the quarks, the state $\epsilon_{ijk} q_i q_j q_k$. A place notation is again used, with quark one in the first slot, *etc.* Now i, j, and k can each be equal to r, g or b, so this is the antisymmetric combination of three colors, analogous to the triple product of three vectors. [It is not identical since the triplet quark is not in the representation equivalent to the $SU(2)$ vector, but the difference is not relevant for our purposes.] These states are formed from three quarks, with no antiquarks. They are the baryons.

To summarize, we have seen that one consequence of color confinement is the remarkable result that two kinds of color singlet, hadronic states made of quarks should exist, mesons formed from $q_i \bar{q}_i$ and baryons formed from $q_i q_j q_k$, as is observed. Of course, states of additional complexity can be formed such as $q\bar{q}q\bar{q}$ or $qqqq\bar{q}$, but they are expected to have higher energy and to be unstable under strong interactions. They are of interest for exploring the "chemistry" of QCD, but we will not consider them further.

So far we have not added the gluons. Since they carry color also, a singlet state can be formed from two gluons by summing over all the colors in a symmetric way [for readers familiar with more group theory, this is the singlet contained in the product 8×8 for the color octet gluons]. These states will also be mesons so the full spectrum should contain mesons from $q\bar{q}$ and from the gluons. The gluon states are called "glueballs". The glueball states will be mixed in with the $q\bar{q}$ mesons, and sometimes have identical quantum numbers. Any particular state could be from either source, and mixing can occur. There are various ways to check the QCD predictions and to distinguish glueballs from $q\bar{q}$ mesons. The clearest test is that the total number of states must be right, regardless of mixing. In particular, more mesons must exist than those expected from $q\bar{q}$ alone.

Another test uses the fact that the unstable mesons decay to lighter mesons. Mesons made of $q\bar{q}$ have decays that depend on the flavor of the quarks (we will discuss this more later) while the gluon couplings are independent of flavor, so the two types of mesons have different decay patterns. Since different quark flavors have different masses, subtleties arise in working out the predictions for experiments, but measurable effects will be present.

We will not study glueball states further. They are, however, as fundamental as $q\bar{q}$ mesons and baryons and it is very important to find the glueball states experimentally and to confirm detailed predictions for their spectrum.

15.3 Quantum Numbers of Mesons and Baryons

Every hadron will be characterized by a set of quantum numbers: its mass, electric charge, baryon number, spin, and it may be an eigenstate of parity, charge–conjugation, *etc.* The mass of the hadron will be determined by several contributions. The quarks in the hadron have masses of their own, which will add to the mass of the hadron. There is a contribution from the interaction energy of a colored particle in the color field of another particle. The problem of calculating the mass of a hadron, and the related problem of the mass of a quark or a gluon in a hadron, are complicated non–perturbative questions that are subjects of considerable research. It is possible to write a phenomenological treatment of the masses, motivated by the structure of QCD but not yet derived from it, that gives a satisfactory description of the theoretical and experimental situation. We will return to consider quark and hadron masses, and particularly to how quark masses are defined when they are tightly bound objects, in Chapter 23.

The other quantum numbers are easier to deal with. The procedure is the same as in atomic and nuclear physics. The $q\bar{q}$ mesons are like positronium. The quark spins can add to spin–zero or spin–one. The orbital angular momentum will be 0, 1, 2, $\cdots$. $\vec{L}$ and $\vec{S}$ will add to give the total spin of the meson. In general the states with $L = 0$ and $S = 0$ or 1 will lie lowest, so they will be the most familiar mesons. Those with $L \geq 1$, and those with zeros in the radial wave function will lie higher, *i.e.* they will be meson resonances. We will enumerate them in the next chapter.

When the q and $\bar{q}$ in a meson have the same flavor, the situation is especially like positronium. Since the fermion and antifermion have opposite intrinsic parity, and inversion of the coordinate system plus a rotation returns the system to its original state, the meson is an eigenstate of parity with eigenvalue

$$P = (-1)^{L+1} \ . \tag{15.1}$$

The $(-1)^L$ comes from the effect of the rotation on the angular wavefuction, since $Y_{LM}(\theta, \phi)$ goes into $(-1)^L$ times itself when $\theta \to \pi - \theta$ and $\phi \to \pi + \phi$. Similarly, if we charge conjugate then $q \leftrightarrow \bar{q}$. The system can be returned to its original state by rotating and interchanging spins, which gives $(-1)^{S+1}$ since spin–zero is antisymmetric and spin–one is symmetric as we saw above. Fermi statistics from interchanging fermions also gives a minus sign. Thus the eigenvalue of C is

$$C = (-1)(-1)^L(-1)^{S+1} = (-1)^{L+S} \ . \tag{15.2}$$

The meson states will be labeled by their total angular momentum J, by P and C, and by the flavor structure of their quarks (there are $u\bar{u}$ mesons, $s\bar{s}$ mesons, $s\bar{u}$ mesons, etc.).

Baryons have three quarks in the wave function. It is necessary to take account of the Pauli principle and ensure that the wave fuction is antisymmetric under the exchange of any two identical fermions. The full spectrum will come from adding orbital angular momentum and the three spins. The simplest states will have $L = 0$ and the spins will add to $1/2$ or $3/2$. Presumably these will be the lightest states. Since we know the lightest baryons (proton, neutron, lambda,$\cdots$) have spin–$1/2$, followed by spin–$3/2$ resonances (Δ, Σ^*, $\cdots$) this is as expected.

One extremely important observation can be made here. Consider the state Δ^{++}, which has spin–$3/2$ and can be made from three u quarks all with spin up, all in orbital s–states. Then the flavor, spin, and angular momentum degrees of freedom are all symmetric under interchange. Without the antisymmetry of the color degree of freedom, the quark interpretation of Δ would be inconsistent with the Pauli principle. The same point can be made from the proton or neutron wave functions, but it is easier to see for the highly symmetric Δ state. Color was introduced because of this argument.

15.4   Comments and Perspective

In this chapter we have argued that the form of QCD implies that colored particles are confined, so that

(i) a colored particle which is given a lot of momentum in a collision will appear as a "jet" of hadrons, and

(ii) meson and baryon states are color singlets.

Suppose a collision occurs. The strong interactions occur on a time scale of 10^{-22} seconds or less, so the hadronization of a quark or a gluon into a jet occurs very rapidly. Binding of quarks into mesons occurs on the same time scale. Some mesons, after formation, can decay by strong interactions, with lifetimes that are often very short. We will see that some of the quarks have weak decays. These take place on time scales on the order of 10^{-12} seconds, much more slowly, so they occur for quarks that have been bound in mesons. What finally emerges from the collision are long-lived and stable states; the presence of the rest has to be deduced.

The arguments in this chapter imply that a certain spectrum of mesons and baryons is expected, although the theory is too complicated to calculate detailed properties. Keeping in mind that the glueball spectrum needs further study, both theoretically and experimentally, we conclude with the following observations:

(1) All the predicted low lying states of $q\bar{q}$ and qqq are observed (in so far as experiments have been done).

(2) No states have been observed that were not predicted. [Extra states have been observed with the quantum numbers expected for glueballs, though the situation is very complicated and may not be untangled without considerably more data.]

(3) All properties of the mesons and baryons are consistent with qualitative (and sometimes quantitative) arguments based on the quark structure picture we have outlined.

This set of results, based on data about dozens of hadrons, acquired over thirty years, is a remarkable achievement. A very large body of data is organized and appears to be understood on the basis of quarks interacting via QCD forces.

Problems

15.1:   We have seen that L and S can be treated as good quantum numbers for hadrons. Since many hadronic states exist, it could have happened that states of the same J but different L and S would mix, and give rise to spectra that do not have a simple interpretation in terms of $q\bar{q}$ quantum numbers. Show that invariance under parity implies that states of even and odd L will not mix. Then show that invariance under charge conjugation implies that there is only one S allowed for each L (demonstrating these is much simpler than finding the actual eigenvalues under P or C). Consequently the number of states that can mix is much smaller, and the simple picture is expected to be useful.

Suggestions for Further Study

Several Scientific American articles give rather detailed discussions of the material in this chapter: K. Johnson, "The Bag Model of Quark Confinement," Y. Nambu, "The Confinement of Quarks," and K. Ishikawa, "Glueballs." The Scientific American article of 't Hooft is also very instructive here. The book of Close and the Les Houches lectures of Quigg both give graduate level discussions of some of the topics of this chapter. T. D. Lee's book has an extensive treatment of the confinement question. Rebbi's Scientific American article, "The Lattice Theory of Quark Confinement," gives a nice view of the lattice approach in general as well as an argument for confinement. A discussion of the subject of Chapters 15–17 at a level similar to ours is given by Gottfried and Weisskopf in Volume I, E1-7.

CHAPTER 16

Light Mesons, Baryons, and Strong Isospin

The one feature we have left out of the picture so far is flavor. There are (at least) six flavors of quarks, and mesons and baryons can be made in all possible ways, *i.e.* $u\bar{u}$, $u\bar{d}$, $u\bar{c}$, $u\bar{b}$, $s\bar{b}$, $s\bar{d}$, uud, usd, $\cdots$.

As described in Chapter 23, and mentioned in Chapter 1, the quarks have an intrinsic mass we have called the free quark mass, the mass they would have even if they were not bound. That mass simply adds to the interaction energies. Mesons with a charmed quark are about 1.3 GeV heavier than the equivalent mesons with only light quarks. Historically, of course, the lightest hadrons were found first, starting with the proton and neutron, because more energetic collisions were required to make heavier ones. The pattern of lighter mesons and baryons was very important in leading to some of the ideas that are part of the Standard Model today. We will not go into much detail about the hadrons, both because they no longer play a major role in the Standard Model, and because they are described in several places. However, it is worthwhile to survey the states and make some observations.

The u and d quarks are very light, with free masses on the order of 10 MeV. Their constituent mass is about 350 MeV, since three quarks make a nucleon mass of about 940 MeV, or quark and antiquark a ρ meson of mass 760 MeV. The constituent mass is mass which results from the binding of massless quarks into a color singlet state. Spin–dependent effects and binding differences in two or three body states imply that simple arguments cannot give too accurate a value for this mass. A strange quark has a mass of about 200 MeV, since strange

185

particles are typically that much heavier than non–strange ones.

16.1   The $L = 0$ Meson States

Here we list some of the lightest states.

<table>
<tr><td colspan="3" align="center">$L = 0$ and $S = 0$ states</td><td colspan="3" align="center">$L = 0$ and $S = 1$ states</td></tr>
<tr><td>Particle</td><td>Quark Content</td><td>Mass (MeV)</td><td>Particle</td><td>Quark Content</td><td>Mass (MeV)</td></tr>
<tr><td>π^+</td><td>$u\bar{d}$</td><td>140</td><td>ρ^+</td><td>$u\bar{d}$</td><td>770</td></tr>
<tr><td>π^-</td><td>$\bar{u}d$</td><td>140</td><td>ρ^-</td><td>$\bar{u}d$</td><td>”</td></tr>
<tr><td>π^0</td><td>$\left(u\bar{u} + d\bar{d}\right)/\sqrt{2}$</td><td>135</td><td>$\rho^0$</td><td>$\left(u\bar{u} + d\bar{d}\right)/\sqrt{2}$</td><td>”</td></tr>
<tr><td>η</td><td>$\left(u\bar{u} - d\bar{d}\right)/\sqrt{2}$</td><td>550</td><td>$\omega^0$</td><td>$\left(u\bar{u} - d\bar{d}\right)/\sqrt{2}$</td><td>780</td></tr>
<tr><td>K^+</td><td>$u\bar{s}$</td><td>494</td><td>K^{*+}</td><td>etc., as above</td><td>890</td></tr>
<tr><td>K^-</td><td>$\bar{u}s$</td><td>494</td><td>K^{*-}</td><td></td><td>”</td></tr>
<tr><td>K^0</td><td>$d\bar{s}$</td><td>498</td><td>K^{*0}</td><td></td><td>”</td></tr>
<tr><td>$\overline{K}^0$</td><td>$\bar{d}s$</td><td>498</td><td>$\overline{K}^{*0}$</td><td></td><td>”</td></tr>
<tr><td>η'</td><td>$s\bar{s}$</td><td>958</td><td>ϕ</td><td></td><td>1020</td></tr>
</table>

The quark content is the same for the spin-zero and spin-one states; that is, a $u\bar{d}$ state with $J = 0$ is a π^+, and with $J = 1$, it is a ρ^+. The $L = S = 0$ states are pseudoscalar mesons, i.e. they have odd parity (see equation (15.1)) and spin-zero. The $L = 0$ and $S = 1$ mesons have $J = 1$ and still odd parity (which depends only on L); they transform under rotations like a vector, and are called vector mesons. The neutral ones ρ^0, ω^0, and ϕ have the same quantum numbers as a photon and can mix with the photon. The neutral pseudoscalars are even under charge conjugation; the neutral vector mesons are odd. [The signs in the quark wave functions for $\pi^0(\rho^0)$ and $\eta(\omega)$ are a little subtle. The minus for the η is meant to mean antisymmetry under exchange of u and d. However, if Clebsch–Gordan coefficients are used with Condon and Shortly phases, an extra minus occurs in the definition of the basic states, and these signs flip; we will not need the signs, so these remarks are just to caution anyone who might use the results. Also, the assignment of $u\bar{u} - d\bar{d}$ for η or ω, and $s\bar{s}$ for η' or ϕ, is very approximate, useful for counting but not for any detailed analysis.]

Except for the pions being unusually light, the masses conform to what might be expected from the free quark masses plus some binding energy, with some variation from spin dependence. The smallness of m_π is understood, as a subtle consequence of the structure of QCD and the smallness of m_u and m_d compared to the scale of QCD, but the explanation involves detailed dynamics and we will not discuss it.

16.2   The $L = 0$ Baryon States

$\underline{L_1 = L_2 = 0 \text{ and } S = 3/2 \text{ states}}$

Particle	Quark Content	Mass (MeV)
Δ^{++}	uuu	1232
Δ^+	uud	"
Δ^0	udd	"
Δ^-	ddd	"
Σ^{*+}	suu	1382
Σ^{*0}	sud	"
Σ^{*-}	sdd	1387
Ξ^{*0}	ssu	1315
Ξ^{*-}	ssd	1321
Ω^-	sss	1672

$\underline{L_1 = L_2 = 0 \text{ and } S = 1/2 \text{ states}}$

Particle	Quark Content	Mass (MeV)
p	uud	939
n	udd	940
Λ	uds	1115
Σ^+	uus	1193
Σ^-	dds	1197
Σ^0	uds	1189
Ξ^0	uss	1315
Ξ^-	dss	1321

For these states only the main quark combinations present in the wave function are listed. The full wave function is rather complicated since it has appropriate symmetrization in flavor, spin and color spaces.

16.3   Decays and Transitions

None of the mesons or baryons—except possibly the proton—are stable. Most of them decay strongly, on a scale of 10^{-23} seconds. Examples are

$$\Delta^{++} \to p\pi^+ \, ,$$
$$\rho^+ \to \pi^+\pi^0 \, ,$$
$$K^{*+} \to K^+\pi^0 \, ,$$
$$\to K^0\pi^+ \, .$$

These occur in a non–perturbative regime, with decay widths on the order of 100 MeV (except occasionally when very small phase space is available). Typically Γ/M is of order 10–20%, rather than of order 1% or less as for most weakly decaying states. Consequently, the widths cannot be calculated perturbatively, and no non–perturbative techniques are yet available.

Some decays are electromagnetic,

$$\pi^0 \to \gamma\gamma \, ,$$
$$\Sigma^0 \to \Lambda\gamma \, ,$$
$$\Delta^+ \to p\gamma \, ,$$
$$\rho^+ \to \pi^+\gamma \, ;$$

some are weak

$$\pi^+ \to \mu^+\nu_\mu \, ,$$
$$K^+ \to \mu^+\nu_\mu \, ,$$
$$\Omega^- \to \Lambda K^- \, ,$$
$$n \to pe^-\nu_e \, .$$

All the weak and electromagnetic decay systematics can be understood by assuming that the transition is at the quark level.

For example, one assumes that the $\Delta^+ \to p\gamma$ and $\rho^+ \to \pi^+\gamma$ decays both occur by a quark spin-flip transition, dominated by $u\uparrow \to u\downarrow +\gamma$ (since the electric charge of the u is largest). The correct angular distributions are obtained and so is the correct ratio of absolute rates. The weak decays can all be described quantitatively by the electroweak theory of Chapter 7 (supplemented by the quark mixing analysis of Chapter 22 in practice).

The systematics of baryon magnetic moments can also be described in terms of quark magnetic moments, with some assumptions.

In general, the hundreds of properties of dozens of light mesons and baryons can be understood qualitatively and sometimes quantitatively in the picture we have described, and there are no discomforting puzzles or problems. It is extraordinary that in spite of the difficulty of carrying out quantitative calculations, good qualitative agreement between experiment and theory holds, as summarized by the statements (1) no states exist that are not expected, (2) no states are expected that do not exist, and (3) all properties appear to be as expected. Considerable work is presently going on to find non–perturbative methods to improve our ability to solve QCD for the behavior and properties of the hadrons.

16.4   The Origin of Strong Isospin Invariance

Our present understanding of the physics of strong isospin invariance is quite different from the interpretation it was given historically. The change in perspective is rather surprising and very instructive.

From its introduction over fifty years ago until the formulation of the Standard Model, the strong isospin invariance of strong interactions was not only a way to help understand data and to relate one process to another, with numerical validity at about the 1% level, it also played a crucial role in the development of ideas and understanding of symmetry principles. As we discussed in Chapter 4, it led to the idea of internal symmetries, and helped provide stimulus to the use of symmetries and group theory in particle physics. It has been of great value in the development of modern ideas.

Now that we have a theory of strong interactions, how is the strong isospin invariance interpreted? If the strong isospin invariance had not been noticed before, the argument might go as follows. The strong force is mediated by gluon exchange, and the gluon interactions are completely flavor–independent. A gluon cannot tell a u from a d; the u and d differ in their electric charge, so electromagnetic interactions allow us to distinguish them. They also differ in their mass, but their masses (of order 10 MeV) are so small compared to the QCD interaction energies (of order 350 MeV) that we can never expect to see effects of the mass difference in strong interactions. Thus all observations are expected to be unchanged to a good approximation if we rotate $u \leftrightarrow d$, which changes $p \leftrightarrow n$ and in general gives all the strong isospin transformations.

Remarkably, we see that the strong isospin invariance is apparently an accident, a consequence of the smallness of m_u and m_d compared to the hadron masses. More precisely, it is the smallness of $m_d - m_u$ that matters. Quark masses satisfy the conditions $m_t \gg m_b \gg m_c \gg m_s \gg m_u$ or m_d. If a large inequality had also occurred with $m_d \gg m_u$, so that $m_d/350$ MeV was not small, we would not have observed strong isospin invariance. Since we do not understand the origin of masses, we can only be grateful that the accident occurred and the associated ideas entered physics and helped its development. Finally, we can turn it around. The observed strong isospin invariance is good evidence that gluon interactions are flavor independent.

Suggestions for Further Study

The book of Close is almost entirely devoted to the interpretation of hadrons in terms of quarks and gluons; it covers not only the spectroscopy, including many complications, but the interactions of hadrons and their scattering. An extensive and modern treatment of the spectroscopy of quarks and gluons is given by Quigg in his 1981 Les Houches lectures.

Heavy
Quarks
$(c, \; b)$

By 1974 some thoughtful physicists had been convinced by theoretical and experimental developments that quarks and gluons were real, and by then the Standard Model existed in its present form. In November 1974 a discovery occurred that in a very short time converted essentially all remaining particle physicists to the point of view that is basically the one described in this book.

At both SLAC and Brookhaven a resonance of mass 3.1 GeV was discovered, with an extremely narrow width, $\Gamma \simeq 70$ KeV. It was named J by the Brookhaven group and ψ by the SLAC group; it is usually referred to as J/ψ. To understand the remarkable impact of the discovery, it helps to see it in context. Many hadron resonances had been discovered in the preceeding two decades. Typically their width was 10–20% of their mass, as expected from decays allowed by strong interactions. At SLAC the J/ψ was produced in $e^+e^- \rightarrow J/\psi \rightarrow$ hadrons. Earlier the ρ^0, ω and ϕ resonances had been observed this way. In $e^+e^- \rightarrow \pi^+\pi^-$, the ρ could be observed by measuring the pion momenta and plotting the number of events versus $M^2 = \left(p_+ + p_-\right)^2$, as described in Chapter 13. A large peak would appear, as in equation (9.21). They were normal hadronic resonances that had the quantum numbers of the photon and the data could be interpreted as in Figure 17.1, with analogous diagrams for ω and ϕ.

191

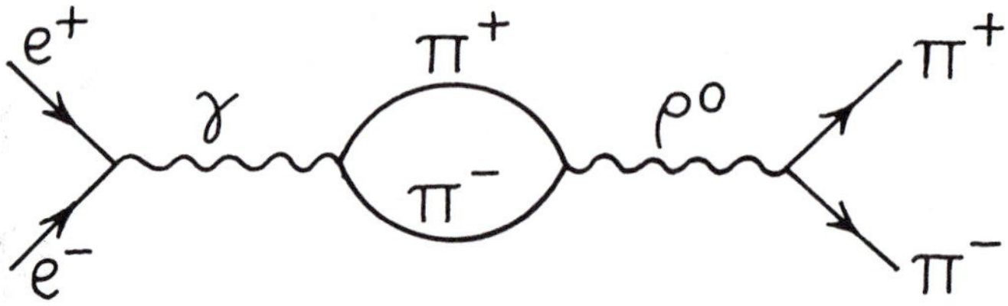

Figure 17.1
Hadronic Resonances.

Their widths were "large," in the sense that the ρ width was 20% of its mass; the ω and ϕ had smaller widths (5–10 MeV) but that was accidental, because their dominant decays were limited by phase space. [The ϕ preferred to decay to $K\overline{K}$ but it was barely above the $K\overline{K}$ threshold, while the ω wanted to decay to $\rho\pi$ but was below the threshold so the ρ was virtual. When the decays were calculated in terms of a coupling strength, that strength was comparable for all three.] If the J/ψ had a "normal" width it would have been over 10^3 times larger than its actual width.

In addition, several theorists had been arguing for the previous year or so that a new heavy quark was required. Although this work was not yet widely accepted, it was becoming known, and contributed significantly to the rapid acceptance of the present picture by theorists. This was particularly true when some theorists who had been thinking along these lines quickly told the SLAC group that, given the interpretation that the J/ψ was a $q\bar{q}$ meson in an orbital s–wave and a spin triplet state, the first radial excitation [which would have the same quantum numbers and therefore could be produced the same way] should be 600 MeV higher in energy. The predicted state, ψ', was immediately found, and the interpretation in terms of a fourth quark became very much the dominant one. The theorists thinking about the J/ψ before it was discovered had realized it would be narrow, though its width turned out to be even smaller than was originally estimated. The flavor of the new quark was called "charm," since it had been named that considerably earlier by the theorists who foresaw it.

If what was being seen was the positronium–like structure of a heavy quark atom, then clearly quarks had to be taken completely seriously as the basis for all particle physics. And if this fundamental, point–like quark had about 1.5 times the mass of the (composite) proton, significant revision in thinking was required.

17.1   Some Charmonium Properties

The J/ψ system is interpreted as the energy levels of the bound system of the charmed quark c and its antiparticle $\bar{c}$. Because the free mass of the c–quark is large compared to its constituent mass, its bound state spectrum is much easier to study than that of the light quarks. It is instructive to see how some of the properties are obtained from data. Similar techniques were used for the $b\bar{b}$ system, and will be used for toponium $(t\bar{t})$ if it is found at an energy accessible to e^+e^- colliders. The $c\bar{c}$ system is called "charmonium." For any heavy quark Q the $Q\bar{Q}$ system is called "quarkonium."

Various processes can be studied. Consider $e^+e^- \to e^+e^-$, $e^+e^- \to \mu^+\mu^-$, and $e^+e^- \to$ hadrons, all in the region of the J/ψ resonance, i.e. at $\sqrt{s} \simeq m_\psi$. Then from equation (9.21), the cross section for scattering through a $J = 1$ resonance is

$$\sigma = \frac{3\pi}{s} \frac{\Gamma_{ee}\Gamma_f}{\left(\sqrt{s} - m_\psi\right)^2 + \Gamma_\psi^2/4} \, . \tag{17.1}$$

For a narrow resonance with a high peak height, study of the resonance shape and measurement of Γ_ψ might not be possible if the width is less than the experimental resolution. To extract information, consider the integrated quantities

$$I_f = \int \sigma d\sqrt{s} \, . \tag{17.2}$$

In particular, using the narrow width approximation (equation (10.4))

$$I_{ee} = \frac{6\pi^2}{m_\psi^2} \frac{\Gamma_{ee}^2}{\Gamma_\psi} \, ,$$

$$I_{\mu\mu} = \frac{6\pi^2}{m_\psi^2} \frac{\Gamma_{ee}\Gamma_{\mu\mu}}{\Gamma_\psi} \, , \tag{17.3}$$

$$I_{\text{had}} = \frac{6\pi^2}{m_\psi^2} \frac{\Gamma_{ee}\Gamma_{\text{had}}}{\Gamma_\psi} \, .$$

Experimentally the I's are proportional to the total number of events collected for each process. Given m_ψ, which is measured from the beam energy to be 3.1 GeV, and data for I_{ee}, $I_{\mu\mu}$, and I_{had}, there are four unknowns. However, $\Gamma_\psi = \Gamma_{ee} + \Gamma_{\mu\mu} + \Gamma_{\text{had}}$ if the detector collects all modes, so the three equations can be solved for Γ_{ee}, $\Gamma_{\mu\mu}$, and Γ_{had}. The results are $\Gamma_{ee} = \Gamma_{\mu\mu} = 5 \pm 1$ KeV, and $\Gamma_{\text{had}} = 70 \pm 10$ KeV. This Γ is indeed narrower than the experimental resolution and cannot be directly detected. The ratio Γ_ψ/m_ψ, which was 10–20% for a typical hadron, is here 2.5×10^{-5}!

A number of other states are also expected, such as the $S = L = 0$ state, which is called η_c. A spin flip transition (like $\rho \rightarrow \gamma + \pi$ or $\Delta^+ \rightarrow \gamma + p$)

$$J/\psi \rightarrow \gamma + \eta_c \qquad (17.4)$$

is expected from the $S = 1$ to the $S = 0$ state, and is observed. Now a complete spectrum and many transitions have been observed, and in all respects the charmonium system has behaved as expected.

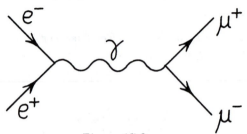

Figure 17.2
Diagram for $e^+e^- \rightarrow \mu^+\mu^-$ away from the J/ψ resonance.

A nice method is available to confirm that the J/ψ indeed has the quantum numbers of the photon. Consider $e^+e^- \rightarrow \mu^+\mu^-$ for example. Away from the J/ψ the amplitude should be given by the photon contribution, as in Figure 17.2, for which the amplitude is, according to the rules of Chapter 7,

$$M_\gamma = \frac{e^2}{s}\overline{u}\gamma_\mu u\overline{u}\gamma^\mu u\ , \qquad (17.5)$$

which is real. The amplitude for the J/ψ contribution is

$$M_{J/\psi} \sim \frac{1}{s - m_\psi^2 - im_\psi\Gamma_\psi} \qquad (17.6)$$

from equation (9.20), and this has real and imaginary parts,

$$M_{J/\psi} \sim \frac{s - m_\psi^2}{\left(s - m_\psi^2\right)^2 + m_\psi^2\Gamma_\psi^2} + i\frac{m_\psi\Gamma_\psi}{\left(s - m_\psi^2\right)^2 + m_\psi^2\Gamma_\psi^2}\ . \qquad (17.7)$$

The real part of $M_{J/\psi}$ will change sign at $\sqrt{s} = m_\psi$, and it will interfere with (the real) M_γ, so if the J/ψ indeed has the same quantum numbers as the photon, there will be an asymmetric interference effect. This is indeed observed.

17.2   The Charmonium Spectrum

The bound state of c plus $\bar{c}$ makes an atomic system analogous to the positronium e^+e^- atoms. As discussed in Chapters 15 and 16, because of the color force the details of the spectrum will be somewhat different. At very small distances, the Coulomb contribution will dominate the potential, while at large distances the rising potential leads to confinement. The detailed shape of the potential cannot be calculated because of the non–linear nature of the gluon contributions, but a number of smooth interpolations between the small and large r regions have been shown to give a good description of the binding potential.

Some of the measured levels of the $c\bar{c}$ charmonium system are shown in Figure 17.3.

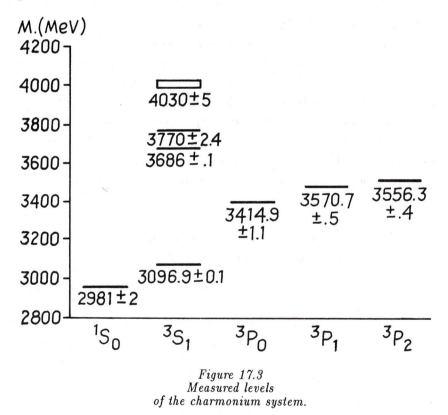

Figure 17.3
Measured levels
of the charmonium system.

Above the threshold for decay into charmed mesons, the widths become relatively large, since a whole set of strong decay modes are present. The D^+ mass is 1869.3 MeV and the D^0 mass 1864.6 MeV, so above a mass of about 3730 MeV any $c\bar{c}$ meson ψ_n has an allowed strong decay $\psi_n \rightarrow D\overline{D}$. We can

picture it as in Figure 17.4,

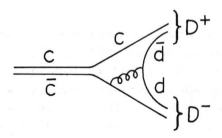

Figure 17.4
Strong decay
of the charmonium system.

though the true process is complicated and involves many gluon contributions
and binding effects. The widths of the $c\bar{c}$ states increase by a factor of 100–1000
above the $D\overline{D}$ threshold. Below the threshold, decays occur by annihilations
through gluons or photons; then the annihilations are at short distances of order
$1/M_c$, and the rates are much smaller. The annihilation rates are also small because of an effect that will be discussed in Chapter 20—that the strong coupling
is decreasing rapidly with the mass scale in this region. For an annihilation the
coupling is needed at a scale of 3 GeV, where its value is $\alpha_3 \simeq 0.15$–0.2, while for
a process such as in Figure 17.4 only long distances are involved and $\alpha_3 \gtrsim 1$. At
least α_3^2 enters in rates, and often a higher power if more gluons are involved.

17.3   Charmed Mesons

In addition to the charmonium states, we expect a complete set of $c\bar{u}$, $c\bar{d}$,
and $c\bar{s}$ mesons. They should be pair produced, with

$$e^+e^- \rightarrow c\bar{c}, \tag{17.8}$$

followed by the separating charmed quarks leading to creation of $u\bar{u}$ and $d\bar{d}$ pairs
(the lightest quarks) as energy is dissipated. Most of the light quarks would end
up in pairs, but some should attach to the c and $\bar{c}$ to make mesons. The names
that the $L = 0$ mesons have been given are, for $S = 0$,

$$\begin{aligned} c\bar{u} &= D^0 \\ c\bar{d} &= D^+ \\ \bar{c}u &= \overline{D}^0 \\ \bar{c}d &= D^- \end{aligned} \tag{17.9}$$

and when a strange quark is involved, less frequently,

$$c\bar{s} = F^+$$
$$\bar{c}s = F^-.$$

(17.10)

The masses are

$$m(D^+) = m(D^-) = 1869.3 \pm 0.6 \text{ MeV}$$
$$m(D^0) = m(\overline{D}^0) = 1864.6 \pm 0.6 \text{ MeV}$$
$$m(F^+) = m(F^-) = 1970.5 \pm 2.5 \text{ MeV}.$$

Note that $m(D^+) + m(D^-) > m(\psi)$, so the ψ is stable against decay into charmed particles; the charmed quarks must annihilate in the decay, which helps explain the relatively long lifetime of the ψ. When $S = 1$ a $*$ is attached, so D^{*0} or $\overline{D}^{*0}$ or $D^{*\pm}$ is formed. D^* is like ρ or K^*, while D and F are like π and K. The notation has now been systematized, with $F^\pm$ being written $D_s^\pm$.

The D mesons would have been stable, but the electroweak theory allows the charmed quark to decay. From the vertices given in Chapter 7, we can draw the diagram (Figure 17.5),

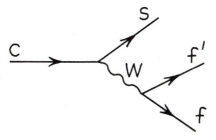

Figure 17.5
Electroweak decay of c to s.

where f' and f are fermions allowed by their coupling to the W and by energy conservation. The possibilities are

$$c \rightarrow se\nu_e$$
$$\rightarrow s\mu\nu_\mu$$
$$\rightarrow su\bar{d}.$$

(17.11)

Because of color, the $u\bar{d}$ channel is three times more probable than the others. The lifetime of the c–quark can be calculated in exactly the way the muon decay was calculated in Chapter 11, giving $\Gamma_c = 5G_F^2 m_c^5 / 192\pi^3$ since there are five channels, and $\tau_c = 1/\Gamma_c \approx 5 \times 10^{-13}$ sec (after some QCD corrections are included). The decay occurs very near to the production point, so only the decay products are directly observed. The path length is long enough, however, for good detectors to observe the separation of production and decay vertices, and that has been done.

The c quark is in a meson and what finally comes out must be hadrons, so more thought is required to understand the decays. We can draw pictures, but it is important to keep in mind that they are pictures rather than Feynman diagrams, as non–perturbative effects are involved, so there are no precise rules for calculations associated with them. One possibility is

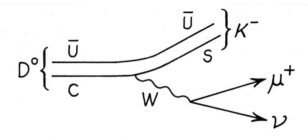

giving $D^0 \to K^- \mu^+ \nu_\mu$. Another is

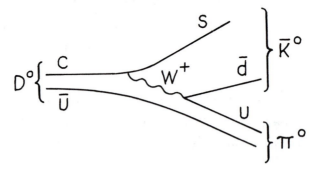

giving $D^0 \to \overline{K}^0 \pi^0$. Note D^0 (and D^+) always give K^- or $\overline{K}^0$, while $\overline{D}^0$ and D^- always give K^+ and K^0. Since the decays are weak ones, the associated widths are extremely small, always smaller then the experimental resolution:

$$\Gamma_D \simeq 5 \frac{G_F^2 m_c^5}{192\pi^3} \simeq 5 \times 10^{-13} \text{ GeV} = 5 \times 10^{-4} \text{ KeV}. \qquad (17.12)$$

17.4   More Leptons and Quarks

After the discovery of the charmed quark, there were two families, and a certain symmetry between leptons (ν_e , e, ν_μ , μ) and quarks (u, d, c, s). No arguments of any kind existed to suggest a need for the discovery of new fermions. Almost immediately the situation changed, as the MARK I group at SPEAR discovered a new lepton, the τ (see Chapter 19). Not long after that the fifth quark, the b, was found at FNAL. Their partners have not yet been directly detected, but there is indirect proof that they exist if the Standard Model correctly describes τ and b interactions, as it seems to; we will describe these arguments in Chapter 25. The t quark could be found at CERN or SLC or TRISTAN or FNAL, depending on its mass (which is unknown, as were all fermion masses before their discovery) and on machine and detector capabilities. The ν_τ could be detected in an appropriate fixed target experiment at CERN or FNAL.

17.5   The b quark

After the b quark was found, studies on its spectrum were carried out at CESR and at DORIS. The energy levels of the $b\bar{b}$ system have been studied and behave as expected. B mesons ($b\bar{u}$, $b\bar{d}$, $b\bar{s}$) have been detected and studied. We will see in the next chapter that the more massive a system, the more its behavior can be expected to be correctly described by perturbative arguments, according to QCD; that expectation is consistent with the behavior of the b system.

Problems

17.1: Suppose the t–quark has a mass in the range of 25–80 GeV. Show its width should be about $\Gamma_t \simeq 9G_F^2 m_t^5/192\pi^3$. Consider a $t\bar{t}$ $^3S_1$ bound state, called θ and named "toponium." If it annihilates to lighter particles, it must do so through three gluons (charge conjugation invariance forbids two and energy and momentum conservation forbid one), so the rate has a factor α_3^3 times a phase space reduction; we might guess

$$\Gamma_\theta \simeq \frac{\alpha_3^3}{32\pi^2} m_\theta .$$

For the ψ and the Υ all decays occur through annihilation. What mass for t would allow both the annihilations and the weak decays to be observed, so that effectively the weak lifetime would be measured, even though the t–quark traveled a distance too short to be observed? What would be expected for the signature of θ decay if all the decays were weak (*i.e.* a t–quark would always decay before the annihilation could occur)? What if the annihilation always occurred before the t had time to decay weakly? How would we know if the weak and annihilation lifetimes were about equal?

17.2:   Explain why the decay $D^0 \rightarrow K^- \pi^+$ is allowed, while $D^0 \rightarrow K^+ \pi^-$ is forbidden. The former is observed and the latter is not, with data showing about a factor of 25 suppression of the forbidden mode at the present level of statistics.

Suggestions for Further Study

The Scientific American articles of Schwitters for the charmed quark, and of Lederman for the b–quark, summarize the discoveries. See also "Quarkonium" by Bloom and Feldman, and the "Personal Accounts" of Richter and of Ting.

CHAPTER 18

Deep Inelastic
Scattering
and Structure Functions

If protons have point–like structure, it ought to be detectable in experiments of the Rutherford type, where the protons are probed with energetic projectiles to see whether any scatter at large angles. That was done with electrons as probes, and results were first reported in 1968. Many further experiments, with electron and neutrino probes, have been carried out since then. The results suggested, remarkably, that the proton was composed of objects that could be the quarks of the spectroscopy we have examined in Chapters 15 to 17, and that in addition almost half of the momentum of a proton was carried by objects that did not have electroweak interactions—the first appearance of gluons. Most surprisingly, it appeared that the quarks behaved as if they were essentially free particles rather than bound so tightly they could not get out.

18.1   Deep Inelastic Scattering

A traditional method to probe for the structure of some target, B, is to shoot a projectile, A, and then study what emerges. If the same things come out, $A + B \rightarrow A + B$, the collision is elastic. Normally the target disintegrates and the collision is inelastic. We will consider electrons as projectiles; similar experiments have been done with neutrino and muon projectiles. All these leptons behave as if they are point–like objects, with no structure, when their own scattering is observed ($e^+e^- \rightarrow e^+e^-$, $e^+e^- \rightarrow \mu^+\mu^-$, and $\nu_\mu e \rightarrow \nu_\mu e$). Therefore any structure effects observed when they are scattered off protons or neutrons can be attributed to the nucleon structure.

201

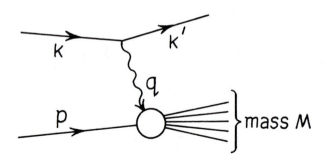

Figure 18.1
Interaction between a probe particle and the target
where the target disintegrates.

Let the initial and final electrons have four–momenta k, k', with energies E, E'. Define a four–momentum transfer $q = k - k'$. In atomic physics electrons might be scattered from an atom to measure the charge distribution. The standard result in atomic and nuclear physics is that

$$\frac{d\sigma}{d\Omega} = \frac{Q^2 \alpha^2 E^2}{4|\overrightarrow{k}|^2 \sin^4\left(\frac{\theta}{2}\right)} |F(q^2)|^2, \tag{18.1}$$

where the target charge is Qe, the electron is scattered through an angle θ, and $F(q^2)$ is a "form factor" that measures the charge distribution. In equation (18.1), the effects of spin have been ignored, *i.e.* it is correct for a spinless projectile, or in the nonrelativistic limit. By the uncertainty principle, the larger q^2 is the better the resolution with which the target is being probed. As Figure 18.1 suggests, for electron and muon projectiles the probes are photons and Z^0's. When the target breaks up, a new variable is needed to describe the total mass of the products, as in Figure 18.1. For elastic scattering, at fixed beam energy, there is only one independent variable, usually taken to be the scattering angle or the momentum transfer. Here for the inelastic collision there are two variables, and M and q^2 can be used. In general, equation (18.1) would be modified to have $F(q^2) \to G(M, q^2)$ where G is a function of the two variables.

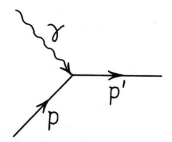

Figure 18.2
Collision of a photon of mass q^2
with a target of momentum p.

It is very instructive to see what we expect to happen if the collision is with a point–like constituent of a proton, as in Figure 18.2, where we imagine a collision of a photon of four–momentum q and mass q^2 with a target of momentum p, making a particle of momentum p' and mass M. The kinematics are

$$p'^2 = M^2 = (p + q)^2 = p^2 + q^2 + 2p \cdot q \qquad (18.2)$$

so we could use as variables q^2 and $p \cdot q$. Since p^2 is a proton mass or a light quark mass (squared), it can be dropped for the large q^2 values of interest to us. This is a situation we have already encountered, with somewhat different kinematics, for production of the W boson in Chapter 10. There we saw we could write the result in terms of some factors times the constituent cross section $\hat{\sigma}$, and for a single particle $\hat{\sigma}$ had a delta function factor. Since we can neglect the mass of the final particle, p^2, here the delta function of Chapter 10 becomes, with $\hat{s} = p'^2$,

$$\delta(\hat{s}) \simeq \delta(q^2 + 2p \cdot q).$$

Thus q^2 and $p \cdot q$ are not independent, so if the scattering were really the sum of scatterings off a set of point–like constituents in the target, the result would be a function of only one variable rather than two.

This is of course what happens. When $2p \cdot q/q^2$ is held constant in plotting data, and the form factor is plotted versus q^2, it is independent of q^2. At large q^2, the proton appears to be a collection of quarks, and the scattering off the proton can be written as a sum of electron–quark scatterings. This phenomenon, that the scattering is only a function of one variable $x = -q^2/2q \cdot p$, is called Bjorken scaling.

When this was first observed, it was rather startling. It quickly led to a considerable increase in the seriousness with which the idea of quark constituents was taken.

18.2   Analysis of the Parton Model

Since the proton contains not only the three quarks we have discussed in a spectroscopic sense in Chapters 15 and 16, but also gluons, and $q\bar{q}$ pairs can always be present in a real situation, a general constituent has been called a "parton" and this picture of the proton is called the "parton model." The quarks that determine the spectroscopic properties of hadrons are called valence quarks, by obvious analogy with the chemical situation, and the quarks that arise from creation of $q\bar{q}$ pairs are called sea quarks, as previously noted. Since a quark is a quark, it is not always possible to maintain the distinction between valence and sea quarks, but their kinematic properties are sufficiently different that the distinction is useful. The valance quarks will be denoted by u_v and d_v.

This approach is made quantitative by assuming that the interaction of the photon probe with the proton is the sum of interactions with individual partons, each parton carrying a fraction of the proton's momentum x, with a probability $f_i(x)$ for the ith parton. Then the scattering will be proportional to constituent cross sections for the scatterings on individual partons, each weighted by a factor Q_i^2 if the parton has an electric charge Q_i .

Since the photon only interacts with quarks and not with (electrically neutral) gluons, the constituent cross section is the same for every parton and factors out. Then the scattering on a proton is proportional to a particular combination of probabilities called $F_2(x)$,

$$F_2(x) = \sum_i Q_i^2 x f_i(x) , \qquad (18.3)$$

which occurs frequently in such a context. $F_2(x)$ is just the sum of structure functions for each quark, weighted by its electric charge squared because of the photon. That our identification of $x = -q^2/2q \cdot p$ is the appropriate one can be checked by noting that the kinematics of equation (18.2) for the parton is $(q + xp)^2 \approx 0$, since the quark carries a fraction x of the proton's momentum. Neglecting the quark mass and also neglecting $x^2 p^2 = x^2 m^2$ gives $x = -q^2/2q \cdot p$. The minus sign is present because q^2 is a spacelike momentum so $q^2 < 0$.

The probabilities $f_i(x)$ are called structure functions. They can be measured in various processes. Once they are measured they can be used, just as we used them in Chapter 10 and as we will use them again. There is one for u quarks in a proton, one for d quarks, one for gluons, etc. They are described further in the next section. Sometimes it is also necessary to specify the hadron, so $f_{u/p}(x)$ is the probability of finding an u–quark in a proton carring momentum fraction x, *etc.*

The preceeding analysis has been presented here heuristically. It can be derived if the resolution with which the proton is probed is sufficiently fine, i.e., for large enough q^2. Then the probe can see the partons one at a time, and quantum mechanical interference effects can be neglected. It turns out that binding effects, which might also be expected, can also be neglected at large enough q^2. That is, the photon interacts with quarks as if they were essentially free particles at large q^2, in spite of naïve expectations and in spite of the fact that quarks are somehow so tightly bound that they are confined. This property can also be derived from QCD. It is called "asymptotic freedom," and when it was derived, thus explaining the apparently contradictory behavior of the quarks, it added greatly to the acceptance of both the reality of quarks and of QCD as the correct theory of strong interactions. We will see the origin of asymptotic freedom in Chapter 20.

18.3   The Structure Functions

From general arguments there are a number of constraints on the structure functions. When these are combined with data, good representations of the structure functions can be obtained over a large region of x. Since the structure functions measure properties that are non-pertubative, they will probably not be calculable from first principles for a long time.

Another interesting and important QCD result affects the structure functions. A quark (or gluon) carrying a momentum fraction x could do so either because it intrinsically had that momentum fraction x or because it originally had a larger momentum fraction but radiated a gluon and dropped to x. There is more phase space available to radiate gluons at larger $|q^2|$, so the structure functions will show a greater probability for carrying small x if $|q^2|$ is larger, and a smaller probability for carrying large x if $|q^2|$ is larger, as in Figure 18.3.

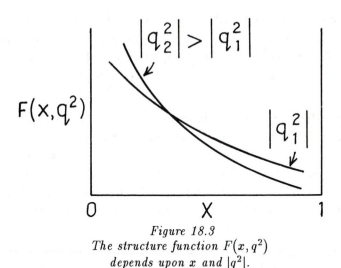

Figure 18.3
The structure function $F(x, q^2)$
depends upon x and $|q^2|$.

Since the probability of gluon emission is calculable at large $|q^2|$ in QCD, if the structure function is known at one q^2 and x, it can be calculated at other q^2 and x. Note that the simplification that occurred by having structure functions depend only on one variable x has partly disappeared, since now they depend on $|q^2|$. But since the $|q^2|$ dependence is calculable the extra complication is minimal. This variation with q^2 is called "scaling violations."

The procedure, then, is to measure the structure functions over a range of x at one q^2, and then calculate them at other q^2. This has been done now, so the reader may use the results, in graphical or numerical form, to calculate any prediction for production from hadrons if the constituent cross section for quarks and gluons is known.

It is worth describing how the various contributions are separated. We will just indicate the steps. Consider the process $\nu_\mu p \rightarrow \mu^- + X$ at large x. Then the interaction is mediated by a virtual W^+, as in Figure 18.4, so the process in the hadron must be $d+W^+ \rightarrow u$, where the d quark is a valence quark carrying a large fraction of the proton momentum. The W^+ does not interact with u quarks or gluons. Thus $F_{d/p}(x)$ can be measured. At smaller x the sea $\bar{u}$'s can get involved giving $\bar{u}+W^+ \rightarrow \bar{d}$. At small x, of course, there is no distinction between sea and valence quarks so the distributions must join. Similarly, $\bar{\nu}_\mu p \rightarrow \mu^+ + X$ measures u quarks at large x, and $\bar{d}$'s at small x. The process $q + g \rightarrow q + \gamma$ is sensitive to gluons in the proton (what is observed is a hard isolated photon plus a jet), and so is $gg \rightarrow gg$ (two jets are observed). So basically what happens is that a small number of structure functions determine a large amount of data; the structure functions are highly overconstrained and a unique set emerges. Once this is done it does not have to be repeated (except perhaps to improve accuracy if better data is available). Since one begins with data that has experimental errors, and uses theory in some approximation to calculate, there is always some error involved in the structure functions. At present there is evidence that hadronic cross sections computed with structure functions can be trusted to about a factor of two, perhaps a little better. Many tests, however, can be defined in terms of ratios that remove much of that uncertainty.

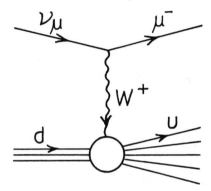

Figure 18.4
The deep inelastic process
$\nu_\mu p \rightarrow \mu^- + X$
mediated by a virtual W^+.

Some structure functions are plotted in Figure 18.5, and Figure 18.6 shows how the momentum of the proton is distributed. For partons that are not shown the equivalent curves can be seen by comparison with those shown; *e.g.* the curve for t lies below that for b, while the curve for c lies above b; $s \approx \overline{d} \approx \overline{u}$, and $d_v \approx u_v$. [In fact, s is somewhat smaller than $\overline{d}$ and $\overline{u}$ because the heavier s–quark is a little harder to pair-produce, and for rather subtle reasons, d_v is somewhat smaller than u_v .]

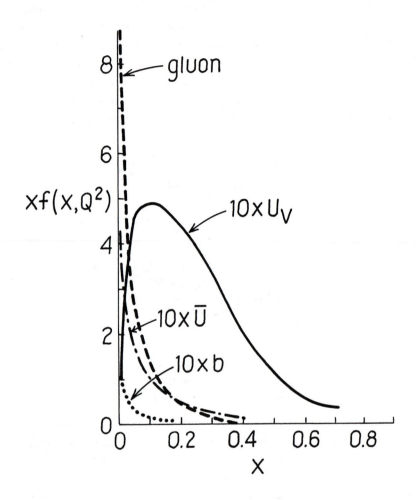

Figure 18.5
Some structure functions and how they depend on x at $Q^2 = -q^2 = 10^4$ GeV^2.
Note that for u_v, $\overline{u}$, and b what is plotted is 10× the structure function.

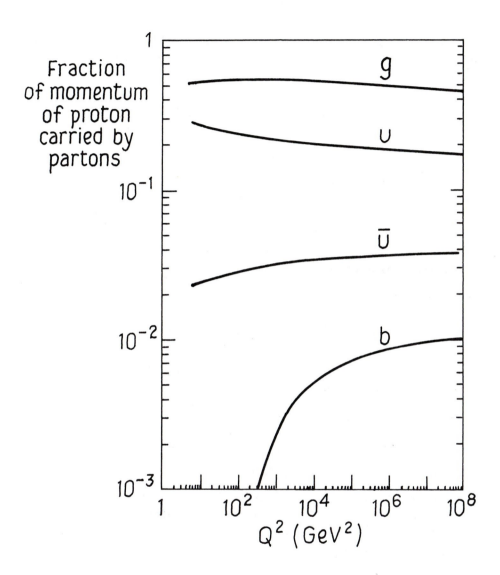

Figure 18.6
The fraction of the momentum of a proton
carried by various partons is shown
as a function of $Q^2 = -q^2$, integrated over x.

Problems

18.1:   When two hadrons collide at high energies, the quarks in them can collide and produce a lepton pair ($\mu^+\mu^-$ or e^+e^-). This is called a Drell–Yan process. For hadrons A and B, we can draw, for example,

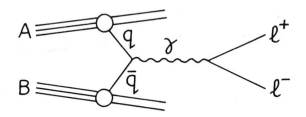

(a) Let $\hat{s} = \left(p_q + p_{\bar{q}}\right)^2$ and $s = \left(p_A + p_B\right)^2$. Show that $\hat{s} \simeq x\bar{x}s$ where x and $\bar{x}$ are the fractions of hadron momenta carried by q and $\bar{q}$.

(b) Explain why the cross section for this process is approximately given by

$$\sigma = \sum_{q,\bar{q}} \int dx\, d\bar{x}\; F_{q/A}(x) F_{\bar{q}/B}(\bar{x}) \hat{\sigma}(q\bar{q} \to \mu^+\mu^-).$$

(c) Argue that the constituent cross section is approximately

$$\hat{\sigma} \simeq \frac{\pi\alpha^2 Q_q^2}{\hat{s}},$$

where Q_q is the electric charge for the quark q in units of e.

(d) Consider two cases for A and B: (i) $A = \pi^+$, with valence quarks $u\bar{d}$, and $B = $ carbon; (ii) $A = \pi^-$, with valence quarks $\bar{u}d$, and $B = $ carbon. The carbon nucleus is isoscalar, *i.e.* it has equal numbers of u and d quarks. Show that the ratio

$$R = \frac{\sigma(\pi^+ \text{carbon} \to \mu^+\mu^- + \text{anything})}{\sigma(\pi^- \text{carbon} \to \mu^+\mu^- + \text{anything})}$$

should be near 1 when $\hat{s}/s$ is small, and that R should approach $1/4$ for $\hat{s}/s \to 1$.

(e) What is the ratio of the Drell–Yan cross section to produce a $\mu^+\mu^-$ pair to that to produce an e^+e^- pair?

18.2:   Interpret the following relations satisfied by structure functions,

$$(a) \quad \int_0^1 x \, dx \, \sum_{i=q,g} f_i(x) = 1 \ ,$$

$$(b) \quad \int_0^1 dx \, \left[f_{u/p}(x) - f_{\overline{u}/p}(x) \right] = 2 \ ,$$

$$(c) \quad \int_0^1 dx \, \left[f_{s/p}(x) - f_{\overline{s}/p}(x) \right] = 0 \ .$$

18.3:   Suppose the structure functions $f_i(x)$ for up and down quarks are given approximately by $u(x) \simeq d(x) \simeq x^{-1/2}(1-x)^3$ for $x \gtrsim 0.05$. In this region what is $F_2(x)$?

18.4:   Consider scattering a neutrino on a fixed lepton or quark target. How does the cross section behave as the ν energy increases (a) at $s \ll M_Z^2$, (b) at $s \sim M_Z^2$? If the target is a quark which is in a proton, how are the answers modified? The experimental data shows that σ grows with E_ν linearly at least up to $E_\nu \simeq 200$ GeV; that result is another strong confirmation of the reality of point–like quarks as constituents of hadrons.

Suggestions for Further Study

The Scientific American article by Jacob and Landshoff may be helpful. Chapters 27-29 and 32 of Dodd covers some of the same material as Chapter 18. The values plotted for the structure functions come from Eichten, Hinchliffe, Lane and Quigg; they give information that allows a computer program to be written to generate the structure functions. Somewhat simpler formulas for structure functions are given by Duke and Owens.

e^+e^- Colliders
and Tests of
the Standard Model

Many Standard Model tests can be performed very cleanly at e^+e^- colliders. Experiments from SLAC and DESY in the decade beginning in 1974 contributed tremendously to the discovery and acceptance of the Standard Model.

The basic quantity we need to organize the data is the cross section for $e^+e^- \rightarrow f\bar{f}$ where f is a point–like spin-1/2 fermion. By comparison of the actual cross section with the point–like one, we want to test whether any given fermion is point–like. We could consider $f = e$, μ, τ, u, d, s, c, b, or t and even f=proton. When $f \neq e$ and we work at $s \ll m_Z^2$ we only need to consider one diagram (Figure 19.1),

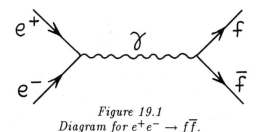

Figure 19.1
Diagram for $e^+e^- \rightarrow f\bar{f}$.

Let f have electric charge $Q_f e$. Then from our rules in Chapter 7, the amplitude is (remembering that $m_\gamma = 0$)

$$M = Q_f e^2 \left(\bar{e}\gamma_\mu e\right) \frac{1}{s} \left(\bar{f}\gamma^\mu f\right) . \tag{19.1}$$

If we work at energies large enough to neglect all the masses of the fermions (for $f = \mu$, u, d, or s this does not require much energy, and for τ and c only a little more), then the factors $\bar{u}\gamma_\mu u$ which have the dimensions of mass can only be given by $\sqrt{s}$, so we have

$$M \simeq Q_f e^2. \tag{19.2}$$

With our usual method we will replace the spin sum by a factor $2^4/2 \times 2 = 4$ since there are a total of $2 \times 2 \times 2 \times 2$ spin projections but we average over the initial ones. That gives

$$\overline{|M|^2} \simeq 4 Q_f^2 e^4. \tag{19.3}$$

Then

$$\frac{d\sigma}{d\Omega} = \frac{\overline{|M|^2}}{64\pi^2 s} = \frac{Q_f^2 e^4}{16\pi^2 s} = \frac{Q_f^2 \alpha^2}{s}. \tag{19.4}$$

Since this has no angular dependence, $\int d\Omega = 4\pi$, so the total cross section is

$$\sigma \simeq 4\pi Q_f^2 \alpha^2 / s. \tag{19.5}$$

The correct answer is $1/3$ of this. (The overestimate arose for a combination of two reasons. First, the photon does not have a longitudinal polarization, so some spin states do not count; that reduces σ by $1/2$. Second, the angular distribution is $1 + \cos^2\theta$, which we have approximated by 2; the $\int d\Omega$ is then reduced by $2/3$.)

Thus we define

$$\sigma_{\text{point}} = \frac{4\pi}{3} \frac{Q_f^2 \alpha^2}{s}, \tag{19.6}$$

as the point–like cross section for producing a $f\bar{f}$ pair with f having electric charge $Q_f e$. For a muon or a tau $Q_f = -1$; for u, $Q_f = 2/3$; and for d, $Q_f = -1/3$.

The angular distribution for $e^+e^- \to f\bar{f}$ is also of interest, as we will see. Because e and f have spin there is no elementary method to get the correct angular distribution, which is

$$\frac{d\sigma_{\text{point}}}{d\Omega} = \frac{Q_f^2 \alpha^2}{4s}(1 + \cos^2\theta). \tag{19.7}$$

Because of its importance, this result is derived in Appendix D. Integrating equation (19.7) over angles gives equation (19.6).

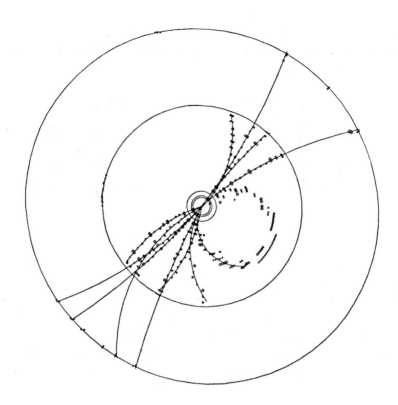

Figure 19.2

This figure shows an event with two jets, from the HRS detector at PEP. The beam axis is into the page. The jets are shown projected onto a transverse plane. Less energetic particles bend more in the magnetic field.

19.1   Are quarks, leptons, and gluons point–like?

How well has the point–like nature of quarks, leptons, and gluons been tested? At PETRA, the processes $e^+e^- \to e^+e^-$, $e^+e^- \to \mu^+\mu^-$, and $e^+e^- \to \tau^+\tau^-$ have been studied for $\sqrt{s} \leq 46$ GeV, and behave to an accuracy of order 10% exactly as expected from equation (19.7), as functions of s and of θ.

The same result holds for $e^+e^- \to q\bar{q}$. It is particularly impressive here, since the quarks are produced as jets, as described in Chapter 15. The jets have the $1 + \cos^2\theta$ expected from equation (19.7) if they are spin-1/2 fermions. For example, the number of jets pointing at $0°$ or $180°$ is twice that pointing at $90°$. The sizes of the cross sections are given correctly for the fractional electric charges normally assigned to the quarks. The c quark and b quark can be identified from their weak decays, so the cross sections for $e^+e^- \to c\bar{c}$ and $e^+e^- \to b\bar{b}$ have been studied to 46 GeV as well, and are point–like. It is not known how to separate u, d, and s, so only the sum of these cross sections has been studied, and it is consistent with point–like behavior.

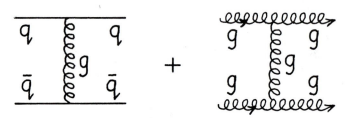

Figure 19.3
Some interactions giving two jets.

At the CERN collider the size and angular distribution of the jet + jet cross section is again as expected for point–like quarks and gluons, up to $\sqrt{s} \simeq$ 150 GeV. The result is basically the Mott cross section times some spin and color factors, since this scattering is due to the processes in Figure 19.3. It is dominated by gluon exchange, giving an angular dependence like photon exchange in electron scattering. Although the various processes cannot be separated directly, the predictions for jet–jet cross sections can be computed as we have described earlier. The agreement is meaningful to about a factor of two accuracy. For comparison, the cross section for $e^+e^- \rightarrow p\bar{p}$ will be about 10^{-9} of the point cross section at $\sqrt{s} = 150$ GeV because the proton is not point–like.

How can these results be interpreted? Historically, structure has always appeared when the available "particles" were probed with projectiles having energies small compared to the mass—for molecules, atoms, nuclei, and nucleons. Here the energies of the probes are two or more orders of magnitude larger than the masses and no evidence for structure has appeared. Ultimately it will remain an experimental question, but it is already clear that quarks and leptons cannot have structure in the same sense that atoms or nuclei or protons had structure. The same result holds for photons and gluons and $W^\pm$ and Z^0 bosons, whose cross sections are all point–like.

19.2   The ratio $R = \sigma/\sigma_{\text{point}}$

A very useful quantity to examine is the ratio

$$R_X = \sigma_{\text{TOT}}(e^+e^- \rightarrow X)/\sigma_{\text{point}} . \tag{19.8}$$

We can examine various contributions to R. If R is at a given value at some $\sqrt{s}$, we can ask how R would increase if some new particle was produced as the energy increased. R is usually defined in a mixed way, so final states that would not trigger the detector can either be corrected for or ignored.

(a) If a new neutrino were produced it would not be detected in a normal collider detector, so

$$\Delta R_{\nu\bar{\nu}} = 0 \ . \qquad\qquad (19.9)$$

(b) A new charged lepton L would be produced with the point cross section, so

$$\Delta R_{L\bar{L}} = 1 \ . \qquad\qquad (19.10)$$

This assumes that all decay modes are detectable, so a correction must be made for any that are not.

(c) A charge $-1/3$ quark would only differ from a lepton in that (i) the coupling at the $f\bar{f}\gamma$ vertex in Figure 19.1 has a factor of $Q_f = -1/3$, which gets squared in the rate, and (ii) there are three colors of quarks, so the rate is enhanced by a factor of 3. The final overall factor relative to a lepton is $1/3$.

(d) For a $Q_f = 2/3$ quark the color factor is still 3 and now $Q_f^2 = 4/9$ so the final factor is $4/3$.

What do we expect to observe then? At low energies the u, d, and s quarks can be produced. It is common to plot R for all hadronic channels, which we abbreviate as R_{had}, subtracting contributions from e, μ, and τ. Then for $\sqrt{s} \lesssim$ 3 GeV, the charm threshold, we expect $R_{\text{had}} = 1/3 + 4/3 + 1/3 = 2$. Once a $q\bar{q}$ pair is produced, all sorts of hadronic states could occur as the quarks hadronize; since the quarks are confined, our lack of knowledge of how hadronization occurs does not matter so long as we sum over all final states, since the quarks always appear in some hadronic channel.

When we cross the $c\bar{c}$ threshold, R should increase by $4/3$, and when we cross the $b\bar{b}$ threshold at $\sqrt{s} \simeq 10$ GeV, R should increase by $1/3$. Local resonance effects can cause fluctuations, but they will be around the expected averages. Additional contributions will occur where a gluon is radiated or exchanged, so R will be somewhat larger than the naïve value. Such corrections should be of order α_s/π, say 5-10%, since that is approximately the probability of emitting a gluon. They have been computed, and there is good agreement between theory and experiment. Up to $\sqrt{s} = 46$ GeV at PETRA, the data is well described by the contributions of the five colored quarks u, d, c, s, and b. One lesson we learn from this is that the Standard Model is well tested here; the result depends on color and the number of colors, on the size of the gluon radiation cross section, on quark electric charges, and on the quarks giving jets with each jet behaving as a single spin-1/2 object. Another lesson is that there is no other electrically charged particle that can be produced as in Figure 19.1 in the energy range $\sqrt{s} \leq 46$ GeV.

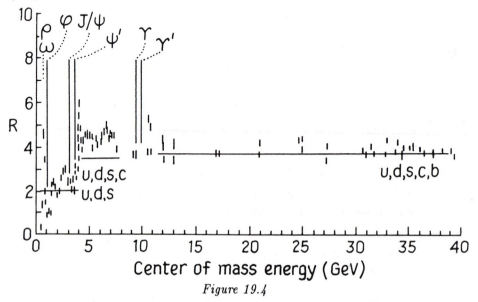

Figure 19.4

Data for R_{had} from the Particle Data Tables. The average behavior of R_{had} is described by the the expected contributions from $e^+e^- \rightarrow q\bar{q}$, summing over all quarks present at a given $\sqrt{s}$.

19.3   The τ and Heavy Leptons

As remarked earlier, the finding of the τ was essentially the only major experimental discovery since the middle 1960's that was unanticipated by theoretical arguments. An active search was made for a heavy lepton with the MARK I detector and the SPEAR data, and one was found.

It is interesting to analyze the τ signatures and how the τ was detected, both for their own sake and because of the generalization to future searches. The production is by

$$e^+e^- \rightarrow \tau^+\tau^-, \tag{19.11}$$

with an expected cross section

$$\sigma_{\text{point}} = \frac{4\pi}{3}\frac{\alpha^2}{s}.$$ (19.12)

To evaluate σ numerically we need $s \geq (2m_\tau)^2 \simeq 16 \text{ GeV}^2$, so near threshold

$$\sigma_{\text{point}} \simeq \frac{4}{(137)^2}\frac{1}{16} \times 4 \times 10^{-28} \text{ cm}^2$$

$$\simeq 5 \times 10^{-33} \text{ cm}^2 .$$ (19.13)

If the experiment had an integrated luminosity of $2 \times 10^{37}/\text{cm}^2$, *e.g.* running for 4×10^6 seconds at $L = 5 \times 10^{30}\text{cm}^{-2}\text{sec}^{-1}$, then 10^5 events would be produced. What would be observed? The decays we expect are

$$\tau^- \rightarrow \nu_\tau \nu_e e^-$$
$$\rightarrow \nu_\tau \nu_\mu \mu^-$$ (19.14)
$$\rightarrow \nu_\tau \bar{u}d,$$

as in Figure 19.5, with analogous decays for τ^+. Since $u\bar{d}$ comes in three colors, the expected branching ratios are 1/5, 1/5, and 3/5 for these modes. (Gluon radiation enhances the $u\bar{d}$ modes by a few percent and reduces the others correspondingly, just as it increases R_{had} a bit.)

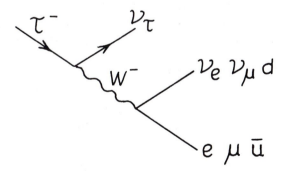

Figure 19.5
Decay of the heavy lepton τ
into lighter fermions.

The events which have the cleanest signature occur when one τ gives a μ and the other an e, and neutrinos escape. Then

$$e^+ e^- \rightarrow e\mu + \text{missing energy} ,$$ (19.15)

and no jets. Each τ has a branching ratio of 1/5 for these modes, so this happens

4% of the time (actually 3% when the corrections are included and squared). Since the basic rate gave $\sim 10^5$ events, a reduction by a factor of 25 or 30 is not a serious problem, and enough events survive.

If one believes that such a signal has been detected, there are a number of consistency checks. For example, if there are N events of the sort of equation (19.15), there must be $3N$ where one τ gives μ and the other gives hadrons (since $\bar{u}d$ must become some hadronic state), $3N$ with e opposite hadrons, and $9N$ with two hadronic jets and missing energy. The events must have the correct dependence on $\sqrt{s}$ to come from production of a spin-1/2 fermion, and the correct $1 + \cos^2\theta$ angular distribution.

It is also necessary to discuss possible backgrounds. The u, d, and s quarks give light mesons and baryons, which rarely decay to $\mu + X$, so such events would not originate from them. The production of $c\bar{c}$ could be worrisome since $c \to s\mu\nu$ or $c \to se\nu$. The production of $c\bar{c}$ is larger than $\tau\bar{\tau}$ by a factor of 4/3 from the charge squared and color factors, and the μ, e branching ratios are naïvely the same. Since many of the $c\bar{c}$ events will have the s quark jet present, and there are only a few more events with $\mu + e$ to begin, the $c\bar{c}$ background cannot fake the $\tau\bar{\tau}$ signal if the detection efficiency for jets is good. This is the kind of analysis that was done for the τ, and a similar approach is made for any hypothetical new signal.

The τ decay provides an independent and very clean check on the color degree of freedom and the number of colors, since the branching ratios are in agreement with the naïve factor of three enhancement of the $u\bar{d}$ channel.

19.4   Observation of Gluons

As described earlier in this chapter, quark jets were clearly observed in $e^+ + e^- \to q + \bar{q}$, with the correct differential cross section. In the late 1970's, first at PETRA, events were observed with three jets, $e^+ + e^- \to j + j + j$. These events were successfully described quantitatively by assuming they were examples of $e^+ + e^- \to q + \bar{q} + g$, with the gluon radiated from either of the quarks. The rate and the angular correlations of the three jets were correctly given by the QCD predictions. An example of a three-jet event is shown in Figure 19.6. These events should be regarded as the discovery of gluons, and the confirmation of their spin and their interactions with quarks.

The results from e^+e^- colliders were not very sensitive to the gluon self-coupling and do not provide a very direct test of that interaction. The behavior of jets at the CERN collider is able to establish the presence of that coupling, because the production of gluon jets from gluons in a proton depends on it rather strongly, for example through the process shown in Figure 11.3d.

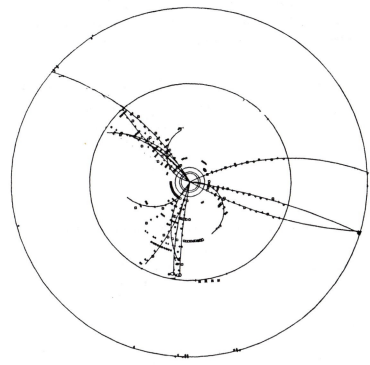

Figure 19.6

An event of e⁺e⁻ → 3 jets is shown, from the HRS detector at PEP. It is interpreted as radiation of a gluon jet by one of the final quarks, as in Figure 11.5. The projection of the event onto a transverse plane is shown, with the beam axis perpendicular to the page.

Problems

19.1: What fraction of events of τ production in $e^+e^- \rightarrow \tau^+\tau^-$ are expected to give purely hadronic final states in the detector? Hadronic plus one charged lepton? How does this prediction depend on the number of colors carried by quarks?

19.2: How much would $R(e^+e^-)$ increase if the following new particles existed and were produced in an experiment as $\sqrt{s}$ increases?

(1) A new lepton $SU(2)$ doublet with electric charges

$$L = \begin{pmatrix} L^{++} \\ L^+ \end{pmatrix}.$$

(2) A color triplet of integer charge states in an $SU(2)$ doublet,

$$Q' = \begin{pmatrix} Q^+ \\ Q^0 \end{pmatrix}.$$

(3) The Standard Model Higgs doublet.

19.3: The measured τ lifetime is 3.1×10^{-13} sec, known to an accuracy of about 10%. How does that compare with the prediction of the Standard Model? Remember the calculation of τ_μ in Chapter 11, and the number of channels.

Suggestions for Further Study

The discovery of the τ is described in the Scientific American article of Perl and Kirk, or the Nature article of Perl. The Scientific American article of Drell may be helpful. For a description of the analysis of the PETRA three–jet data and the relation to gluons see Söding and Wolf.

CHAPTER 20

Coupling Strengths Depend on the Momentum Transfer

We are used to thinking of the fine structure constant as a number, but in fact it is not constant. As the momentum transferred in a scattering changes, or as the distance scale probed in an interaction changes, α varies. Similar behavior holds for the other couplings α_2 and α_3.

A photon exchange diagram, such as Figure 20.1,

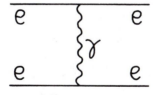

Figure 20.1
Feynman diagram of a
photon exchange.

between electrically charged particles, gives the Rutherford scattering cross section, and Coulomb's Law, when calculated to order α^2 in the cross section. But in a quantum field theory there are higher order corrections, such as Figure 20.2,

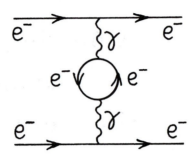

Figure 20.2
Feynman diagram for an
electron loop in a photon exchange.

A virtual e^+e^- pair forms and then disappears. Clearly any number of such lines can be added. A full treatment of the implications of adding such loops for all processes is beyond the scope of this book; among other things it leads to the theory of renormalization. To evaluate the contribution of Figure 20.2, we have to integrate over all possible momenta for the electron in the loop, since the virtual momenta could take any value. The loop contribution is just an intermediate state, so our rules for writing vertices plus the normal quantum theory results for inserting a set of intermediate states tell us how to write down the contribution from Figure 20.2. When we do so, we find that this contribution is not finite, as we will see below. One of the great accomplishments of past decades is learning how to eliminate those infinite contributions in a consistent way in gauge theories, a process called renormalization. Our treatment will suggest how this happens. Our main purpose in this chapter is to explore some of the more practical consequences for particle physics of the existence of these loops. Most significantly, we will find that a calculable variation is introduced in coupling strengths.

20.1   QED

We can proceed by studying an even simpler process, the vertex of Figure 20.3. The momenta are labeled in the figure, and $k = k' + q$. Using the

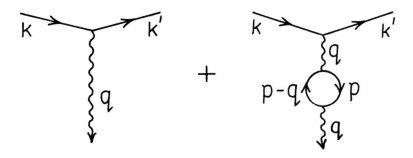

Figure 20.3
Electron–photon vertex
with one loop correction.

rules from earlier chapters, we can write the two contributions approximately as

$$e\bar{u}(k')\gamma^\mu u(k)\epsilon_\mu - \int \frac{d^4p}{(2\pi)^4} \left[e\bar{u}(k')\gamma^\mu u(k) \right] \times$$

$$(20.1)$$

$$\frac{1}{q^2} \frac{\left[e\bar{u}(p)\gamma_\mu u(p-q) \right] \left[e\bar{u}(p-q)\gamma^\lambda u(p) \right]}{(p^2 - M^2)\left[(p-q)^2 - M^2 \right]} \epsilon_\lambda$$

where a sum over the spins of the electrons in the loop is implied. The photon polarization vector is ϵ_μ and M is the electron mass. The integration over p is present because the electron in the loop can have any momentum, and all contributions will add. We can take a common factor out of the two terms, so the sum is

$$e\bar{u}(k')\gamma^\mu u(k) \times$$

$$(20.2)$$

$$\left[\epsilon_\mu - \frac{e^2\epsilon^\lambda}{q^2} \int \frac{d^4p}{(2\pi)^4} \frac{\left[\bar{u}(p)\gamma_\mu u(p-q)\right]\left[\bar{u}(p-q)\gamma_\lambda u(p)\right]}{(p^2 - M^2)\left[(p-q)^2 - M^2\right]} \right]$$

which we can write as

$$e\bar{u}(k')\gamma^\mu u(k) \left[\epsilon_\mu + \epsilon^\lambda T_{\mu\lambda} \right].$$

$$(20.3)$$

Complete evaluation of T requires a lengthy calculation. In the evaluation of $T_{\lambda\mu}$, terms proportional to q_λ or q_μ can be dropped, since the photon polarization satisfies $\epsilon_\mu q^\mu = 0$; the electron current $j^\mu = \bar{u}(k')\gamma^\mu u(k)$ also satisfies $q_\mu j^\mu = 0$. The result is

$$T_{\mu\lambda} = g_{\lambda\mu} I(q^2) \,, \tag{20.4}$$

$$I(q^2) = \frac{\alpha}{3\pi} \int\limits_{M^2}^{\infty} \frac{dp^2}{p^2} - \frac{2\alpha}{\pi} \int\limits_0^1 dx \; x(1-x) \ln\left(1 - \frac{q^2 x(1-x)}{M^2}\right). \tag{20.5}$$

The integral in equation (20.2) appears to have a quadratic divergence, since each $\bar{u}u \sim p$, but the leading piece does cancel when the algebra is done. There is still a logarithmic divergence, the infinite piece mentioned above, which appears as the first term in equation (20.5). Note the crucial point, however, that the infinite piece is independent of q^2. That the answer has the form of equation (20.4) is clear, since $T_{\mu\lambda}$ has two indices and can only depend on $g_{\mu\lambda}$ and on the four–vector q_μ, and we have just seen that any term proportional to q_μ drops out. The precise form of equation (20.5) only follows after a careful and long calculation.

The finite contribution in equation (20.5) is an integral which can be done analytically. We are most interested when $|q^2|$ is large, in order to study short distance collisions, so it is worthwhile making the approximation $-q^2/M^2 \gg 1$. Then

$$\ln\left(1 - \frac{q^2 x(1-x)}{M^2}\right) \approx \ln\left(-q^2/M^2\right) \tag{20.6}$$

in which case, putting the infinity in the form of an upper limit Λ,

$$I(q^2) \simeq \frac{\alpha}{3\pi} \ln \frac{\Lambda^2}{M^2} - \frac{\alpha}{3\pi} \ln\left(\frac{-q^2}{M^2}\right), \tag{20.7}$$

since $\int dx \; x(1-x) = 1/6$. The two terms combine, so

$$I(q^2) \simeq \frac{\alpha}{3\pi} \ln \frac{\Lambda^2}{(-q^2)} \, . \tag{20.8}$$

The M^2 has dropped out, as is reasonable when we are considering a high energy limit. Putting equations (20.4) and (20.8) into equation (20.3) gives, for the amplitude describing Figure 20.3,

$$ie\bar{u}(k')\gamma^\mu u(k)\epsilon_\mu \left[1 - \frac{\alpha}{3\pi} \ln \frac{\Lambda^2}{(-q^2)}\right]. \tag{20.9}$$

Since the virtual photon will have to be absorbed, we also attach an electron line at the bottom, as in Figure 20.4.

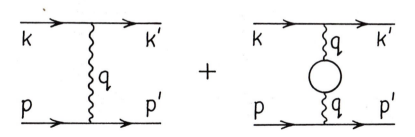

Figure 20.4
Electron scattering
including one loop correction.

Then the amplitude for Figure 20.4 is

$$e^2 \left[1 - \frac{\alpha}{3\pi} \ln \frac{\Lambda^2}{(-q^2)} \right] (\bar{u}(k')\gamma^\mu u(k)) \, (\bar{u}(p')\gamma_\mu u(p)) \ . \tag{20.10}$$

So far there has been some complicated algebra. Now the physics considerations enter. First a technical point. We could have included several loops in a chain. Since each intermediate photon has momentum q, the loops obviously factor, and we would get a series of the form $1 - \epsilon + \epsilon^2 - \ldots$, which sums to $1/(1 + \epsilon)$. Thus the full coefficient of the spinors in equation (20.10) should be

$$\frac{e^2}{\left[1 + \frac{\alpha}{3\pi} \ln \left(\frac{\Lambda^2}{(-q^2)} \right) \right]} \ . \tag{20.11}$$

Now the main physics point enters. We have been assuming that $e^2/4\pi = \alpha = 1/137$. But what we measure as α necessarily includes the contributions of Figure 20.4, with any number of loops along the chain, so we should get the measured value of $1/137$ only after including all of the loop corrections. The answer depends on q^2. So in practice we must measure α at some particular q^2. Let us call the particular choice μ^2, *i.e.* we measure $\alpha = 1/137$ at $q^2 = -\mu^2$. [Remember, q^2 is space–like so it is negative; μ^2 is positive.] Call the bare coupling that goes at each vertex in Figure 20.4 e_0, and $\alpha_0 = e_0^2/4\pi$. Then the physical amplitude is given by a sum as in Figure 20.5,

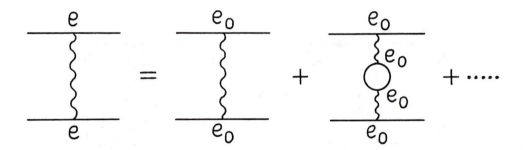

Figure 20.5
The physical amplitude given by
a sum of terms
involving the bare coupling.

and equations (20.10) and (20.11) tell us that the value measured at μ^2 is

$$\alpha(\mu^2) = \frac{\alpha_0}{1 + \frac{\alpha_0}{3\pi}\ln\left(\frac{\Lambda^2}{\mu^2}\right)} . \qquad (20.12)$$

Finally, we can calculate α at any q^2, from equation (20.11)

$$\begin{aligned}
\alpha(q^2) &= \frac{\alpha_0}{1 + \frac{\alpha_0}{3\pi}\ln\left(\frac{\Lambda^2}{(-q^2)}\right)} \\
&= \frac{\alpha_0}{1 + \frac{\alpha_0}{3\pi}\ln\left[\left(\frac{\Lambda^2}{(-q^2)}\right)\left(\frac{\mu^2}{\mu^2}\right)\right]} \\
&= \frac{\alpha_0}{1 + \frac{\alpha_0}{3\pi}\left[\ln\frac{\Lambda^2}{\mu^2} + \ln\left(\frac{\mu^2}{(-q^2)}\right)\right]} . \qquad (20.13)
\end{aligned}$$

Now solve equation (20.12) for $1 + \frac{\alpha_0}{3\pi}\ln\frac{\Lambda^2}{\mu^2} = \frac{\alpha_0}{\alpha(\mu^2)}$ so equation (20.13) gives

$$\alpha(q^2) = \frac{\alpha_0}{\frac{\alpha_0}{\alpha(\mu^2)} + \frac{\alpha_0}{3\pi}\ln\frac{\mu^2}{(-q^2)}} . \qquad (20.14)$$

The α_0 in the numerator and denominator cancels, and so

$$\alpha(q^2) = \frac{\alpha(\mu^2)}{1 + \frac{\alpha(\mu^2)}{3\pi}\ln\frac{\mu^2}{(-q^2)}} . \qquad (20.15)$$

Equation (20.15) is the result of interest, and it is rather remarkable. There is no dependence on Λ or α_0. Only finite quantities enter. The coupling strength

$\alpha(q^2)$ is expressed in terms of only physical quantities, $\alpha(\mu^2)$ which is a measured value at some particular $q^2 = -\mu^2$, and q^2 itself. α depends on q^2. If we found $\alpha = 1/137$ at $q^2 = -\mu^2$, then α is different from $1/137$ at other q^2.

$\alpha(q^2)$ is called a "running" coupling strength (often it is called a coupling constant, or running coupling constant, but since it is not constant these are unfortunate names). We calculated with electrons in the loop. If we had put muons or taus or quarks in the loop we would have an equally valid contribution. Thus the correction terms should be multiplied by a factor

$$n_\ell + 3\left(\frac{4}{9}\right)n_u + 3\left(\frac{1}{9}\right)n_d \qquad (20.16)$$

where n_ℓ is the number of charged leptons, n_u the number of $Q = 2/3e$ quarks, n_d the number of $Q = -1/3e$ quarks, and a factor of three is included for quark color. Each contribution enters with its electric charge squared since it couples to a γ at each side of the loop. If $|q^2|$ is small then some heavy fermion might give a reduced effect, since the fermion mass in the propagators will suppress the integral. Consequently a full calculation will include threshold effects as $|q^2|$ increases. If quarks and leptons occur only in families, then N families contribute $N[1 + 4/3 + 1/3] = 8N/3$. Loops with $W^\pm$ should be included as well, if $|q^2| \geq M_W^2$.

The sign between the two terms in the denominator of equation (20.15) is extremely important, and has a simple physical explanation. At higher $|q^2|$, $\alpha(q^2)$ is larger, since the denominator gets smaller. This corresponds physically to a screening effect. Imagine a negative charge at the origin. Lots of charged pairs emerge from the vacuum. [The fermion loop in Figure 20.2 can be thought of as a particle–antiparticle pair emerging from the vacuum and annihilating back into the vacuum, but having a net physical effect that is observable.] For each pair the positive charge is attracted to the negative charge at the origin, the negative charge repelled. So a probe at some distance sees the negative charge at the origin shielded by a net positive charge. If a probe gets closer to the origin (higher $|q^2|$) it sees less charge shielding and so a net larger negative charge; α is larger.

The effect is not negligible. Suppose $\alpha = 1/137$ at $\mu^2 = 4M_e^2$. The coefficient has the factor from equation (20.16); if we compute $\alpha(M_W^2)$ we could use $n_\ell = n_u = n_d = 3$. Then we multiply the $\alpha \ln(-q^2/\mu^2)$ term by a factor of 8, so

$$\frac{\alpha(M_W^2)}{\alpha(4M_e^2)} \simeq \frac{1}{1 - \frac{8}{3\pi \times 137}\ln\frac{M_W^2}{4M_e^2}} \simeq 1.1 \ . \qquad (20.17)$$

This could be overestimated if the t quark contribution is suppressed until larger $|q^2|$, and because of threshold effects, *etc.*, but it indicates the effect.

The general correction we have discussed is observable in other ways as well. For small q^2 the analysis would be relevant in the atomic physics domain, and then the result from equation (20.5) contributes to the Lamb shift.

20.2   QCD

A similar effect occurs for QCD, but a new feature enters, with remarkable consequences. We will just consider the changes that occur here relative to the QED case, and discuss the implications. The equivalent of Figure 20.3 is now Figure 20.6 where the lines represent quarks and gluons. The new feature is the presence of the third diagram. It occurs for QCD because gluons interact with themselves, while photons do not. The loop of quarks in the second figure clearly gives the same contribution for every flavor of quark, since the quark-gluon coupling is independent of flavor. To include the second diagram in a QCD calculation only a color factor has to be changed relative to what happened in the QED case; the factor $\alpha(\mu^2)/3\pi$ is replaced by $\alpha_3(\mu^2)/6\pi$ for each flavor.

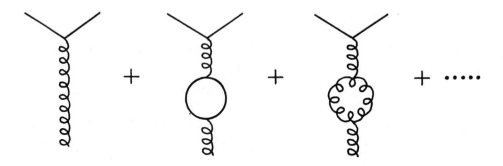

Figure 20.6
The QCD Feynman diagram for a gluon exchange
with the first order quark and gluon loop corrections.

The third diagram has a similar space-time structure, and gives only a numerical factor as well, but a very important one. First, since eight gluons contribute, and the color charge of a gluon is larger than that of a quark, the third diagram contributes more than the second. More importantly, its contribution is of the opposite sign. This is what we should expect qualitatively, because of the self-interaction of the gluon. To understand that, consider a heavy blue quark at the origin. Sometimes a dissociation occurs, with $q_b \rightarrow q_r + g_{b\bar{r}}$, as in Figure 20.7. Then a probe would not see the blue quark at the origin but instead would see the blue color charge on the gluon moved out into the gluon cloud, rather than being more concentrated at the origin. Thus there is an antiscreening effect because the radiated gluon can carry the color charge. As $|q^2|$ gets larger, the probability of radiation gets larger, so less and less color charge is concentrated at the original quark, and it behaves more and more like a free particle!

This property is called "asymptotic freedom." The reader may recognize that it is the behavior that was observed in scattering electrons (and neutrinos) from quarks in hadrons in Chapter 18. Finding that asymptotic freedom emerged from QCD was of considerable significance in gaining rapid acceptance for QCD and for the idea that quarks were real.

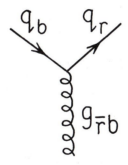

Figure 20.7
Color flip of a quark
due to spontaneously radiating a $b\bar{r}$-gluon.

The result of combining the two corrections is the replacement

$$\alpha(\mu^2)/3\pi \rightarrow \frac{\alpha_3(\mu^2)}{4\pi}\left(\frac{2}{3}n_f - 11\right) \tag{20.18}$$

in equation (20.15) where n_f is the number of quark flavors. Thus

$$\alpha_3(q^2) = \frac{\alpha_3(\mu^2)}{1 + \frac{\alpha_3(\mu^2)}{12\pi}(33 - 2n_f)\ln(-q^2/\mu^2)}. \tag{20.19}$$

There is a renormalization as before. The value of $\alpha_3(\mu^2)$ is determined by a measurement at $q^2 = -\mu^2$, and then α_3 can be calculated at any other q^2.

As long as $33 - 2n_f$ is positive (remember, n_f is the number of flavors, presently $n_f = 6$), when $|q^2|$ is increased the denominator increases, and $\alpha_3(q^2)$ decreases, the behavior of asymptotic freedom. For very large $|q^2|$ the result becomes independent of $\alpha_3(\mu^2)$.

At the other end, for $|q^2|$ small the two terms in the denominator have opposite signs, so $\alpha_3(q^2)$ gets large. At some $q^2 = -\Lambda_{QCD}^2$ the denominator can vanish so the QCD force at that q^2 has become very strong. (Our approximate calculation will not be correct there since when $\alpha_3(q^2)$ is larger lots of other diagrams will be important, but the qualitative effect that the QCD force is large there will be correct.) Solving for Λ_{QCD}^2,

$$\ln \Lambda_{QCD}^2/\mu^2 = -\frac{12\pi}{\alpha_3(\mu^2)}\frac{1}{33 - 2n_f}$$

$$\Lambda_{QCD} = \mu \exp\left\{-\frac{6\pi}{(33 - 2n_f)\alpha_3(\mu^2)}\right\}. \qquad (20.20)$$

Suppose at some large μ^2, e.g. $\mu = 10$ GeV, that $\alpha_3 = 0.2$, and that $n_f = 5$ there. Then

$$\Lambda_{QCD} \simeq 166 \text{ MeV}. \qquad (20.21)$$

We expect QCD to get strong enough to bind quarks and gluons into hadrons on a scale of a few times Λ_{QCD}, just where it happens.

QCD is a remarkable theory. It can confine colored particles into color singlet hadrons on a scale of ≤ 1 GeV, and account for why strong interactions occur on this scale. At the same time, it can provide interactions that become weak if short distances are probed—a quark struck by a large $|q^2|$ probe behaves as if it is essentially free.

An interesting insight emerges if α, α_2, and α_3 are sketched vs. $|q^2|$ as in Figure 20.8. We have looked at α, which grows, and α_3, which decreases with $|q^2|$. For α_2 the result is like α_3, since the gauge boson loops will still dominate because the electroweak charge of the W is larger than that of the fermions. The result is the same as that of equation (20.19), with 33 replaced by a somewhat smaller number.

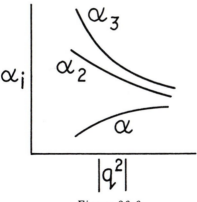

Figure 20.8
Qualitative variation of α_i
with $|q^2|$.

The structure of the theory implies that the force with the largest symmetry group will become strongest at low energies, and all of the theories with non–Abelian groups will get stronger at low energies and be asymptotically free. On the other hand, the force with a $U(1)$ symmetry gets weaker at low energies and at the same time is not confining. Thus it is no accident that the weak, long-range electromagnetic force was the first one discovered, while the confining force is strong and leads to hadrons. The behavior of the forces at much larger $|q^2|$ will be considered in Chapter 27. The behavior sketched in Figure 20.8 suggests that the forces have strengths that are more similar at larger $|q^2|$ than at the $|q^2|$ we are familiar with.

Problems

20.1:   Show that any coupling strengths $\alpha_i(q^2)$ satisfy an equation of the form

$$\frac{\partial \alpha_i(q^2)}{\partial \left[\ln\left(\frac{-q^2}{\mu^2}\right)\right]} = B\alpha_i^2 \ .$$

What does B depend on?

20.2:   If Λ_{QCD} is measured to be 225 MeV, what is α_3 at $-q^2 = M_\Upsilon^2$?

20.3:   If there were twice as many light quarks and leptons as is observed, what would $\alpha(M_W^2)$ be approximately?

CHAPTER 21

Production
and Detection
of a Higgs Boson

The Higgs "sector" of the Standard Model is of particular importance. As we have emphasized in Chapter 8 in the discussion of spontaneous symmetry breaking, where masses were introduced into the theory, the underlying physics of the Higgs mechanism is not understood. In the Standard Model the simplest form of the Higgs mechanism requires a neutral particle, the Higgs boson, to exist. So far the Higgs boson has not been found; in fact, through 1986 no experiment had ever been published that could have detected the Standard Model Higgs boson even if it had been present in the region where the experiment looked, and no experiment has been able to look in the most likely regions.

The difficulty in finding the Higgs boson H^0 arises, in part, because its couplings are proportional to mass, as was derived in Chapter 8, so they are small for the light particles that are most copiously available. Another reason is that the mass of H^0 is unknown; as we saw in equation (8.21) and the discussion below it, M_H depends on the coefficient λ of the Higgs self interaction in the Higgs potential. Since there is no understanding of the physical origin of λ, its numerical value is not known. Nor does any other observable depend on λ in a way that allows λ to be extracted.

Since M_H is unknown, searches have to be planned for all M_H, which is much more difficult than designing an experiment to look for H at a specific mass. As we will see, different techniques are required for different mass ranges. Even if a particle is found at one mass, it will be necessary to examine other regions to gain full understanding, because approaches beyond the minimal Standard

Model require additional Higgs bosons, and because the spectrum of scalars can be quite different if they are fundamental point–like objects or if they represent artifacts of other interactions. Such broad searches are possible, and we can see what is required by analyzing the situation for all M_H.

There are no general limits that can be set on M_H either. Constraints do exist, but they only become limits with assumptions added. For example, it is often stated that there is a lower limit of about 10 GeV on M_H. But that is only correct if no fermions exist on the scale of M_W. If M_t is above about 60 GeV, or if a fourth family exists, then M_H has a lower bound that is only calculable when all of the masses of heavy fermions are known; the lower bound could be close to zero. Effectively there is no lower limit. At the other end it is often stated that $M_H < 1$ TeV. But that is also only a qualitative guide. What it means is that as M_H is imagined to be larger than about 1.5 TeV, some processes that have Higgs boson contributions in internal lines will begin to get loop contributions and higher order contributions that are as important as the tree level ones that are usually calculated. Then the measured values should deviate from the perturbative predictions. This happens for two classes of observables. One class is a few low energy quantities that can be measured very precisely. Unfortunately, numerical studies suggest that effects are very small if M_H is less than several TeV, while if M_H is larger than that it is not known how to calculate what deviations are expected; so far a clear result has not emerged from these studies. If extremely precise measurements are made at future colliders, they may be sensitive to M_H. The second is the scattering of $W^\pm$ and Z bosons, which gets contributions from H exchange. When $M_H \gg 1$ TeV, the perturbative contributions to $W^+W^- \rightarrow W^+W^-$, for example, reach the limit allowed by unitarity when $\sqrt{s} \approx 1.8$ TeV, which means the scattering must deviate from the perturbative predictions. Although it has been learned in recent years how to interpret production of W pairs in terms of collisions of beams of W's, calculations again indicate that the size of the effects is small until $\sqrt{s} \gtrsim 2$ TeV. Events at that scale will be produced at the planned SSC, but luminosity and detector capabilities will have to be optimum to allow sufficient data to be obtained. A machine that can study WW interactions in the region with $M_{WW} \lesssim 3$ TeV could be built with electron or proton beams, but that will not happen for some years.

21.1   Higgs Couplings

We saw in Chapter 8 that at any fermion–Higgs vertex there was a factor

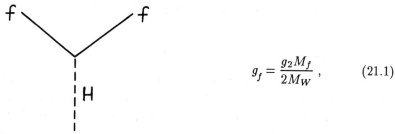

$$g_f = \frac{g_2 M_f}{2 M_W} , \qquad (21.1)$$

and at a Higgs–gauge boson vertex, a factor (problem 2, Chapter 8)

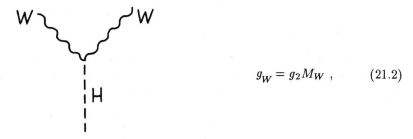

$$g_W = g_2 M_W , \qquad (21.2)$$

where $W = W^\pm$ or Z^0. No ggH or $\gamma\gamma H$ vertex occurred at the tree level, though one does occur via loops of fermions or $W^\pm$. From these contributions, we have to build up all the Higgs production and decays.

21.2   Higgs Decays

From these couplings, we can calculate the relevant Higgs decays. Consider $H \to f\bar{f}$. The matrix element is

$$M = g_f \bar{u} u . \qquad (21.3)$$

Let us assume $M_H \gg M_f$. Then $\bar{u} u$ can be replaced by M_H. By our counting, there are four final spin states and one initial one so

$$\overline{|M|^2} \simeq 4 g_f{}^2 M_H^2 . \qquad (21.4)$$

Then from equation (9.6)

$$\Gamma_{Hf\bar{f}} \simeq \frac{\overline{|M|^2}}{16\pi M_H} = \frac{1}{4\pi} M_H \frac{g_2^2 M_f^2}{4 M_W^2} = \frac{\alpha_2}{4} \frac{M_f^2}{M_W^2} M_H . \qquad (21.5)$$

The correct answer is 3/2 of this. The main point to note is that the naïve width

$\Gamma \sim \alpha_2 M_H$ is modified by a factor $(M_f/M_W)^2$. Thus H decays dominantly to the heaviest fermion allowed by energy conservation (when $2M_f \sim M_H$ there is a phase space factor that should modify our approximate formula).

Next consider $H \rightarrow WW$ (remember, $W = W^\pm$ or Z). Now the matrix element is

$$M = g_W \epsilon \cdot \epsilon' \qquad (21.6)$$

where ϵ and ϵ' are the polarization four–vectors of the two W's, and let the W's have four–momenta $k_\mu = (k_0; \vec{k})$ and $k'_\mu = (k'_0; \vec{k}')$. From equation (21.2), $g_W = g_2 M_W$. For massive spin–one particles, our simple method of calculating widths and cross sections sometimes breaks down. The reason is related to the physics of longitudinal vector bosons, which originated in the theory of Goldstone bosons, so it is no accident the simple method breaks down. Fortunately in this case, it is not hard to do the calculation.

We only need to know the longitudinal polarization vectors, and it is easy to write them. For a W at rest, the tranverse (x,y) and longitudinal (z) polarization states can be chosen to be the obvious ones for any spin–one state,

$$\left. \begin{array}{l} \vec{\epsilon}^{(x)} = (1, 0, 0), \\ \vec{\epsilon}^{(y)} = (0, 1, 0), \\ \vec{\epsilon}^{(z)} = (0, 0, 1). \end{array} \right\} \qquad (21.7)$$

Now we need the longitudinal polarization vector for a moving W with four–momentum k^μ. It has the form

$$\epsilon_\mu^{(z)} = \left(\epsilon_0^{(z)}; \vec{\epsilon}^{(z)} \right). \qquad (21.8)$$

As always, if we can write a four–vector that reduces to $(0; 0, 0, 1)$ in the rest frame, and satisfies the condition $\epsilon_\mu k^\mu = 0$, we have the correct and unique answer. [The condition $\epsilon_\mu k^\mu = 0$ can be thought of as the only Lorentz invariant way to guarantee that the number of spin projections remains three, even though the particle is moving. In principle, there are four independent four–vectors, but imposing one condition reduces the number of independent ones to three.] Since only k^μ is available to construct ϵ_μ, we see the result must be

$$\epsilon_\mu^{(z)} = \frac{1}{M_W} (|\vec{k}|; k_0 \hat{z}). \qquad (21.9)$$

The same result follows by a standard Lorentz transformation on the third vector of equation (21.7) since $\gamma = E/M_W$ and $\beta\gamma = |\vec{k}|/M_W$. In the rest frame (21.9) clearly must reduce to (21.7), and ϵ_μ is orthogonal to k^μ as required. As the energy of a W gets large, $|\vec{k}| \simeq k_0$, and

$$\epsilon_\mu^{(z)} \to \frac{k_\mu}{M_W} \, , \tag{21.10}$$

so

$$\epsilon \cdot \epsilon' \simeq \frac{k \cdot k'}{M_W^2} \, . \tag{21.11}$$

To evaluate $k' \cdot k$, note that $M_H^2 = (k + k')^2 = 2M_W^2 + 2k \cdot k'$. If we work in the approximation that $M_H^2 \gg M_W^2$, then $k \cdot k' \simeq M_H^2/2$, so

$$\overline{|M|}^2 \simeq \frac{g_2^2 M_H^4}{4 M_W^2} \, . \tag{21.12}$$

Then

$$\Gamma_{HWW} \simeq \frac{1}{16\pi M_H} \frac{g_2^2 M_H^4}{4 M_W^2} = \frac{\alpha_2 M_H^3}{16 M_W^2} \, . \tag{21.13}$$

This is *larger* than a naïve estimate $\alpha_2 M_H$ by a factor M_H^2/M_W^2; the simple calculations break down here for a deep reason, the presence of the $1/M_W^2$ in equation (21.11) and therefore in Γ_H . This is the correct rate for $H \to W^+ W^-$; including $H \to ZZ$ multiplies it by 3/2, in our approximation where $M_H^2 \gg M_W^2$ and M_Z^2 . When $M_H \simeq 2M_W$, there is a phase space factor that has to be included.

A remarkable consequence of equation (21.13) is that Γ_H grows rapidly with M_H. If everything is put in TeV units, then

$$\Gamma_H \simeq \frac{1}{2} M_H^3 \, , \tag{21.14}$$

so $\Gamma_H \simeq M_H$ when $M_H \simeq 1.4$ TeV. By then, presumably, it is appropriate to think of M_H as a parameter of the theory rather than the mass of a particle.

21.3   Ways to Search for Higgs Bosons

Several useful ways to search for H^0 have been found.

(a) The quarkonium states ψ, Υ and toponium (which we will call θ) will have decays $\psi \to \gamma H^0$, $\Upsilon \to \gamma H^0$, and $\theta \to \gamma H^0$. We can estimate the rate as follows. For the decay of any resonance into a weakly coupled final state, we can picture the process as in Figure 21.1.

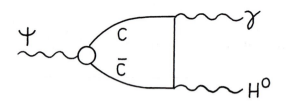

Figure 21.1
Decay of the quarkonium state ψ
into $\gamma+$ Higgs boson.

At the $c\bar{c}\gamma$ vertex, we expect a factor $\frac{2}{3}e$, so in the width this will give $\frac{4}{9}\alpha$. At the $Hc\bar{c}$ vertex there is a factor $g_2 M_f/2M_W$, so in rate a factor $\alpha_2 M_f^2/4M_W^2$. Since here the masses of ψ and of H^0 might not be very different, we have to be careful about the phase space; equation (9.6) says there is a factor of the final momentum, which is $p' = \left(M_\psi^2 - M_H^2 \right)/2M_\psi$. Then there will be factors associated with the bound state wave function, exactly as in atomic physics or positronium decay. We can bypass all those factors by a comparison. Consider also the decay $\psi \to \mu^+\mu^-$, as in Figure 21.2. By the same sort of arguments, this has factors $\frac{4}{9}\alpha^2$, and we can neglect m_μ so no additional phase space factors occur. The bound state factors will be the same, so if we take a ratio they cancel. Thus using $M_\psi \simeq 2M_c$,

$$\frac{\Gamma(\psi \to \gamma H)}{\Gamma(\psi \to \mu^+\mu^-)} \simeq \frac{\alpha_2 M_\psi^2}{8\alpha M_W^2} \left(1 - \frac{M_H^2}{M_\psi^2} \right) . \tag{21.15}$$

The final momentum p' for $\psi \to \mu\mu$ is just $M_\psi/2$, so the last factor is the ratio $p'(\psi \to \gamma H)/p'(\psi \to \mu\mu)$. A final factor of two has been introduced since there are two orderings for emitting the γ and H, *i.e.* two diagrams.

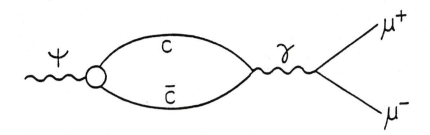

Figure 21.2
Decay of the quarkonium state ψ
into muons.

The best way to use equation (21.15) is to put in data for $\Gamma(\psi \to \mu^+\mu^-)$. A similar equation holds for Υ and for θ, with $M_c \to M_b$ or M_t . If numerator and denominator are divided by Γ_ψ ,

$$\mathrm{BR}(\psi \to \gamma H) \simeq \frac{\alpha_2 M_\psi^2}{8\alpha M_W^2}\left(1 - \frac{M_H^2}{M_\psi^2}\right)\mathrm{BR}(\psi \to \mu^+\mu^-) . \qquad (21.16)$$

$\mathrm{BR}(\psi \to \mu^+\mu^-)$ is about 7%, and $\mathrm{BR}(\Upsilon \to \mu^+\mu^-) \simeq 3\%$. The factors in front are given in Table 21.1 .

Table 21.1
Mass ratios for decay rates
in GeV

m(GeV)	$\alpha_2 m^2/2\alpha m_W^2$
1.5	7.6×10^{-4}
4.5	6.9×10^{-3}
45	0.69

We see that for the ψ the branching ratio is about 5×10^{-5}, so well over 10^6 ψ's would be needed to see an effect; no experiment has yet run long enough to get that sample. For Υ the situation is an order of magnitude better, and at CESR enough events could be obtained by the CUSB detector during 1987-88 to see

$\Upsilon \to \gamma H$ if $M_H \lesssim$ 5–6 GeV. For toponium the γH branching ratio is over a percent (assuming $\theta \to \mu^+ \mu^-$ has a branching ratio of several percent as is expected), so toponium will be an extremely valuable Higgs boson factory if it occurs at a mass where study is possible. The experiments proceed by searching for a monochromatic photon opposite a decaying state, since a photon of energy $E = \left(M_V^2 - M_H^2\right)/2M_V$ emerges from the two-body decay (for $V = \psi$, Υ, or θ). Then the particles opposite the γ can be studied to determine whether the decaying particle has spin–zero and decays into various channels in proportion to their mass.

[There is a numerical caveat. The higher order diagrams for $\psi, \Upsilon, \theta \to H\gamma$, such as those with an extra internal or soft gluons, have been shown to reduce the expected rate by nearly a factor of two, so somewhat more events will be needed than suggested above.]

(b) At e^+e^- colliders with $\sqrt{s} \geq M_Z$, the process of Figure 21.3 can give a detectable rate for Higgs boson production if the luminosity is large enough. For $\sqrt{s} \simeq M_Z$ the cross section is enhanced by sitting on the Z resonance, but the final $\ell^+\ell^-$ are from a virtual Z so a careful calculation is needed to evaluate the rate. If machines such as SLC and LEP that will have $\sqrt{s} \simeq M_Z$ reach their design luminosity, they can search for a Higgs boson for $M_H \lesssim$ 40 GeV, depending somewhat on how well background separations can be performed. Lower luminosity reduces this sensitivity; higher luminosity will be needed to establish the behavior of an effect if it is found.

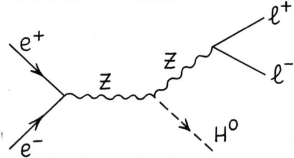

Figure 21.3
Scheme for an e^+e^- collider
to produce the Higgs particle.

The $\ell^+\ell^-$ are detected. Although they are not monoenergetic, the $\ell^+\ell^-$ mass peaks at the high end, as near to M_Z as it can get. For events with a high mass $\ell^+\ell^-$, the accompanying particles can be examined to see if they could be from H^0.

(c) If $M_H \gtrsim 40$ GeV, no machine under construction as of 1986 can find it. If LEP II is constructed at some $\sqrt{s}$, it can search almost up to $M_H \lesssim \sqrt{s} - M_Z$, except that if $M_H \gtrsim 85$ GeV there is a background problem which probably renders the search insensitive. Once $\sqrt{s} > 2M_Z$, the rate for $e^+e^- \rightarrow ZZ$ with one $Z \rightarrow \ell^+\ell^-$ and the other decaying hadronically is considerably larger than the signal for $e^+e^- \rightarrow ZH$, and the signature is very similar. Above about 40 GeV or 85 GeV (depending on LEP II), of currently planned facilities only the SSC has a possibility of providing direct experimental information on Higgs physics; e^+e^- colliders at higher energy and very high luminosity could do so also.

A machine such as the SSC will open a huge window to study Higgs physics, using a variety of techniques, if it can run at its planned large luminosity.

(a) When $M_H < 2M_W$, the dominant decays are $H \rightarrow t\bar{t}$ or $b\bar{b}$. These are very difficult modes to search for because many more $t\bar{t}$ or $b\bar{b}$ pairs are produced from QCD processes such as $gg \rightarrow t\bar{t}$. Instead, rare decays have to be used. Fortunately, decays such as $H \rightarrow \gamma\gamma$, $H \rightarrow Z\gamma$ with $Z \rightarrow \ell^+\ell^-$, $H \rightarrow \theta\gamma$ with $\theta \rightarrow \ell^+\ell^-$, $H \rightarrow \Upsilon\gamma$, and $H \rightarrow \psi\gamma$ all occur via fermion and W loops with large enough branching ratios (10^{-3} to 10^{-5}) to allow searches. This approach is crucially dependent on the high luminosity of the SSC. In this mass range, 10^6 or more H^0 would be produced in a year of 10^7 seconds, so a clean branching ration of 10^{-5} is as low as one could hope to go. Combining all of the above modes, it appears possible to search from some minimum value of order 30 GeV up to $2M_W$. All these channels have significant backgrounds, so it may take a very high luminosity to get a large enough signal to noise ratio.

(b) When $M_H > 2M_W$, the H decays are dominated by $H \rightarrow W^+W^-$ and $H \rightarrow ZZ$, as is clear by comparing equations (21.5) and (21.14). When $M_H \lesssim 500$ GeV, so many H^0 are produced that the mode $H \rightarrow ZZ$ followed by both $Z \rightarrow e^+e^-$ or $\mu^+\mu^-$ can be looked for; it gives a very clean detector signal and there are no backgrounds that mimic the signal. Other modes of W^+W^- and ZZ can also be looked for. Γ_H is narrow enough that H will appear as a resonance in the W^+W^- and ZZ channels.

At larger M_H two qualitatively important things happen to change the nature of the search. First, as Γ_H increases, the Higgs width gets so large that it becomes difficult to see a resonant shape in the WW mass spectrum; for $M_H = 1$ TeV, equation (21.14) tells us that $\Gamma_H \approx 500$ GeV, and beyond that Γ_H rapidly becomes so large that it is not directly meaningful to talk of a Higgs particle.

The second problem arises because the rates get so small at larger M_H that it is necessary to look in the channel where one W decays to $q\bar{q}$. Then events from other Standard Model processes look a lot like the signal one is searching for. The signal of interest is an excess of W^+W^- events. But a common process is the production of $W + q + g$, from a $q + g$ initial state. Whenever the mass of the $q + g$ jet pair is near the W mass, this background mimics the signal. The relevant rates can be calculated since all the processes are perturbative Standard Model ones, and the background is larger than the signal. Since the background is not identical to the signal, careful study of events will allow the signal to be detected, but only if extremely good detectors with the right properties are built.

21.4   Large M_H

As M_H increases, one might imagine that the rate gets so small that no signal could be detected, and we cannot get positive information about Higgs physics. That turns out not to be true, for very important and basic reasons in the Standard Model. We can understand it as follows. If $M_H \gtrsim$ few hundred GeV, the dominant mechanism to produce a Higgs boson is by the process of Figure 21.4, where f and f' are any fermions (qq, $q\bar{q}$, or e^+e^-). This process can be interpreted as having beams of W's which scatter, so we are studying WW scattering. The W's are emitted from the fermions with a calculable probability of carrying a given fraction of the fermion's momentum; the situation is like the structure function analysis of Chapters 10 and 18, but here we are dealing with point–like fermions and W's so all the structure functions are calculable. We will not give the details, but only assure the reader that the techniques needed in order to think in terms of scattering with real W's are known.

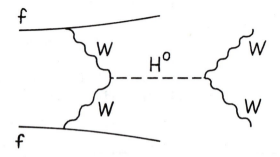

Figure 21.4
The mechanism to produce
a heavy Higgs boson.

The diagram for $WW \to H \to WW$ in Figure 21.4 is essentially one we have calculated. To be specific, consider the process $W^+ + W^- \to Z + Z$. Let the W polarization vectors be ϵ_W and ϵ_W' in the initial state and ϵ_Z and ϵ_Z' in the final state. Then the HWW and HZZ vertices are just $g_2 M_W \epsilon \cdot \epsilon'$, so the matrix element for $WW \to H \to ZZ$ is

$$M = \frac{g_2^2 M_W^2 (\epsilon_W \cdot \epsilon_W')(\epsilon_Z \cdot \epsilon_Z')}{M_H^2 - \hat{s}} . \tag{21.17}$$

In Section 21.2, we showed that for a longitudinal W, which is the most interesting case, $\epsilon \cdot \epsilon' \simeq k \cdot k'/M_W^2$ at high energies. Here let the initial W's have momenta p and p', and final Z's momenta k and k'. Then $\epsilon_W \cdot \epsilon_W' = p \cdot p'/M_W^2$ and $\epsilon_Z \cdot \epsilon_Z' = k \cdot k'/M_W^2$. Also, $\hat{s} = (p + p')^2 = (k + k')^2$, so if $\hat{s} \gg M_W^2$, then $p \cdot p' \simeq \hat{s}/2 \simeq k \cdot k'$. Thus

$$M \simeq \frac{g_2^2}{4 M_W^2} \frac{\hat{s}^2}{M_H^2 - \hat{s}} . \tag{21.18}$$

Since $\sigma(WW \to WW) = |M|^2/16\pi\hat{s}$, we see that σ grows as a power of $\hat{s}$, which is a meaningless result since cross sections cannot grow in an unbounded power law way. [Technically this is referred to as a violation of conservation of probability, or a violation of unitarity.] In fact, before we even calculated we should have realized that other diagrams contribute to the process $WW \to WW$. Besides the subprocess shown in Figure 21.4, there are contributions shown in Figure 21.5.

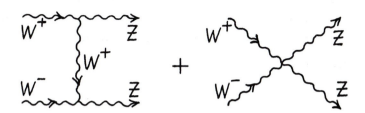

Figure 21.5
Additional processes contributing
to $W^+ W^- \to ZZ$.

The vertices for these (the $W^+ W^- Z$ and $W^+ W^- ZZ$ vertices) are implicitly written in Section 2.8e, but we do not need the details to work out the most important consequence. Since the contributions of Figure 21.5 do not depend on M_H, they give some function of $\hat{s}$ and M_W^2. Then the full amplitude is the sum of Figure 21.4 and Figure 21.5,

$$M = \frac{g_2^2}{4M_W^2} \frac{\hat{s}^2}{M_H^2 - \hat{s}} + f(\hat{s}, M_W^2) \qquad (21.19)$$

where we have just written $f(\hat{s}, M_W^2)$ for the contributions of Figure 21.5. The implications of this are very important.

(a) It is indeed true, and we could show it by a somewhat lengthly explicit calculation to obtain $f(\hat{s}, M_W^2)$, that from equation (21.19) (as must happen) $M \to$ const as $\hat{s} \to \infty$, so the full cross section does not grow as a power of $\hat{s}$. In particular, suppose $\hat{s} \gg M_H^2$. Then we see that $f(\hat{s}, M_W^2)$ grows proportional to $\hat{s}$ in order to cancel the first term in equation (21.19).

(b) Even more important, suppose we want to understand what happens if M_H is very large, *i.e.* if somehow there is not a Higgs boson in the theory at masses of order the weak scale. Then when $M_H \to \infty$, the first term in equation (21.19) gets very small for $M_H^2 \gg \hat{s}$, so in this region $M \simeq f(\hat{s}, M_W^2)$. But now as $\hat{s}$ gets large, f grows like $\hat{s}$ because, as we showed in (a) $f \sim \hat{s}$, and f does not depend on M_H so its behavior does not change as M_H gets large. Thus, if M_H is large, $\sigma(W^+W^- \to ZZ)$ will apparently grow like $\hat{s}^2$, contrary to what is allowed. We see that *even for $M_H \to \infty$ an experimental signal appears in WW scattering.* We have not given the actual amplitudes and cross sections here, but a calculation shows that the effect of the growth with $\hat{s}$ is in principle experimentally detectable at a machine with the luminosity and energy of the SSC. Detection requires not only the production of W pairs at sufficient rate by the machine, but detection of W pairs with sufficient precision to establish that the W's are longitudinally polarized, since that is where the physics we have discussed appears.

21.5   Comments

What we have seen in this chapter is that even though no one knows what the physics of the Higgs sector will turn out to be, we do know how to do a complete set of experiments to untangle what is there, and the experimental facilities needed are within our reach (though at a large economic cost). In particular, we have seen in the last section that effects of a Higgs boson cannot be pushed to arbitrarily high mass; if $M_H \to \infty$, detectable effects appear in the production rate and properties of pairs of W bosons.

If Higgs fields correspond to point-like scalars that are to be included in the fundamental Lagrangian of the theory, the Higgs bosons might be expected to have masses of order M_W (within a factor of a few either way). Some approaches along these lines imply the existence of several scalars. If instead Higgs physics is a representation of some underlying dynamics, very heavy states could occur.

Some approaches to constructing a fundamental theory that includes the Standard Model lead to a particular variation on the form the Higgs sector will take. For technical reasons, in such theories, which include supersymmetric theories, there have to be two doublets of Higgs fields instead of one. It still happens that three of the Higgs degrees of freedom are Goldstone bosons that become the longitudinal states of $W^\pm$ and Z so these gauge bosons can get mass. Then, because we began with eight degrees of freedom instead of four, there are five physical scalar bosons rather than one. Three are neutral, and there is a charged pair $H^\pm$. One of the neutral states is very much like the single H^0 of the minimal Standard Model. Although many details are different, the basic point that this system of scalar bosons could be found experimentally and studied if it existed is still true. And there are a variety of ways to distinguish this kind of Higgs sector from the minimal Standard Model Higgs sector, including the presence of charged scalars in one case and not in the other, different couplings to various channels, and so on. The experimental unraveling of the physics of the Higgs sector could begin as early as 1988, and could conceivably be finished before 2000.

Problems

21.1:   Calculate, approximately, the branching ratio for $H^0 \rightarrow \tau^+ \tau^-$ if

(a) $M_H = 8$ GeV,

(b) $M_H = 15$ GeV.

Suggestions for Further Study

One view of the difficulties associated with understanding the Higgs sector of the Standard Model is provided in Veltman's Scientific American article. A thorough treatment of the properties of Higgs bosons in the minimal Standard Model is given in Chapter 24 of Okun (1982).

CHAPTER 22

Quark
(and Lepton)
Mixing Angles

Some readers will have noticed that, with the Feynman rules we have written so far, s and b quarks are stable. They are not coupled to any lighter quarks. That occurred because of a subtle assumption that we did not make explicit; we assumed that the quarks which went into the left–handed electroweak doublets were the quarks of definite mass. That is, we assumed that the eigenstates of the electroweak Hamiltonian were the eigenstates of the mass Hamiltonian. Since we do not understand the origin of mass, we have no reason to make that assumption, and in fact it is wrong. For simplicity, first suppose there are only two families of quarks. Then we could write the charged current of Chapter 7, to which the W couples, as

$$
\begin{aligned}
J_{\mathrm{ch}}^{\mu} &= (\,\overline{u} \quad \overline{c}\,)\, \gamma^{\mu} P_L \begin{pmatrix} d \\ s \end{pmatrix} \\
&= \overline{u}\gamma^{\mu} P_L d + \overline{c}\gamma^{\mu} P_L s\,,
\end{aligned}
\tag{22.1}
$$

where we have used row and column vectors in a flavor space. By u, c, d, and s, we mean the mass eigenstates, the energy levels of the system. But then we should call the weak eigenstates something different; it is an experimental question as to whether they are equal [they are not]. So we replace $\begin{pmatrix} d \\ s \end{pmatrix}$ by $\begin{pmatrix} d' \\ s' \end{pmatrix}$ where the q' states are defined to be the weak interaction eigenstates. One set of eigenstates can be expanded in terms of another, so we write

$$
\begin{pmatrix} d' \\ s' \end{pmatrix}_L = V \begin{pmatrix} d \\ s \end{pmatrix}_L
\tag{22.2}
$$

where V must be a unitary 2×2 matrix. As shown in Appendix B, the most general unitary 2×2 matrix can be written with three angles θ, α, and γ,

$$
V = \begin{pmatrix} \cos \theta e^{i\alpha} & \sin \theta e^{i\gamma} \\ -\sin \theta e^{-i\gamma} & \cos \theta e^{-i\alpha} \end{pmatrix},
\tag{22.3}
$$

so

$$
\begin{aligned}
d' &= \cos \theta e^{i\alpha} d + \sin \theta e^{i\gamma} s \\
&= e^{i\alpha} \left(d \cos \theta + s \sin \theta e^{i(\gamma - \alpha)} \right)
\end{aligned}
\tag{22.4}
$$

$$
\begin{aligned}
s' &= -\sin \theta e^{-i\gamma} d + \cos \theta e^{-i\alpha} s \\
&= e^{-i\gamma} \left(-d \sin \theta + s \cos \theta e^{-i(\alpha - \gamma)} \right).
\end{aligned}
\tag{22.5}
$$

We can redefine the relative phases of the quark states without changing any observables, so we can multiply d' by $e^{-i\alpha}$, s' by $e^{i\gamma}$, and s by $e^{-i(\gamma - \alpha)}$. If mass terms are present, these phases can be absorbed by similar transformations of s_R and d_R. With these replacements,

$$
d' = d \cos \theta + s \sin \theta,
\tag{22.6}
$$

$$
s' = -d \sin \theta + s \cos \theta,
\tag{22.7}
$$

and

$$
V = \begin{pmatrix} \cos \theta & \sin \theta \\ -\sin \theta & \cos \theta \end{pmatrix}.
\tag{22.8}
$$

Then the form of equation (22.1) that should have been used from the beginning is

$$
J_{\text{ch}}^{\mu} = (\overline{u} \quad \overline{c}) \gamma^{\mu} P_L \begin{pmatrix} d' \\ s' \end{pmatrix}
\tag{22.9}
$$

$$
= (\overline{u} \quad \overline{c}) \gamma^{\mu} P_L V \begin{pmatrix} d \\ s \end{pmatrix}
\tag{22.10}
$$

$$
\begin{aligned}
= &\overline{u} \gamma^{\mu} P_L d \cos \theta + \overline{u} \gamma^{\mu} P_L s \sin \theta \\
&- \overline{c} \gamma^{\mu} P_L d \sin \theta + \overline{c} \gamma^{\mu} P_L s \cos \theta.
\end{aligned}
\tag{22.11}
$$

There are two new terms, both multiplied by $\sin \theta$, and the old terms are reduced by $\cos \theta$. The angle θ is called the Cabibbo angle and has been measured to be

$\theta \approx 13°$. If θ had come out to be zero, then equation (22.11) would have reduced to equation (22.1) and the s quark would have been stable. Now it can decay via the coupling to u from the second term above. The electroweak vertices are now as in Figure 22.1.

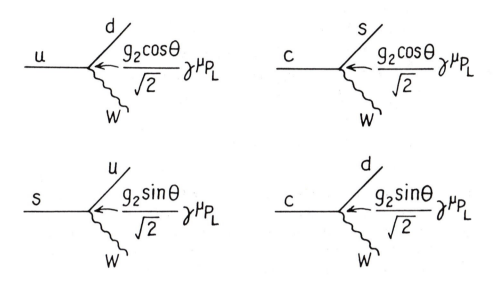

Figure 22.1
The electroweak vertices.

Note that we chose, as is conventional, to rotate the down–type quarks, but that is entirely general. If we had rotated both $(\overline{u}\ \overline{c})$ and $\left(\begin{smallmatrix} d \\ s \end{smallmatrix}\right)$, we would have had a current of the form $(\overline{u}\ \overline{c})\,\gamma^\mu P_L\, V_{\rm up}^\dagger V_{\rm down}\left(\begin{smallmatrix} d \\ s \end{smallmatrix}\right)$, but the product of two rotations is a rotation so we can replace $V_{\rm up}^\dagger V_{\rm down}$ by a single rotation matrix V.

Next we have to check whether there is any change in the structure of neutral currents due to this rotation. The Z couples to (still working with two families)

$$J^\mu_{\rm neu} = \sum_{f=u,c,d,s} \left(\overline{f}_L\gamma^\mu\left[T_3^L - Q\sin^2\theta_w\right]f_L + \overline{f}_R\gamma^\mu\left[-Q\sin^2\theta_w\right]f_R\right), \quad (22.12)$$

so now we replace d and s by d' and s'. Then the terms that could change are

$$
\begin{aligned}
&\left(\bar{d}_L\cos\theta + \bar{s}_L\sin\theta\right)\gamma^\mu\left[T_3^L - Q\sin^2\theta_w\right]\left(d_L\cos\theta + s_L\sin\theta\right)\\
&\quad +\left(-\bar{d}_L\sin\theta + \bar{s}_L\cos\theta\right)\gamma^\mu\left[T_3^L - Q\sin^2\theta_w\right]\left(-d_L\sin\theta + s_L\cos\theta\right)\\
&\quad +(L\to R)\\
&= \bar{d}_L\gamma^\mu\left[T_3^L - \sin^2\theta_w\right]d_L\left(\cos^2\theta + \sin^2\theta\right)\\
&\quad +\bar{s}_L\gamma^\mu\left[T_3^L - \sin^2\theta_w\right]s_L\left(\sin^2\theta + \cos^2\theta\right)\\
&\quad +\bar{d}_L\gamma^\mu\left[T_3^L - \sin^2\theta_w\right]s_L\left(\cos\theta\sin\theta - \cos\theta\sin\theta\right)\\
&\quad +\bar{s}_L\gamma^\mu\left[T_3^L - \sin^2\theta_w\right]d_L\left(\cos\theta\sin\theta - \cos\theta\sin\theta\right)\\
&\qquad +(L\to R)\\
&= \sum_{d,s}\left(\bar{f}_L\gamma^\mu\left[T_3^L - Q\sin^2\theta_w\right]f_L +(L\to R)\right)
\end{aligned}
$$

$$(22.13)$$

so we are back to the original result; the neutral current is diagonal in mass eigenstates or weak eigenstates. That is called the GIM mechanism. It has very profound consequences for decays, since the Standard Model theory has no vertices of the form $\bar{s}dX$ where X is a gauge boson or Higgs boson. It is indeed observed that kaon decays involving an $s \to d$ are much smaller than $s \to u$ decays. As a consequence, decays involving $s \to d$ (called flavor changing neutral currents) are very interesting because they are possible probes of new interactions.

Finally, the results have to be generalized to the three families of quarks we know about. It is clear that the way to proceed is by analogy with equation (22.10). The full charged current is

$$
J_{\mathrm{ch}}^\mu = (\bar{u}\ \ \bar{c}\ \ \bar{t})\gamma^\mu P_L V \begin{pmatrix} d \\ s \\ b \end{pmatrix},
$$

$$(22.14)$$

and V is a 3×3 unitary matrix. It is shown in Appendix B that an $n \times n$ unitary matrix has n^2 independent real parameters, nine here. We can redefine the phases of five quark states; the sixth would amount to an overall phase for all the states so it does not help. That leaves four parameters to describe the matrix. It is also shown in Appendix B that an orthogonal $n \times n$ matrix which describes rotations has $n(n-1)/2$ real parameters, three here. Thus, one of the parameters in V must enter as a relative phase. Then the terms in the Hamiltonian$\sim W_\mu J_{\mathrm{ch}}^\mu$

can be complex, and we know from quantum theory that the theory will not be invariant under transformations involving time reversal (or equivalently, CP; see Chapter 24).

The matrix V is called the Kobayashi–Maskawa matrix. From the Particle Data Tables, its entries are presently measured to have values (magnitudes)

$$V = \begin{pmatrix} 0.9742{-}0.9756 & 0.219{-}0.225 & 0{-}0.008 \\ 0.219{-}0.225 & 0.973{-}0.975 & 0.037{-}0.053 \\ 0.002{-}0.018 & 0.036{-}0.052 & 0.9986{-}0.9993 \end{pmatrix}. \qquad (22.15)$$

Ordering the families by mass, we see that transitions by one unit are small, and transitions by two units are very small. [If decays $b \to u+$ anything are observed, the upper right–hand element will be measured; some have been reported but not confirmed.]

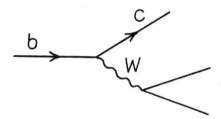

Figure 22.2
Decay of a b–quark.

Note that now the b–quark can decay as well. One of the terms in equation (22.14) is

$$\bar{c}\gamma^\mu P_L V b$$

with coefficient $V_{23} \approx 0.045$. Then b decay goes as in Figure 22.2 where W can go into $\mu\nu_\mu$, $e\nu_e$, $\tau\nu_\tau$, $u\bar{d}$, and $c\bar{s}$. At the $\bar{b}cW$ vertex there is a factor of V_{23} in addition to the usual coupling $g_2/\sqrt{2}$. Then the width of the b is approximately

$$\Gamma_b \simeq \frac{9V_{23}^2 G_F^2 M_b^5}{192\pi^3} \qquad (22.16)$$

where 9 comes from the nine open channels (neglecting phase space corrections for M_c and M_τ). This gives

$$\frac{\Gamma_b}{\Gamma_\tau} \simeq \frac{9}{5} V_{23}^2 \left(\frac{M_b}{M_\tau}\right)^5 \simeq 0.4 \qquad (22.17)$$

so we expect the b to live about $2\frac{1}{2}$ times longer than τ because it can only decay via the rotation from the mass eigenstates to the weak eigenstates. [In practice τ_b was first measured and V_{23} was deduced from the observation of the b lifetime].

At present the elements of V are parameters that have to be measured, just as fermion masses have to be measured. Eventually, it is hoped that the relation between the weak eigenstates and the mass eigenstates will be calculable, so the elements of V can be expressed in terms of ratios of masses.

We could have carried out a similar procedure for the leptons. If, however, any pair of the quarks are degenerate in mass, then we cannot tell the mass eigenstates apart, so we could perform a rotation and make the relevant angle or element of V become zero. Thus, if neutrinos have zero mass, there is no need to have lepton mixing angles; the weak eigenstates and mass eigenstates can be the same. In Chapter 29, we discuss briefly the situation if neutrino masses are nonzero.

Problems

22.1:   Check the unitarity of the Kobayashi–Maskawa matrix (22.15). Which elements are not constrained by the unitarity requirement, given the numerical ranges allowed there?

CHAPTER 23

Quark
and Hadron
Masses

The purpose of this chapter is to clarify a little the problem of mass in particle physics. Although the subject is rather complicated, and the most fundamental parts are not understood, some organizing principles can be introduced.

There are basically three sources of the mass of hadrons. The most familiar is the mass quarks and gluons have because they are confined objects. If a particle is confined to a region on the order of one fermi in size, it has, from the uncertainty principle, $\Delta p \sim 1/\Delta x \sim 200$ MeV. Converting this to an energy is not a unique procedure, since we do not know what mass to use, but one can see that a hadron containing quarks and gluons might have a total mass of a few times 200 MeV, as is observed. The average mass per constituent of a bound quark or gluon is called its constituent mass. Because the constituent mass involves bound state and non–perturbative interactions, it can not be calculated very well.

As we saw with the Higgs mechanism, quarks and leptons and W and Z get mass in some fundamental way, which we describe by an interaction with a scalar field. Such mass would add to the mass of a hadron containing a quark. This second kind of mass we have called the free quark mass, or for historical reasons, the current algebra mass.

If we scatter a probe off a hadron and the probe has a wavelength of the order of the hadron size, it will sense the entire hadron mass, including the constituent mass. If a probe has a wavelength small compared to the hadron size, it can scatter off a single constituent, a quark or a gluon. Then it can sense the free particle mass, and the quark or gluon can recoil as if it were free. Thus

the various masses can in principle be measured. We have given a set of values
for the masses in Chapter 1.

The situation here is not, in principle, different from the rest of physics.
All masses are measured by some interaction. For particles that are not color
singlets and are not too massive, it is hard to construct a probe with an energy
large enough to have a wavelength small compared to the mass of interest. Thus
technically it is hard to measure quark masses. Indirect methods have been used
to arrive at the numbers given in Chapter 1.

One of the classic "mass" problems in physics is the neutron–proton mass
difference $\Delta M = M_n - M_p$. It gets an electromagnetic contribution, since n and
p have different constituents. A little thought shows ΔM_{EM} is negative, since
the proton constituents have the larger charge; the data, of course, has $\Delta M > 0$.
That $\Delta M > 0$ is now interpreted as being due to $M_d > M_u$. The old problem
has been reinterpreted, but will not be solved until $M_d - M_u$ is calculable. A small
consolation is that the values given in Chapter 1 for the free mass of M_d and M_u
come from a technical analysis that does not include the n–p mass difference,
but is consistent with the observed $M_n - M_p$, which indicates that the present
view is consistent.

Finally a third contribution to the mass comes from interactions that are
calculable, such as perturbative interactions with γ, W, and Z. The electromagnetic contribution to the n–p mass difference mentioned above is of this type,
as is the shift of a few percent in the W and Z masses. Such contributions are
called radiative corrections, since they often involve radiation and reabsorption
of a gauge boson. Because these are calculable, they help test the structure of
the theory. Measurements of M_Z and M_W at SLC and FNAL will provide the
first tests of these corrections in the late 1980's.

CHAPTER 24

CP
Violation

A small but very interesting effect was observed in the middle 1960's. It is called CP violation; we will explain what that means in this chapter. As with other phenomena, it appears to be possible to accommodate CP violation straightforwardly in the Standard Model, though there is no necessary physical reason why it needs to be present. No one knows whether it is an effect whose origin is profound, or one that is essentially an accident. If there is to be a possibility of deriving the net baryon number of the universe, an effect like the observed CP violation must be present in the interactions of the gauge bosons and/or Higgs bosons with quarks and leptons in the very early universe, but so far there is no known logical connection between the CP violation that has been observed and that needed for the early universe.

Some background is necessary to explain what CP violation is. Parity, P, is a symmetry operation. If a system is described by a quantum mechanical wavefunction, $\psi(\overrightarrow{x})$, then the transformed wavefunction $P\psi(\overrightarrow{x})$ has all coordinates inverted through the origin. If the system is invariant, $P\psi(\overrightarrow{x}) = \pm\psi(\overrightarrow{x})$. Interactions can be classified according to their transformations under parity. Since particles can be created or absorbed, intrinsic parity can also be assigned to particles. The overall parity of a state is its parity under space reflection times the intrinsic parities of the particles in the state.

The Standard Model Hamiltonian (Lagrangian) for a charged current interaction is of the form $\overline{f}\gamma^{\mu}P_{L}fV_{\mu}$, where V_{μ} is a gauge boson. Since $P_{L} = (1 - \gamma^{5})/2$, this is a sum of two terms, one transforming as a vector (γ^{μ}) and one as an axial vector ($\gamma^{\mu}\gamma^{5}$), with opposite parity. If one term or the other were present, or if they could not interfere, parity would "be conserved" since only squares enter into observables. Since the two terms do interfere, a system of quarks or leptons can change its parity. Since the two terms are of the same strength and can completely interfere, it is said that parity is violated maximally. We can say the same thing in quantum mechanical language. For a transition from a state ψ to a state ψ' via a Hamiltonian H, the matrix element is $\langle\psi'|H|\psi\rangle = \langle P\psi'|PHP^{-1}|P\psi\rangle$. Invariance requires $PHP^{-1} = \pm H$, which will not hold if H is a sum of two terms that transform oppositely under P.

Another symmetry operation is charge conjugation, C. If it were a good symmetry, then whenever a particle could undergo certain interactions, so could its antiparticle. By turning all particles in a process into their antiparticles, we would get another process that would happen with equal probability. The Standard Model also "maximally violates" C invariance, since (for example) processes occur involving left–handed neutrinos, but not left–handed antineutrinos.

However, if we operate with the product CP, we turn a left–handed neutrino into a right–handed antineutrino. That is because under P we expect $\overrightarrow{x} \rightarrow -\overrightarrow{x}$ and $t \rightarrow t$ so $\overrightarrow{p} \rightarrow -\overrightarrow{p}$ and $\overrightarrow{\sigma} \sim \overrightarrow{x} \times \overrightarrow{p} \rightarrow \overrightarrow{\sigma}$. Then the helicity $\overrightarrow{\sigma} \cdot \overrightarrow{p}$ changes sign so right–handed turns into left–handed. And the Standard Model does have electroweak interactions of right–handed antineutrinos as well as left–handed neutrinos. It is CP invariant, in the sense that if a process occurs, so does the CP transformed process. CP changes particles moving in one direction into antiparticles moving in the opposite direction with opposite helicity.

This part of the Standard Model was formulated in the mid–1950's, and it was hoped that even though P and C were separately not conserved by the weak interactions, CP was a valid symmetry. In the mid–1960's, however, it was found that that was not quite true. Instead, although processes and their CP conjugates occur, their probabilities to occur are not identical but differ a little, about one part in a thousand. It is this small difference in probabilities which is called CP violation.

To understand how it could arise in the Standard Model, first note that another symmetry operation is time reversal, T. It is known that the combined operation of CPT is a good symmetry for all quantum field theories. Thus a violation of CP invariance implies a violation of T or vice versa. The quantum mechanical transformation gives $\langle\psi'|H|\psi\rangle = \langle T\psi|THT^{-1}|T\psi'\rangle$. If the theory is T invariant these two quantities are equal. The time reversal operation can be written $T = UK$, where K complex conjugates and U is a matrix in the same space as H. In particular, if H is not real the complex conjugation will mean that $THT^{-1} \neq H$ and time reversal invariance (and CP) is violated.

We saw in Chapter 22 that the charged current could be written

$$(u \quad c \quad t) \gamma^\mu P_L V \begin{pmatrix} d \\ s \\ b \end{pmatrix}$$

where V was a 3×3 matrix. The most general V could be made to depend on three real quark mixing angles and a phase, *i.e.* V is allowed to be complex. Then the Hamiltonian is complex if that phase angle is non–zero, and T and CP–invariance are violated. Thus the Standard Model can easily incorporate CP violation.

It is hoped that eventually the matrix V will be calculable since it provides the transformation between the mass eigenstates and the weak eigenstates. Until then we will not be sure if the phase angle in V is the source of CP violation.

One interesting point to note is that we also saw in Chapter 22 that if there were only two families of quarks, then the 2×2 matrix V could, with full generality, be made real. Thus only if there are three or more quark flavors can CP violation occur through the complexity of V. Whether this observation is fundamental or not is not known.

So far, CP violation has been observed only in the decays of kaons. The neutral kaons are K^0 and $\overline{K}^0$, particle and antiparticle. Since CP is almost a good symmetry we can form CP eigenstates called $K_L = K_0 - \overline{K}_0$ and $K_S = K_0 + \overline{K}_0$; obviously K_S is even under CP and K_L is odd. [The neutral kaon system has a number of interesting properties which we will not discuss.] One decay mode possible for a neutral kaon is $K \rightarrow \pi\pi$, and another is $K \rightarrow \pi\pi\pi$. The system of two pions is an even eigenstate of CP. Then K_S can decay into $\pi\pi$, but K_L cannot. Since $\pi\pi\pi$ has a much smaller phase space than $\pi\pi$, the width of K_L is much smaller than the width of K_S, and its lifetime much longer. This is indeed observed, and shows that CP is approximately a good symmetry.

To measure the violation of CP, we can compare two CP conjugate modes. The data is

$$\frac{\Gamma(K_L \rightarrow \pi^- e^+ \nu_e) - \Gamma(K_L \rightarrow \pi^+ e^- \overline{\nu}_e)}{\Gamma(K_L \rightarrow \pi^- e^+ \nu_e) + \Gamma(K_L \rightarrow \pi^+ e^- \overline{\nu}_e)} = 0.00333 \pm 0.00014 .$$

The effect is small but many standard deviations from zero. For the decay $K_L \rightarrow \pi\mu\nu$ the equivalent number is 0.00319 ± 0.00038. This basically determines the size of the CP violation.

In spite of much effort, additional measurements of CP violation effects have been hard to carry out. Further measurements in the kaon system are in progress. For technical reasons the effects are expected to be larger in the mesons with b–quarks in the Standard Model, but small for charmed mesons. These predictions need testing. Current experiments have only been able to look for effects in heavy quark decays at the 10% level (and observed none), which is much larger than the expected effects. Although CP violation, as we know it today, can be incorporated in the Standard Model, as discussed above, the validity of this description is not tested, and it should be considered as less well founded than the rest of the Standard Model.

Problems

24.1:   Show that a state of two π^0's has even CP. Which of the K_L or K_S could decay to $\pi^0 \pi^0$?

Suggestions for Further Study

Volume I, Chapter E11 of Gottfried and Weisskopf, and Chapter 8 of Hughes, contain introductory surveys of the fascinating behavior of the K^0 and $\overline{K}^0$ system and of the symmetries discussed in this chapter. The Nobel Lectures of Cronin and of Fitch are instructive.

Why the t–Quark and the τ–Neutrino Must Exist

So far the t–quark and the τ–neutrino have not been seen directly, not because of any problem but (presumably) because appropriate experiments or machines and detectors to observe them have not been available. The t–quark is assumed to be too heavy to have been found so far; at PETRA a t–quark would have been detected if $M_t \lesssim 23$ GeV. As TRISTAN, SLC, and FNAL analyze data at reasonable luminosities, a t–quark will be observed if it is lighter than about 30 GeV, 48 GeV, and 120 GeV respectively.

In this chapter we explain that there is already compelling evidence that the t–quark exists if the Standard Model is a valid description of b–quark interactions. That is of interest not only for its own sake, but as an illustration of the way the Standard Model works, and as an application of the Standard Model techniques. It will be instructive even after the t–quark is found. Similar remarks apply to the τ–neutrino.

25.1 Forward–Backward Asymmetries

Consider $e^+e^- \rightarrow b\bar{b}$, at PEP and PETRA energies, such as $\sqrt{s} \simeq 30$ GeV. There are two contributions, from γ and from Z^0, as shown in Figure 25.1.

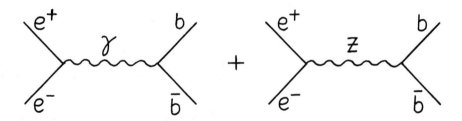

Figure 25.1
Feynman diagrams contributing
to $b\bar{b}$ production.

An angular distribution that is symmetric about 90° is obtained from either contribution, but when both are present they can interfere and an asymmetric distribution results. If $d\sigma/d\Omega$ is the differential cross section, the forward–backward asymmetry is defined by

$$A_{FB} = \frac{\int_0^1 d\Omega \frac{d\sigma}{d\Omega} - \int_{-1}^0 d\Omega \frac{d\sigma}{d\Omega}}{\int_0^1 d\Omega \frac{d\sigma}{d\Omega} + \int_{-1}^0 d\Omega \frac{d\sigma}{d\Omega}} \qquad (25.1)$$

where the limits are on $\cos\theta$. Even at $\sqrt{s} = 30$ GeV the contribution of the Z is not negligible; it is suppressed by $\Gamma_Z/(\sqrt{s} - M_Z)$ relative to the large peak at $\sqrt{s} = M_Z$, but the Z couplings are larger than the photon couplings since $g_2 = e/\sin\theta_w$, and the result is an interference term of order 10–20%.

The key point is that the coupling of the b to the Z is proportional to $\left(T_3^b + \frac{1}{3}\sin^2\theta_w\right)$ as we have seen in Chapter 7. This is about $+0.07$ if $T_3^b = 0$ (i.e. if there is no t quark), while it is -0.43 if $T_3^b = -1/2$ (i.e. if the b is in a doublet with another quark, the t–quark). Thus T_3^b is measurable. The data gives $T_3^b = -0.49 \pm 0.10$ at present, so the b–quark is in a doublet and there has to exist a heavier quark to be its partner. The heavier quark is, by definition, the t–quark.

The experiments separate b–quarks from all the other quarks that appear in $e^+e^- \rightarrow q\bar{q}$ by a variety of techniques depending on the kinematics of a heavier quark decaying. Most importantly, consider the decay $b \rightarrow \mu\nu X$ where X represents a quark jet. In the rest frame of the b, the most energy the muon can have is $M_b/2$. For lighter quarks the same argument holds. Thus, for muon transverse momenta relative to the b jet between $M_c/2 \simeq 0.9$ GeV and $M_b/2 \simeq 2.4$ GeV, all events should be due to b–quark decays. The same argument applies to the ν which gives missing momentum. Putting several such arguments together gives a good sample of *b*–quarks without much background from other lighter quarks.

25.2 *b*–Quark Decays

The same result can be obtained in another rather instructive way (which occurred earlier historically). It is known that the *b*–quark decays. In the Standard Model, the decays occur through quark mixing (Chapter 22), and the allowed vertices are $b \rightarrow c + W^-$ and $b \rightarrow u + W^-$. They are proportional to the elements V_{bc} and V_{bu} of the CKM matrix (Chapter 23). But this picture requires that *b* be in a doublet, so *t* must exist, although its mass is unknown.

Suppose instead that *b* was a singlet so no *t*–quark existed. Could we account for *b*–decays? Since *b* would have no interactions with $W^\pm$, and in the Standard Model nothing can decay to lighter states via its interactions with *Z*, *b* would have to decay by mixing with the lighter quarks *s* and *d*, which are in doublets and can interact as usual. For example *b* could decay as shown in Figure 25.2.

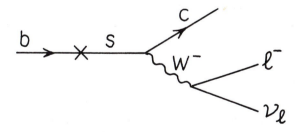

Figure 25.2
b-quark decaying by mixing
with lighter quarks.
(This hypothetical decay does not occur
for the real b-quark, but would occur if
the b were not a doublet.)

But if that process can occur, so can the decay of Figure 25.3,

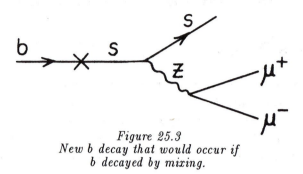

Figure 25.3
New b decay that would occur if
b decayed by mixing.

which is forbidden in the Standard Model. Further, the relative rates of Figure 25.3 and Figure 25.2 must be determined by the ratio of the Z couplings given by the factors $g_2(T_3 - Q\sin^2\theta_w)/\cos\theta_w$, to the W couplings $g_2/\sqrt{2}$. A few lines of calculations are needed to get the precise answer since these coupling factors occur separately for left–handed and right–handed fermions, whose contributions must be summed. The answer is

$$\frac{\Gamma(b \to \ell^+\ell^- X)}{\Gamma(b \to \ell^- \nu_\ell X)} \approx 0.11 \tag{25.2}$$

for $\ell = \mu$ or e. Such unusual decays with $\ell^+\ell^-$ pairs have been looked for and not found, with a limit several times below the minimum allowed in equation (25.2). Thus, for b decays to be allowed, b must be in a doublet and a t–quark must exist.

25.3   The τ–Neutrino

Similar arguments hold for ν_τ . If τ was not in a doublet, it could not decay except by mixing, and then decays such as $\tau \to \mu\mu\mu$ would occur; they do not. And T_3^τ is measured to be $-1/2$ with about 15% accuracy. Thus ν_τ must exist. Further, from the agreement of the τ lifetime with the Standard Model calculations, to about 10% accuracy, the expected mechanisms must operate, as discussed in Chapter 19; M_{ν_τ} must be small enough not to show up as a phase space suppression. Somewhat better limits can be obtained from the kinematics of special decays; for example, the energy of the pion in $\tau^\pm \to \nu_\tau\pi^\pm$ can be measured. From the kinematics of Appendix C, $E_\pi = \left[M_\tau^2 - M_{\nu_\tau}^2 + M_\pi^2\right]/2M_\tau$. The current direct limit is $M_{\nu_\tau} < 70$ MeV. Cosmological arguments suggest $M_{\nu_\tau} \lesssim 100$ eV; we will not go through these arguments here.

Direct detection of ν_τ will require a fixed target experiment of some kind. The main sources of ν_τ will be τ and $D_s^\pm$ decays (the D_s^+ is the $c\bar{s}$ meson), since $\tau \to \nu_\tau X$ and $D_s^\pm \to \tau \nu_\tau$; the latter is expected to have a branching ratio of order 1%. When ν_τ interact (assuming they are stable or long-lived), they will produce τ's which can then be observed. Since other neutrinos do not produce τ's, an appropriate arrangement will be able to establish the existence of ν_τ.

25.4   The Mass of the t–Quark

Although the theory plus some data can tell us that the t and ν_τ exist, they do not tell us their masses. As we just described for ν_τ, since it is lighter than τ there is evidence about its mass. For t there is less direct evidence at present; as mentioned before, PETRA data requires $M_t \gtrsim 23$ GeV, and if $M_t \lesssim 120$ GeV it will be observed at the Tevatron collider before about 1990.

A knowledge of M_t is important for several tests of the Standard Model, and for extracting other information. An example of the latter (see problem 1) is that when M_t is not small compared to M_Z, the $t\bar{t}$ mode of the Z decay does not contribute the full amount of a charge 2/3 quark, so a correction must be made to interpret Γ_Z in terms of restrictions on the number of neutrino families or other invisible particles to which Z might couple. More subtle possibilities occur where a t–quark can contribute in loop diagrams, such as those of Chapter 20 that determine how couplings change, and a knowledge of M_t is necessary to fully test the theory.

Problems

25.1:   Suppose the luminosity at SLC is not large enough to discover the t–quark and determine whether $M_t < M_Z/2$, but is large enough to measure Γ_Z to an accuracy of 60 MeV. What is the effect of a t–quark mass on interpreting Γ_Z in terms of the existence of the t–quark or the existence of additional neutrino families? The contribution to the Z width from a massive fermion f, relative to its contributions where $M_f = 0$, is

$$\frac{\Gamma_Z(M_f)}{\Gamma(0)} = \frac{\beta^3 + \left(1 - 4Q_f \sin^2 \theta_w\right)^2 \beta \frac{(3-\beta^2)}{2}}{1 + \left(1 - 4Q_f \sin^2 \theta_w\right)^2},$$

where Q_f is the electric charge of the fermion and $\beta = \left(1 - 4M_f^2/M_Z^2\right)^{1/2}$. What other measurements could allow a separation of the effects from $Z \rightarrow \nu\bar{\nu}$ and $Z \rightarrow t\bar{t}$?

25.2:   Neglecting the masses of all final particles, write the ratio of partial widths for $b \rightarrow s\mu^+\mu^-$ to $b \rightarrow c\mu^-\nu$ by the mechanism of Figures 25.2 and 25.3. Note that all unknown quantities and phase space factors cancel in the ratio. Compare with the predictions of the Standard Model. About how many b–quarks have to be produced to find out whether this decay is occurring at the expected level?

Suggestions for Further Study

The data on forward–backward asymmetries of b and τ and on b decays was taken from Saxon's review; references to the original data and theory can be traced from there.

CHAPTER 26

Open
Questions

We have often emphasized that although the Standard Model gives a description of all of the phenomena that are known in particle physics, it is conceptually incomplete. There are a variety of indications that more fundamental physics remains to be discovered. Nevertheless, the Standard Model will describe phenomena on the scale of interactions below 100 GeV or so, and perhaps much higher. The new physics will extend and strengthen the foundations of the Standard Model, but because the Standard Model is already a relativistic quantum field theory it will remain the effective description, whether it is a fundamental theory or not.

The new physics questions can be put into three categories.

(1) The Higgs physics of the Standard Model has to be understood. That is the central problem facing particle physics today. It is a part of the Standard Model, but the form of the solution may be part of the transition to new physics. If Higgs bosons exist as light fundamental scalars (with masses of order M_Z or less) we will be led to one kind of world view, the kind suggested by the paths being pursued with names like supersymmetry or superstrings. If, on the other hand, Higgs physics does not exist in the form of discrete particle states (one or more) then a very different and presently unclear approach will be needed. Whatever the outcome, at present it appears to be essentially an experimental question. As discussed in Chapter 21, we know how to obtain data to answer the question, but it will be the late 1990's at the earliest before the full answer is in.

(2) It could happen that the gauge theory of the Standard Model gets extended. A large number of possibilities have been considered.

(a) More families could exist, interacting according to the Standard Model. There are no limits that can be set on the numbers or masses of quarks and leptons—it is an experimental question. [Sometimes it is said that there are such limits, but what is meant is that given certain assumptions in addition to the Standard Model, such as the absence of supersymmetry, a limit can be set; if the limit is wrong it implies the assumptions are wrong, which is well worth knowing, so testing the limits experimentally is valuable.] If lots of families existed, we would probably expect an explanation in terms of the energy levels of some system of more fundamental objects, while if only a few families exist, we might expect an explanation in terms of the representations of some larger symmetry.

(b) The electroweak and QCD forces, and the quarks and leptons, could be unified into a simpler structure. This possibility, one hinted at by several patterns and properties of the Standard Model, is discussed in Chapter 27. Theories that include such unification are called "Grand Unified Theories".

(c) The Standard Model treats left–handed and right–handed fermions very differently, putting them into $SU(2)$ doublets and singlets, as we have seen. It could be that at a higher scale they are treated symmetrically, and that what we see is only a low energy limit. Future experiments will clarify this question. Approaches having this property are called "Left–Right Symmetric Theories".

(d) Although so far no transitions have been observed between the weak eigen-states of families, such as $\mu \rightarrow e$, it could be that additional gauge bosons exist to generate such transitions but they are much heavier. Then the associated interactions will be very weak, and such effects could be seen only in very rare decays or at energies comparable to the masses of the heavy bosons. Gauge symmetries that connect families are called "Horizontal" ones.

(e) There could be a larger group that contains part or all of the Standard Model groups, and has representations including the fermions we know and others. New kinds of fermions could occur that, for example, are like quarks but have both left–handed and right–handed states as $SU(2)$ singlets, or carry both quark and lepton number. Models have been constructed that include such states.

(f) A symmetry, called supersymmetry, which relates fermions and bosons has been suggested. It has a number of attractive theoretical features. In Chapter 28, we will describe how to test experimentally whether such a symmetry is occurring at a given energy scale; the analysis there provides a good example of how the Standard Model has given us tools to evaluate such questions.

(3) In the first area, Higgs physics, we saw that we can calculate all the alternatives even though we do not know the outcome, and, in principle, experiments can be performed to find out about Higgs physics. In the second case, new physics would appear as extensions of the gauge theory structure that has provided the beautiful Standard Model theory. The third area is more vague. Some unanticipated, strong interaction could enter, accounting for composite quarks and leptons and gauge bosons through a complicated dynamics. The Standard Model would be an effective theory at its energy scale, but its form would be only incidentally connected to an underlying theory. Once experiments get to the scale of the new interactions, quite a different world might appear. This is a logical possibility, though there are no indications at present that it might happen.

Whatever happens in the future, a real break with the past has occurred. Previously there was no mathematically consistent theory, and it was not possible to extrapolate to a new energy or intensity. Now, because we have the Standard Model we can calculate the Standard Model predictions for higher energy or greater luminosity or for any perturbative situation. In addition, as the next three chapters will illustrate, for any hypothetical extension of the Standard Model we can calculate cross sections and experimental signatures. We do not know which new physics will occur, if any, but we do know that it will first show up as small deviations from Standard Model behavior, that we can test any particular hypothesis, and that we can decide whether any particular question can be answered at a given present or future facility.

CHAPTER **27**

Grand
Unification

One way to extend the Standard Model, and perhaps embed it in a more fundamental theory which answers some of the open questions, is to try for further unification. Remarkably, it is possible to construct models which unify quarks and leptons and which also unify the electroweak and strong forces! At the same time some fundamental questions can indeed be answered. The approaches which have been attempted are called grand unified theories (GUT). So far none of them have emerged as being particularly likely to actually describe nature, but several are consistent with data. Some of the main results of these theories can be understood as an application of Standard Model techniques.

27.1   Unifying Quarks and Leptons;
Electric Charge and the Number of Colors

The many similarities between quarks and leptons suggest putting them into representations of a larger symmetry group, so that the relations between them are consequences of the theory. It turns out that a variety of ways to do that have been found. We will look at the simplest.

Just as $SU(2)$ representations begin with (apart from the singlet) the doublet, $SU(5)$ representations begin with a 5-component object. An $SU(5)$ multiplet can be assigned as

$$\left(\begin{pmatrix} \nu_e \\ e^- \end{pmatrix} \\ \begin{pmatrix} \overline{d}_r \\ \overline{d}_g \\ \overline{d}_b \end{pmatrix} \right)_L .$$

(27.1)

The top two are the $SU(2)_L$ doublet. The next three states are the color triplet $\overline{d}_L$. It must be a $\overline{d}_L$ since if we are to have only three states, given the three colors, an $SU(2)$ singlet is needed, while only a left–handed fermion can go with the left–handed lepton doublet since angular momentum and this internal symmetry group must commute. Put differently, there will be raising and lowering operators that move up and down in the above multiplet, and they must not change a spin projection. Recall that right–handed fermions and left–handed antifermions are $SU(2)$ singlets, so $\overline{d}_L$ is an $SU(2)$ singlet and a color triplet. Simply by considering the above representation, we can find several interesting results.

As shown in Appendix B, the generators of $SU(n)$ transformations are traceless. Imagine that the generators are represented by matrices. Diagonal generators have their eigenvalues as the diagonal elements, so their trace is just the sum of their eigenvalues, which must consequently be zero. This is familiar for angular momentum, where the diagonal generator is J_Z. The eigenvalues of J_Z always sum to zero $(1/2, -1/2; 1, 0, -1; etc.)$. This mathematical property has important consequences because we want the $SU(2)$ and $U(1)$ invariances of the theory to be part of the $SU(5)$ invariance. We saw in Chapter 7 that the electric charge operator was a linear combination of the diagonal $SU(2)$ and $U(1)$ generators, $Q = T_3 + Y/2$. Thus we want to require that the sum of the eigenvalues of the electric charge be zero.

For the above representation, this implies

$$Q(\nu_e) + Q(e^-) + 3Q(\overline{d}) = 0$$

(27.2)

so we have derived $Q(\overline{d}) = -\frac{1}{3}(0-1) = \frac{1}{3}$! The fractional charge of quarks is related to the number of colors, and the relation comes out right. Such an embedding of quarks and leptons into one simple group explains why $Q(e^-) = -Q(p)$, $i.e.$ why charge is quantized, and why atoms are neutral. Just from this result the reader can see why it is difficult to believe there is not some validity to the GUT approach.

27.2   Unification of Forces

In Chapter 20, we learned that the couplings α_1, α_2, and α_3 satisfy, approximately, an equation

$$\frac{1}{\alpha_i(M^2)} = \frac{1}{\alpha_i(\mu^2)} + \frac{b_i}{4\pi} \ln \frac{M^2}{\mu^2} , \tag{27.3}$$

where M is the mass scale (or momentum transfer scale) at which we want to calculate α_i, μ is a scale where the coupling is measured, and b_1, b_2, and b_3 are calculated for the $U(1)$, $SU(2)$, and $SU(3)$ interactions. In Chapter 20, we looked in some detail at how the equivalent b for the electromagnetic interaction arose.

We also wrote the coefficients b_2 and b_3. They are

$$b_3 = 11 - \frac{4n_F}{3} , \tag{27.4}$$

$$b_2 = \frac{22}{3} - \frac{4n_F}{3} . \tag{27.5}$$

Here n_F is the number of families with mass $M_F \lesssim M$ so that they enter the loop in Figure 20.2 (M_F is a typical family mass). One point has to be clarified in order to write the correct equation for α_1. There really are three closely related $U(1)$ groups that get discussed. One is the $U(1)$ of electromagnetism. One is the $U(1)$ of Chapters 6 and 7 with the associated boson B_μ. The third differs from the second only in normalization. That difference arises because the hypercharge generator (Y) was normalized in Chapter 7 by a particular argument, and it happens not to be a normalization that is the same as the one for the $SU(2)$ and $SU(3)$ generators. When we put all the groups into $SU(5)$, we have to correct the normalization for the $U(1)$ part to make it consistent with the others. We will distinguish the couplings with the following notation; all of the couplings vary with q^2. We write $\alpha = e^2/4\pi$, $\alpha_1 = g_1^2/4\pi$ where g_1 is the coupling of Chapter 7, $g_1 = e/\cos\theta_w$. And we define $\alpha_1' = \frac{5}{3}\alpha_1$; α_1' is the coupling that is properly normalized for $SU(5)$. The factor $\frac{5}{3}$ is calculated in Section 3. The coefficient b_1' is then

$$b_1' = \frac{-4n_F}{3} . \tag{27.6}$$

The negative contributions in equations (27.4)–(27.6) arise from fermion loops and the positive contributions from the gauge boson loops (gluons for $SU(3)$, W's for $SU(2)$).

If it makes sense to discuss unifying the forces, their strengths have to be related in some sense. Since the α_i vary differently with mass scale, we can ask if at some mass they are equal. Define

$$\alpha_5 = \alpha_1' \left(M_G^2 \right) = \alpha_2 \left(M_G^2 \right) = \alpha_3 \left(M_G^2 \right) \tag{27.7}$$

to be the value of the couplings when they come together (if they do), at a mass M_G.

Using two of the three equations (27.3), we have

$$\frac{1}{\alpha_2(\mu^2)} + \frac{b_2}{4\pi} \ln \frac{M_G^2}{\mu^2} = \frac{1}{\alpha_3(\mu^2)} + \frac{b_3}{4\pi} \ln \frac{M_G^2}{\mu^2} \ . \tag{27.8}$$

Rearranging gives

$$\frac{1}{\alpha_2(\mu^2)} - \frac{1}{\alpha_3(\mu^2)} = 2 \frac{b_3 - b_2}{4\pi} \ln \frac{M_G}{\mu} \tag{27.9}$$

and

$$b_3 - b_2 = 11 - \frac{22}{3} = \frac{11}{3} \ , \tag{27.10}$$

(note this is independent of n_F so only the gauge boson loops directly enter). Solving,

$$\ln \frac{M_G}{\mu} = \frac{6\pi}{11} \left(\frac{1}{\alpha_2(\mu^2)} - \frac{1}{\alpha_3(\mu^2)} \right) \tag{27.11}$$

which determines M_G. This equation is independent of μ in principle; if μ changes so do α_2 and α_3 and the result for M_G will not vary. In practice because of measurement errors, there is some sensitivity, which is enhanced because M_G depends exponentially on the result. If we take $\mu = 10$ GeV, $\alpha_2(\mu) = 1/30$, and $\alpha_3(\mu^2) = 0.10$, we have

$$\ln \frac{M_G}{M_W} \simeq 34.25 \ ,$$

so

$$M_G \simeq 7.5 \times 10^{15} \ \text{GeV} \ . \tag{27.12}$$

The precise number should not be taken too seriously since it depends exponentially on what we put in for α_2 and α_3. The main point is if the couplings come together it is at a very large mass scale, or at a very short distance. Only there (about 10^{-29} cm) are the forces of particle physics finally the same strength.

Equating α_1', α_2, and α_3 at M_G gives two equations. We have put in two pieces of measured data, $\alpha_2(\mu^2)$ and $\alpha_3(\mu^2)$, and then solved one equation for the value of M_G. Now we could solve the other equation for $\alpha_1(\mu^2)$ and check that it is consistent. In practice, the numerical results are so sensitive to effects we have left out (such as Higgs bosons and threshold effects in the loop diagram calculations for the b_i and the precise numerical values of $\alpha_2(\mu^2)$ and $\alpha_3(\mu^2)$) that we would only obtain approximate agreement. When the precise calculation is done M_G comes out to be $3.6 \pm 3 \times 10^{14}$ GeV, and the consistency check for α_1 agrees with its measured value within errors.

Graphically we have approximately the situation in Figure 27.1, since each $1/\alpha_i$ is a linear function of $\ln(M/\mu)$:

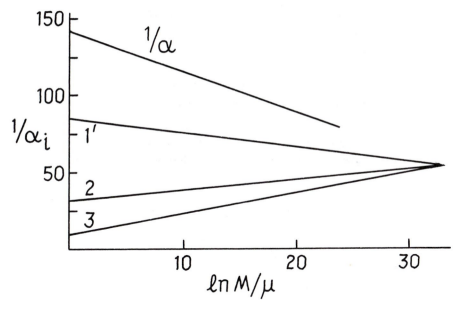

Figure 27.1
Behavior of the coupling strengths with mass scale.

The slopes are determined by the b_i, *i.e.* the gauge fields and the particular group structure. If the intercepts for $M \simeq \mu$ are right, the lines for $1/\alpha_i$ can come together.

27.3   Calculation of $\sin^2 \theta_w$

If $SU(2)$ and $U(1)$ are subgroups of a larger group, then the ratio of the couplings g_1 and g_2 is determined. That means that the weak mixing angle θ_w of Chapter 7 is no longer a free parameter to be fixed by experiment. Since θ_w is well measured [$\sin^2 \theta_w \simeq 0.23$ as discussed in Chapter 11], obtaining the correct value is a strong constraint on any attempt at unifying $SU(2)$ and $U(1)$. There are two parts to the calculation. First, the symmetry structure enters, but that determines $\sin^2 \theta_w$ where the symmetry holds, at a mass scale M_G. Then to see what $\sin^2 \theta_w$ is at our mass scale ($\sim M_W$), we have to express $\sin^2 \theta_w$ in terms of the couplings and use equation (27.3).

For the symmetry part, the important step is to write the electric charge in terms of the $SU(5)$ generators T_a. We know we can write

$$Q = T_3 + cT_1 , \qquad (27.13)$$

where T_3 is the $SU(2)$ diagonal generator and T_1 is the $U(1)$ generator, and both are consistently normalized $SU(5)$ generators; c is a constant to be determined, which is not necessarily unity because the normalization of the $U(1)$ generator relative to the $SU(n)$ generators was not fixed before. For the $SU(5)$ generators, the covariant derivative is (by analogy with what we have written earlier for $SU(2)$ and $SU(3)$)

$$\partial^\mu - ig_5 T_a V_a^\mu = \partial^\mu - ig_5(T_3 W_3^\mu + T_1 B^\mu + \ldots) \qquad (27.14)$$

where V_a^μ are the $SU(5)$ gauge bosons. Note there is only one coupling, g_5, for all interactions.

Using equations (7.12), (7.13), and (7.19),

$$\begin{aligned} B^\mu &= A^\mu \cos \theta_w + Z^\mu \sin \theta_w , \\ W_3^\mu &= -A^\mu \sin \theta_w + Z^\mu \cos \theta_w ; \end{aligned} \qquad (27.15)$$

then equation (27.14) gives as the coefficient of A_μ

$$\begin{aligned} -g_5 T_3 \sin \theta_w &+ g_5 T_1 \cos \theta_w \\ &= -g_5 \sin \theta_w (T_3 - \cot \theta_w\, T_1) \\ &= eQ . \end{aligned} \qquad (27.16)$$

Thus we can identify

$$e = g_5 \sin \theta_w \tag{27.17}$$

and

$$c = - \cot \theta_w \ . \tag{27.18}$$

Finally, we have to calculate c from the properties of $SU(5)$. We use a theorem, that for any representation R of a simple group,

$$\text{Tr}_R T_a T_b = N_R \delta_{ab} \ . \tag{27.19}$$

This says take any two generators, T_a and T_b. Take the trace of their product over a particular representation R. The answer is zero unless $a = b$, in which case it is a number N_R that depends on the representation but not on a or b. [For ordinary angular momentum this says $\text{Tr } J_x^2 = \text{Tr } J_y^2 = \text{Tr } J_z^2$ and $\text{Tr } J_x J_y = 0$, etc.   For spin $\text{Tr } J_i^2 = 1/2$ for the $J = 1/2$ representations, $\text{Tr } J_i^2 = 2$ for $J = 1$ representations, etc., so $N_{1/2} = 1/2$ and $N_1 = 2$.] We apply this to Q^2,

$$\text{Tr } Q^2 = \text{Tr}(T_3 + cT_1)^2 = \text{Tr } T_3^2 + c^2 \text{Tr } T_1^2 \tag{27.20}$$

where $\text{Tr } T_3 T_1 = 0$ has been used. Putting $\text{Tr } T_1^2 = \text{Tr } T_3^2$ and solving gives

$$1 + c^2 = \frac{\text{Tr } Q^2}{\text{Tr } T_3^2} \ . \tag{27.21}$$

For our fundamental representation, the 5–plet of equation (27.1),

$$\text{Tr } Q^2 = 0 + 1 + 3\left(\frac{1}{9}\right) = \frac{4}{3} \ , \tag{27.22}$$

$$\text{Tr } T_3^2 = \frac{1}{4} + \frac{1}{4} + 0 + 0 + 0 = \frac{1}{2} \tag{27.23}$$

so

$$1 + c^2 = \frac{8}{3} \ . \tag{27.24}$$

Using equation (27.18) then gives

$$\sin^2 \theta_w = \frac{1}{1 + c^2} = \frac{3}{8} \ . \tag{27.25}$$

Thus it is an $SU(5)$ prediction that $\sin^2 \theta_w = 3/8$.

This can not yet be compared to experiment since it is a prediction for the mass scale where the $SU(5)$ symmetry is good, while we know the $SU(5)$ symmetry is badly broken at the weak scale or at low energies, where the forces do not have the same strengths. To see how to calculate the prediction at lower scales, we just have to express $\sin^2 \theta_w$ in terms of the couplings.

The Lagrangian of Chapter 7 had a term

$$g_1 \overline{\psi} \gamma^\mu \frac{Y}{2} \psi B_\mu \qquad (27.26)$$

which becomes

$$g_1 \overline{\psi} \gamma^\mu (Q - T_3) \psi B_\mu \qquad (27.27)$$

using $Q = T_3 + Y/2$. Putting $T_1 = (Q - T_3)/c$ from equation (27.13) into equation (27.14) gives a term

$$\frac{1}{c} g_5 \overline{\psi} \gamma^\mu (Q - T_3) \psi B_\mu \qquad (27.28)$$

so by comparison

$$\left. \begin{array}{c} g_5 = c g_1 \; , \\[2mm] \alpha_5 = c^2 \alpha_1 \; . \end{array} \right\} \qquad (27.29)$$

Thus, at the unification scale, we have $\alpha_1 = \alpha_5/c^2$, while $\alpha_2 = \alpha_5$, the difference only being due to the historical normalization choice for α_1 made in Chapter 7, as discussed in the first section of this chapter.

From equation (7.19), evaluating $\sin^2 \theta_w$ and all the couplings at any given scale,

$$\sin^2 \theta_w = \frac{g_1^2}{g_1^2 + g_2^2} = \frac{\alpha_1}{\alpha_1 + \alpha_2} = \frac{1}{1 + \dfrac{\alpha_2}{\alpha_1}} \; . \qquad (27.30)$$

Now we are ready to evaluate $\sin^2 \theta_w$ at low energies, because we have it expressed as a function of the couplings and we know how they vary. First, to check, at the unification scale using equation (27.29) in equation (27.30) gives $\sin^2 \theta_w = 1/(1 + c^2)$ as in equation (27.25). Now to evaluate α_1 and α_2 at low energies, we should put in α_5 and calculate with equation (27.3) plus equation (27.4) to low energies. Since the reverse calculation already has been done, we know the answers are that (*e.g.* from Chapter 7)

$$\left.\begin{array}{l} \alpha_1(M_W^2) \simeq 0.010 \\ \alpha_2(M_W^2) \simeq 0.032 \end{array}\right\} \tag{27.31}$$

so we predict

$$\sin^2 \theta_w \simeq \frac{1}{1 + \frac{0.032}{0.010}} \approx 0.23 . \tag{27.32}$$

We see that $\sin^2 \theta_w$ has changed from 3/8 at the unification scale to about 0.23 at the weak scale. Note that this is a rather large change, *i.e.* $(3/8 - 0.23)/(3/8) = 0.39$. $\text{Sin}^2 \theta_w$ will also vary between M_W and a few GeV, but not very much so we will not work at the level of precision needed to calculate that variation.

To put it differently, equation (27.30) shows how to calculate $\sin^2 \theta_w$ at any scale, given α_1 and α_2 at that scale. There is a normalization difference between α_1 and α_2 because of the arbitrary normalization of the $U(1)$ part of the Lagrangian for the Standard Model. At the unification scale, we have

$$\alpha_2(M_G^2) = \frac{\alpha(M_G^2)}{\sin^2 \theta_w(M_G^2)}$$

and

$$\alpha_1'(M_G^2) = \frac{5\alpha(M_G^2)}{3\cos^2 \theta_w(M_G^2)} = \frac{5}{3}\alpha_1(M_G^2) ,$$

the 5/3 coming from equations (27.29) and (27.21). To calculate $\sin^2 \theta_w$ at any other scale we have to use equation (27.30) and equation (27.3) for α_1 and α_2, putting in M_G. Since we already know that what comes out is consistent with the values of α_1 and α_2 at low energies, we can use equation (27.31) without further arithmetic. When everything is done with proper care and precision, for the simplest model $\sin^2 \theta_w$ comes out to be about 0.215.

To understand why $\sin^2 \theta_w$ changes, *i.e.* why the couplings change differently, it is necessary to draw the diagrams [like Figure 20.2 that led to $\alpha(q^2)$] that lead to $\alpha_1(q^2)$ and $\alpha_2(q^2)$. $SU(5)$ has $5^2 - 1 = 24$ gauge bosons that occur in the loops. Since only eight gluons $+ W^\pm + Z + \gamma = 12$ of them are observed, the others must be heavy and as $-q^2$ decreases they no longer contribute importantly to the loops. Because the couplings have different strengths, the loop contributions that become unimportant have different numerical values for α_1,

α_2, and α_3, giving different values for the α_i. To understand quantitatively why α_1 decreases while α_2 increases, we would have to examine the calculations in more detail than is appropriate here.

The history is amusing. By 1974, when grand unified theories were first emerging, $\sin^2 \theta_w$ had been measured only crudely and had a value of about 0.35 with large errors, so 3/8 was considered a significant success. Then in 1975, it was realized the prediction must be corrected for the variation in coupling constants, and that the correct prediction was about 0.20–0.23 depending on what q^2 in the low energy region was used, in some disagreement with experiment. As data improved from 1975–1985, the reported values steadily decreased toward the theoretical prediction, until at present they agree to within about ±0.01 in $\sin^2 \theta_w$. The prediction is a strong constraint on possible grand unified theories; unfortunately it is not unique to the $SU(5)$ example we have considered, but can emerge from a variety of grand unification models that differ in the unification group and possibly other ways.

27.4   Proton Decay

When up and down quarks are put in $SU(2)$ doublets, the gauge bosons $W^{\pm}$ act as "raising" or "lowering" operators causing transitions from $u \rightarrow d$, $W^+ + d \rightarrow u$ or $W^- + u \rightarrow d$. Similarly, when quarks and leptons are put into an $SU(5)$ multiplet and the $SU(5)$ theory is made a gauge theory [i.e. one invariant under local $SU(5)$ phase tranformations], $SU(5)$ gauge bosons arise. Some of them are the familiar ones γ, $W^{\pm}$, Z, and g but the rest can cause transitions quarks↔leptons. If the quarks in a nucleon can turn into leptons, then the proton can decay, and proton number (baryon number in general, but proton number in practice since it is the only stable hadron) is not conserved. We expect, of course, quarks to turn into leptons in a way that conserves color since $SU(3)$ is a part of the $SU(5)$ invariant theory, and electric charge will be conserved just as it is in the Standard Model.

To find the quantum numbers of the new bosons, we can proceed by analogy with smaller groups. For $SU(2)$, for fermions in doublets to couple to gauge bosons, we have $2 \times \overline{2} = 1+3$ and the W's are in the triplet. For $SU(3)$, it becomes $3 \times \overline{3} = 1+8$ and the gluons are in the octet. For $SU(5)$, it is $5 \times \overline{5} = 1+24$. We can trace the quantum numbers by remembering that 5 contains $(2,1)+(1,3)$ where the quantities in brackets are $(SU(2)$ multiplicity, $SU(3)$ multiplicity). Then $\overline{5}$ contains $(\overline{2},1)+(1,\overline{3})$ so $5 \times \overline{5}$ contains $(2 \times \overline{2},1)+(1,3 \times \overline{3})+(2,\overline{3})+(\overline{2},3)$. Multiplying out gives $(1,1)+(3,1)+(1,1)+(1,8)+(2,\overline{3})+(\overline{3},2)$. The singlet under both is the 1 in $1+24$, and changes no quantum numbers. The

$(3, 1)$ and $(1, 1)$ are the W^i_μ and B_μ, and the $(1, 8)$ are the gluons. The remaining states, an $SU(2)$ doublet of color triplets and their antiparticles, are the new bosons. They are usually denoted as

$$\begin{pmatrix} Y^\alpha \\ X^\alpha \end{pmatrix},$$ (27.33)

with electric charges $Q_Y = -1/3$, $Q_X = -4/3$.

The new vertices in the theory are shown in Figure 27.2. As always the lines can be reversed by putting particle↔antiparticle. Now any process can occur that can be drawn with the vertices of Chapter 7 plus these. In particular, one possible transition is shown in Figure 27.3. It gives rise to the decay $p \rightarrow e^+\pi^0$, which is not allowed in the Standard Model because the diagrams cannot be constructed, but is allowed here.

Our usual technique to estimate the width is not expected to apply here, because hadronic binding effects are involved. Since we expect M_X to be large, and presumably of the order of the grand unification scale we found in Section 1, we only need a very crude answer and we can proceed as follows. The matrix element must have a factor g_5^2/M_G^2, so the width has a factor g_5^4/M_G^4. By dimensions, the width must be proportional to a mass, and the only mass that could be relevant is the proton mass, M_p. Thus, up to a numerical factor, the width must be

$$\Gamma_{p \rightarrow e^+\pi^0} \simeq \frac{\alpha_5^2 M_p^5}{M_G^4}.$$ (27.34)

The numerical factor will be proportional to the probability of two quarks being in the same place so they can annihilate, which is significantly less than unity. [Such factors are measured by decays of mesons like $\pi^\pm \rightarrow \mu^\pm \nu$, $\rho^0 \rightarrow e^+e^-$, etc. For mesons in the one GeV range the region over which there is a large probability of finding the two quarks is on the order of 10^{-3} GeV^3.] So the lifetime $\tau \simeq 1/\Gamma$ will be even longer than indicated by equations (27.34). Careful and precise calculations, valid to better than an order of magnitude, have been done for $\Gamma_{p \rightarrow e^+\pi^0}$ and confirm our qualitative arguments.

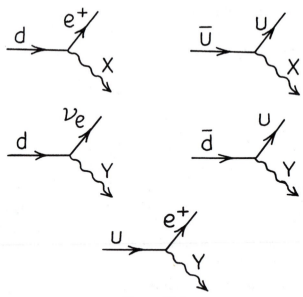

Figure 27.2
New interactions
with the X and Y particles.

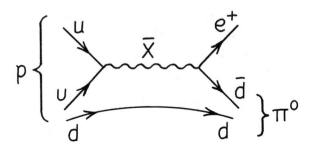

Figure 27.3
Feynman diagram
that allows the proton to decay.

Numerically, since τ_p varies as M_G^4, it is very sensitive to the value of M_G. If M_G increases by a factor of 3, τ_p increases by about 81 (α_5 would also change, as would various corrections, so the precise factor would have to be calculated). Using $M_G = 3.6 \times 10^{14}$GeV, we get $\tau_p \sim 3.6 \times 10^{38}$ sec, or about 10^{31} years. The universe has a lifetime of about 10^{10} years, so although protons would not be stable they would appear very stable on the scale of the lifetime of the universe. That is consistent with our perception of them as stable.

To look for proton decay, then, it is necessary to get together a large number of them and carefully observe them. One cm^3 of H_2O contains about 6×10^{23}

nucleons, so a cube 10 meters on a side contains almost 10^{33} nucleons. That is about $10,000$ tons of water!

The signature of the $p \rightarrow e^+ \pi^0$ decay is photons, since $\pi^0 \rightarrow \gamma\gamma$ and e^+ gives a Čerenkov cone. Thus the detector must be sensitive to very few photons, and the water must be so pure that a decay anywhere inside the chamber is visible to phototubes anywhere around the walls. Other decays are possible as well; different grand unified theories predict different patterns of decays.

The experiment has been underway since about 1982, and so far there is no evidence for proton decay. When careful calculations are carried out for the simplest $SU(5)$ grand unified model that can be constructed, with minimal structure, it predicts a lifetime of about 10^{30} years, and that is now excluded by experiment. Supersymmetric grand unified theories tend to predict longer lifetimes, 10^{32}–10^{33} years, and are at present not inconsistent with the data.

Experiments will eventually achieve a level of over 10^{32} years, either finding a signal or setting limits. Beyond that level intrinsic backgrounds that look just like the signal are too frequent to allow a signal to be seen. Such backgrounds arise from, for example, an upper atmosphere cosmic ray collision which produces pions, which then decay to neutrinos, followed by neutrino interactions that produce the same final signature as a real decay and cannot be separated by analysis.

27.5   The Baryon Asymmetry

There is one piece of evidence that has tentatively convinced many people that baryon number is indeed not conserved. That is the apparent fact that the universe has an asymmetry, with the number of baryons minus the number of antibaryons, n_B, relatively large. The direct evidence for that result is good at least to scales of the size of the cluster of galaxies we are in; otherwise both cosmic ray antimatter and x–rays from matter–antimatter annihilation would have been observed. Indirect arguments lead to the result $n_B/n_\gamma \simeq 10^{-10}$, $i.e.$ there is one baryon for each 10^{10} photons in the universe. This number is about eight orders of magnitude larger than has been obtained in any theory without violation of baryon number conservation.

A result such as $n_B/n_\gamma \sim 10^{-10}$ can be obtained, starting from the beginning of the big bang with no net baryon number, if three conditions are satisfied. It is necessary that (i) the basic vertices from the Lagrangian violate conservation of baryon number, (ii) the basic processes have to violate CP invariance (see Chapter 24), and (iii) the universe cannot be in thermodynamic equilibrium while

(i) and (ii) hold. We have seen that the first of these occurs in the theories that unify quarks and leptons. The second was described in Chapter 24; it is observed at today's energies, so it is very plausible that it holds at a fundamental level of the theory. However, there is not yet any unique connection known between the interactions of the heavy gauge bosons and Higgs bosons of $SU(5)$ at the high temperatures of the early universe, and the interaction of the W's at the weak scale. The third condition is necessary because if the system were in equilibrium, then every process would go in both directions, so a process creating net baryon number (*e.g.* $e^+\overline{d} \rightarrow u+u$) would be balanced by one going the other way, unless the universe had cooled before the second one could occur, so the second one became less probable. The standard big bang theory of the expanding universe provides what is needed for point (iii).

There are many subtle points that have to be included before a convincing case can be made that the baryon asymmetry did arise from an initial symmetry. If physics is to be able to include the origin of the universe in its domain of explanation, probably an explanation of the baryon asymmetry is required, so it is encouraging that the technical possibility exists. Finding a signal for proton decay would be of great importance in achieving these goals.

Suggestions for Further Study

Several Scientific American articles, Unified Theory of Elementary Particle Interactions by Georgi, The Decay of the Proton by Weinberg, The Search for Proton Decay by LoSecco, Reines, and Sinclair, and The Cosmic Asymmetry Between Matter and Antimatter by Wilczek, cover various aspects of the material of this chapter. The $SU(5)$ example of Section 27.1 is further described in the Physics Today article by Georgi and Glashow. The calculation of $\sin^2\theta_w$ of Section 27.2 can be followed in the paper of Georgi, Quinn, and Weinberg. The idea that a baryon asymmetry could arise by the mechanism advocated here was first discussed as a hypothetical possibility by Sakharov, before any theory in which the effect could occur existed.

CHAPTER 28

Supersymmetry

Supersymmetry is the name given to a hypothetical symmetry of nature. Basically it is a symmetry which relates fermions and bosons. Just as there are operators that change neutron → proton, or $e^- \rightarrow \nu_e$, we can postulate the existence of operators that change bosons into fermions,

$$Q\,|b\rangle = |f\rangle \,, \tag{28.1}$$

with a conjugate operator going the opposite way. Q leaves all quantum numbers unchanged except spin. It has been shown that mathematically consistent, supersymmetric, quantum field theories can be constructed. The motivations for studying supersymmetric theories, and for hoping that nature utilizes them, are quite strong. However, at the present time there is no experimental evidence that nature is supersymmetric. In this chapter we will discuss a few aspects of supersymmetry, partly for its own sake, and partly because it is a typical example of how the Standard Model gives us the tools to quantitatively test whether additional physics is present. A chain of arguments very much like the ones we will write can be made for any hypothetical gauge theory extension of the Standard Model.

If the Standard Model were part of a supersymmetric theory, with the symmetry not broken at all, it would be very obvious. Every one of the quarks, leptons, and gauge bosons would have a partner, generated by using equation (28.1) or its equivalent for fermions, that differed in spin but was otherwise identical. Some of the states are listed in Table 28.1. Supersymmetric partners are denoted by a ~. They are usually named by attaching -ino for a gauge boson, or s- for a fermion.

If there were an unbroken supersymmetry, then many phenomena would occur. There would be a super-hydrogen atom with $\widetilde{e}$ bound to a proton. The chemistry of multiselectron atoms, with bosons rather than fermions bound to the nucleus, would be very different. There would be additional weak interactions, with $\widetilde{W}$ and $\widetilde{Z}$ exchanged, and so on. Clearly none of these things happen, and nature does not have an unbroken supersymmetry.

Table 28.1
Supersymmetric States

particle	supersymmetric partner	spin of partner	name
γ	$\widetilde{\gamma}$	1/2	photino
e_L	$\widetilde{e}_L$	0	selectron
u_R	$\widetilde{u}_R$	0	up squark
g	$\widetilde{g}$	1/2	gluino
ν_μ	$\widetilde{\nu}_\mu$	0	muon sneutrino
$\vdots$	$\vdots$	$\vdots$	$\vdots$

Since we know of the broken symmetry of the electroweak theory, perhaps it is reasonable to also assume that the supersymmetry is broken. Just as with the fermion masses in the Standard Model, a supersymmetric theory can be written that allows the superpartners to have arbitrary masses, but no one has found a way to calculate the masses. At present one can only search for the superpartners in whatever mass range is accessible to experiment. Just as in the Standard Model, once one assumes mass values for the superpartners, the theory is fully predictive; all rates can be calculated.

To calculate in the supersymmetric Standard Model, we need the Feynman rules. It is clear what they are. We just take the rules for the Standard Model and replace the particles by their partners in pairs, keeping the coupling strengths the same. The replacement has to be in pairs since otherwise the number of half-integral spin particles would be odd, and it would be impossible to conserve angular momentum in a transition. Then we see, for example, that the full theory has vertices as in Figure 28.1.

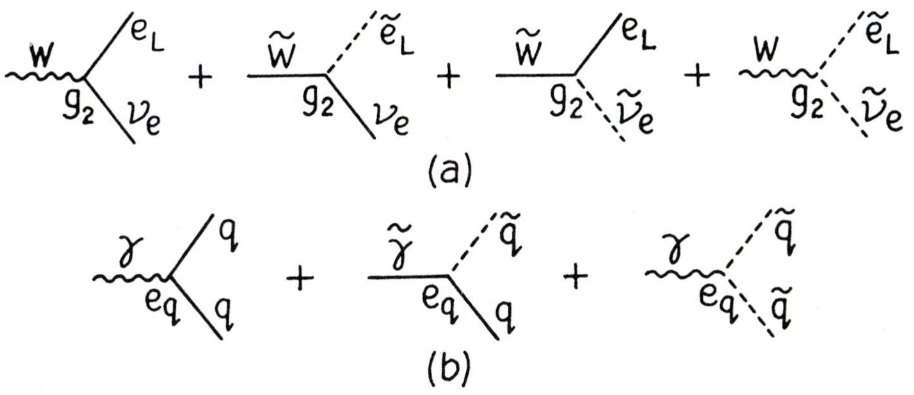

Figure 28.1
Vertices in the supersymmetric Standard Model.

In addition to the interaction of a photon with quarks, there is a quark-squark-photino interaction and a photon-squark-squark interaction. The strengths of all the gauge couplings are just the measured ones we already know, because the measured couplings would know about the existence of the supersymmetric theory even if we don't. [Because the couplings change with momentum transfer, as described in Chapter 20, if the superpartners were very much heavier than M_W there would be differences in the couplings. They could be calculated by an analysis like that of Chapter 20.] We have skipped over the space–time dependence of the vertices in Figure 28.1, which changes as the spin changes, because it is not needed to obtain our semi–quantitative estimates of rates. If it were necessary to know the space–time dependence one would have to go back and construct the full Lagrangian, which would then generate the appropriate space-time dependence; it is usually the simplest possibility that occurs.

Since all the vertices involve superpartners in pairs, we can draw three important conclusions for a normal supersymmetric theory,

(*i*) supersymmetric partners will be produced in pairs starting from normal particles,

(*ii*) the decay of supersymmetric partners will contain a supersymmetric partner,

(*iii*) the lightest supersymmetric partner will be stable.

28.1   Production and Detection
of Supersymmetric Partners

Starting from beams of quarks and leptons, we can draw a variety of diagrams to produce superpartners. Some are shown in Figure 28.2.

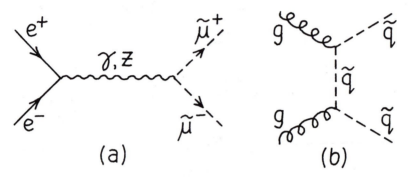

Figure 28.2

Some ways to produce supersymmetric partners,
(a) at an e^+e^- collider, (b) at a hadron collider.

The production cross sections involve the same couplings we are used to, so the cross sections are typical of production rates for W's, quarks, etc., except that there is phase space suppression if the superpartners are heavy. Next we have to ask how the partners would act once they are produced. For simplicity let us assume that gluinos are heavier than squarks, so the decay $\tilde{q} \to q\tilde{g}$ is not allowed by energy conservation, and that photinos are lighter than squarks and than zinos and winos. Then the dominant decays for any sfermion with electric charge will be $\tilde{f} \to f+\tilde{\gamma}$; e.g. $\tilde{\mu} \to \mu+\tilde{\gamma}$, or $\tilde{d} \to d+\tilde{\gamma}$. As we have learned, typical decay widths for a superpartner of mass $\widetilde{M}$ will be $\widetilde{\Gamma} \simeq \alpha\widetilde{M}$. With $\widetilde{M} \sim$ tens of GeV, $\widetilde{\Gamma}$ is of order 0.1-1 GeV, so the associated lifetimes are short compared to 10^{-20} sec, and only the decay products emerge into the detector.

To complete the analysis, it is necessary to decide which will be the lightest supersymmetric particle (LSP) since all the others will decay into it. There are several possibilities; we will assume it is the photino for simplicity. If some other superpartner were lighter than the photino we could go through a similar analysis; details change, but qualitative conclusions do not.

Since all the superpartners that are produced will decay in a very short time, only normal particles plus the photino will enter the detector. To detect the presence of supersymmetry we must be able to detect the $\tilde{\gamma}$. To see how to do that we have to study how it interacts.

The $\tilde{\gamma}$ will interact by hitting a quark in the detector, as in Figure 28.3, and exciting a $\tilde{q}$.

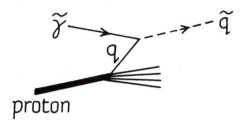

Figure 28.3

A photino interacting.

The $\tilde{q}$ could be real or virtual depending on the available energy. For simplicity we assume the $\tilde{q}$ is real. The cross section for this is one we have already learned to write in Chapters 10 and 18. It is

$$\sigma = \sum_q \int dx \; q(x) \; \hat{\sigma}(\hat{s}) \tag{28.2}$$

where x is the fraction of the proton's momentum carried by the quark, $q(x)$ is the quark structure function already discussed in Chapters 10 and 18, and $\hat{\sigma}$ is the constituent cross section for $\tilde{\gamma} + q \rightarrow \tilde{q}$. There is a sum over all the quarks in the proton. The square of the center of mass energy of the $\tilde{\gamma}$ and the q is $\hat{s}$, so $\hat{s} = \widetilde{M}^2$ where $\widetilde{M}$ is the squark mass. Also, $\hat{s} = xs$, where s is the square of the center of mass energy of the photino and proton.

The matrix element is approximately

$$M \simeq e_q e \bar{u} u \tag{28.3}$$

where e_q is the quark charge ($2/3$ or $-1/3$). As usual we can replace the spinors by the appropriate mass, $\bar{u}u \simeq \widetilde{M}$. Then proceeding just as for W production in Chapter 10,

$$\hat{\sigma} \simeq \pi \, e_q^2 \, e^2 \, \delta(\hat{s} - \widetilde{M}^2) \, . \tag{28.4}$$

Writing $\hat{s} = xs$, this is $\hat{\sigma} = \pi \, e_q^2 \, e^2 \, \delta(x - \widetilde{M}^2/s)/s$. Inserting $\hat{\sigma}$ in (28.2), the delta function allows the integral to be done, so

$$\sigma \approx \frac{4\pi^2 \alpha}{\widetilde{M}^2} \sum_q e_q^2 \, x \, q(x) \, , \tag{28.5}$$

where we have replaced s by $\widetilde{M}^2/x$. The factor $\sum\limits_{q} e_q^2\, x\, q(x)$ is just the structure function $F_2(x)$ from Chapter 18, so

$$\sigma(\widetilde{\gamma}p) \approx \frac{4\pi^2\alpha}{\widetilde{M}^2} F_2(\widetilde{M}^2/s)\;. \tag{28.6}$$

Note that although we are working in a hypothetical theory, we have calculated the photino interaction cross section in terms of familiar quantities, plus an assumed squark mass. To estimate $\sigma(\widetilde{\gamma}p)$ numerically, we need to pick a value for $\widetilde{M}$. Analyses such as the one we are going through have been done, and currently imply that a signal for a squark would have been seen if $\widetilde{M} \leq 70$ GeV, so for a qualitative estimate we assume $\widetilde{M} \simeq M_W$. Looking up F_2 in the Particle Data Tables we find that over a range of x in the region of $x \sim 0.1$, $F_2 \simeq 0.15$. Then $\sigma(\widetilde{\gamma}p) \simeq 2.5 \times 10^{-33}$ cm^2. This is typical of a neutrino cross section, about 10^{-7} of a pion cross section. A typical $\widetilde{\gamma}$ will not interact in a detector— it will escape, carrying away momentum. Thus the experimental signature of supersymmetry is an event where apparently momentum is not conserved. Such events can also occur if neutrinos are produced, for example in decays of W's or of heavy quarks, but then a charged lepton is also produced. If events are ever discovered with apparent failure of conservation of momentum, and no charged leptons, they could be the signal of supersymmetry. Then detailed analysis can establish whether they could in fact come from production of superpartners. The relative rates for various processes, the distribution of missing momentum from large to small, and a number of other quantitative predictions can all test whether a supersymmetric interpretation is possible.

At present, searches for superpartners have been carried out at PETRA, PEP, and the CERN collider. No signals have been seen. Analyses indicate that a signal would have been seen if $\widetilde{M}_q \leq 70$ GeV, $\widetilde{M}_g \leq 65$ GeV, $\widetilde{M}_e \leq 65$ GeV, $\widetilde{M}_\mu \leq 22$ GeV, $\widetilde{M}_\tau \leq 22$ GeV, $\widetilde{M}_\nu \leq 1.5$ GeV; in some cases there are minor caveats. Basically these numbers can be taken as lower limits on the masses superpartners could have. New searches will be done at TRISTAN, SLC, and FNAL during 1987-1988.

The absence of a signal is not surprising even if nature were supersymmetric, since the existing limits are still below M_W, where a naïve guess might have expected superpartners to appear. We would have been very lucky if a detectable superpartner had been light enough to be found in searches before now. The main lesson to learn, from the point of view of this book, is that we were able to develop quantitative tests for the presence of supersymmetry, even though it is a hypothetical theory. The logic is very general. New particles can be produced in detectable quantities only if they carry at least one of the charges that allows them to couple to the particles of the Standard Model. They must carry electric charge, or color charge, or weak charge. If they do, we can compute their production rates and decay properties with Standard Model techniques, and make quantitative tests of new ideas.

28.2   The Lightest Supersymmetric Particle and Dark Matter

At the end of the introductory section of this chapter we noted that the LSP would normally be a new stable particle. It often happens that an extension of the Standard Model will lead to a new stable particle. In this section, we will examine some of the consequences of the existence of such an object; we will focus on the LSP, but the kinds of consequences obtained may be rather general. [It is technically possible to construct a theory where the LSP is not absolutely stable, but it is rather unnatural, and we want to concentrate on what happens if there is a new stable particle, so we will assume the LSP is stable].

In the early universe, at temperatures where all the particles were relativistic, the LSP would be in equilibrium with the other particles. There would be as many of them as there were quarks, etc. As the universe cooled, especially if the LSP has a mass in the several GeV range, as might be expected if the general supersymmetry mass scale were of order M_Z, many LSP's would be left over. Some would annihilate, but a number perhaps of order the number of baryons would be left. If the mass of the LSP were several times a baryon mass, they might provide significantly more of the total mass of the universe than baryons do. In a complete theory, there would presumably be a relation between the mass of the LSP and the proton mass, so the amounts of mass in each kind of particle would not be arbitrary.

The LSP's would behave very differently from baryons. Since their interactions are typically electroweak, with low energy cross sections in the 10^{-38} cm^2 range, they would not undergo nuclear reactions and form luminous stars—they would be "dark matter." They would still have normal gravitational interactions, of course, so they would concentrate near galaxies, but because their interactions are so weak they would also not lose energy easily, and would probably be spread out in spherical halos.

Given this chain of reasoning, it is exciting that astronomers have in recent years found evidence, from the behavior of galaxies, that galaxies contain several times more matter than the amount we can see in the form of stars. It is far from established that such matter must be a new form of weakly interacting particle, but that is a possible interpretation of the data. In addition, if it is believed that the universe is flat, *i.e.* just closed, then considerable matter is required that cannot be in the form of baryons.

If the universe were full of such particles, it is interesting to see how we could observe them in the laboratory. First, they could be observed in collider experiments where sleptons or squarks or gluinos were directly produced, and the existence and mass of the LSP were inferred from the event structure.

Second, the dark matter particles are spread more or less uniformly throughout our galaxy, and we are moving through them as the earth moves. A quark in a nucleus in a detector could collide with an LSP and the energy transferred to the nucleus could be detected. Let us have a velocity v as we move through the galaxy, and let an LSP of mass $\widetilde{M}$ collide with a nucleus of mass M. Then the momentum transfer could be as much as $\Delta p \leq 2\widetilde{M}v$, and the energy transfer $\Delta E \leq 4\widetilde{M}^2 v^2/2M$. Our velocity relative to the galaxy is $v \sim 10^{-3}c$. To get a feeling for the numbers, assume $\widetilde{M} \sim M$ and ask for $\Delta E > 1$ KeV. Then we need $\widetilde{M} > 1/2$ GeV, which is indeed the range of interest.

To see an effect it is not only necessary to transfer enough energy, but to have events occur at a reasonable rate. The interaction rate will be the product of three factors, $R = \text{flux} \times \sigma \times N$, where the flux is the number density of LSP's times the relative velocity v, σ is the cross section for an LSP to interact with a nucleus, and N is the number of target nuclei in a detector; N is of order Avagadro's number.

We can estimate the number density from the astronomical numbers. Newton's law gives for particles moving in the galaxy that $v^2/r = \widetilde{M}G_N/r^2$ so the mass density is $\rho = \widetilde{M}/(\frac{4\pi}{3}r^3) = v^2/(\frac{4\pi}{3}r^2 G_N)$. As above, $v \simeq 10^{-3}c$ and we are at a radius in the galaxy of $r \simeq 10$ kpc $\simeq 3 \times 10^{22}$ cm. Then $\rho \simeq 2 \times 10^{-24}$ gm/cm^3, *i.e.* a density of about 1 GeV/cm^3. If the matter were in protons there would be about 1 proton per cubic cm; we can put approximately $n = \rho/\widetilde{M} \simeq 1/\text{cm}^3$ if $\widetilde{M}$ is of order a GeV.

Finally, σ is the interaction cross section. Just on dimensional grounds it must be about

$$\sigma \simeq G_F^2 \, \widetilde{M}^2 \, A^2$$

where the last factor occurs because the interaction is very nonrelativistic, so the wavelength is of order the size of the nucleus and the interaction can be coherent over A nucleons in a nucleus. If we multiply all these factors a number of order 10^{-2}/sec interactions emerges, which is large enough to allow some optimism. A number of people are currently trying to make detectors to find dark matter.

Problems

28.1:   Draw one or two diagrams that lead to gluino production at a hadron collider. Estimate, qualitatively, the cross section for gluino production. Draw a diagram for gluino decay. The diagrams can be constructed using Standard Model vertices and replacing pairs of particles by their supersymmetric partners, as in the first section of this chapter. What would events of gluino production look like in a detector?

28.2:   What normal Standard Model processes have the same signature as supersymmetry? Find at least two, one involving $W^{\pm}$ production and $\tau^{\pm}$ decay, and other involving Z^0 production. How could a signal for supersymmetry be detected in spite of them?

28.3:   Estimate the effect on Γ_Z if sneutrinos were light, $\widetilde{M}_\nu \ll \frac{1}{2}M_Z$.

Suggestions for Further Study

There is a Scientific American article on the motivation for supersymmetry by Freedman and van Nieuwenhuizen, and one along the lines of this chapter by Haber and Kane.

Neutrino

Masses?

The neutrinos (ν_e, ν_μ, and ν_τ) have masses small compared to those of the other fermions. Only upper limits are presently known. Current measurements give $m_{\nu_e}/m_e \leq 4 \times 10^{-5}$, $m_{\nu_\mu}/m_\mu < 2.4 \times 10^{-3}$, and $m_{\nu_\tau}/m_\tau < 4 \times 10^{-2}$; any or all of the neutrinos could have zero mass.

There is no understanding as to why neutrino masses should be small. As we saw in Chapter 7, the photon mass is zero because there is an unbroken $U(1)$ symmetry, associated with the linear combination of generators $T_3 + Y/2$. In Chapter 8 we saw that a broken global symmetry could give rise to massless Goldstone bosons. For neutrinos there is no known reason to think either of these mechanisms operates.

In Chapter 7 we saw that a fermion mass term could be written using the Higgs doublet if left-handed and right-handed fermions of a given flavor existed; $g\overline{f}_L \phi f_R$ becomes a mass term when ϕ gets a vacuum expectation value. A mass obtained from such an interaction is called a Dirac mass. For this mechanism to apply, there must be right–handed neutrinos; none are presently known to exist. Detecting or producing a right-handed $SU(2)$ singlet neutrino is not easy, since it has no interactions with $W^\pm$ (because $W^\pm$ only interacts with members of a doublet), it has no interactions with Z or γ (since Q and T_3 are both zero), and it has no interactions with gluons (since it is uncolored). If no right-handed neutrinos existed, neutrinos would not get Dirac masses. We still would not understand, however, why right-handed fermions exist for the other fermions but not for neutrinos.

There is another kind of mass term that can be written for neutrinos since they are electrically neutral. The term $m\overline{\nu}_L \nu_L^c$, where ν^c is the charge-conjugate state, satisfies the conditions to be a mass term in the Lagrangian. A fermion that has the property that its charge conjugate is equal to itself is called a Majorana fermion. If neutrinos are Majorana particles they could also have masses described by such a term. There is no known reason why they should not. As remarked in Chapter 24, the charge conjugate of a left–handed fermion is right–handed.

Thus, with or without right-handed neutrinos, it is surprising that neutrinos do not have masses like those of the other fermions. Many models have been constructed in which neutrinos do have mass. None of them are yet theoretically compelling, so we will not examine any of them. Rather, we will study what might be observed if neutrinos did have mass. A variety of experiments are in progress or planned to detect neutrino masses if they are non–zero, so some effect could be discovered in the next few years.

29.1   If $m_\nu \neq 0$; Neutrino Oscillations

There is no sure way to detect neutrino masses. However, one method is very promising and also interesting to examine, so let us consider it. Suppose some or all of the neutrinos do get mass by some mechanism. For simplicity, consider only ν_e and ν_μ. Just as for quarks in Chapter 22, there is no apparent reason why the weak interaction eigenstates should be identical with the mass eigenstates.

If the weak eigenstates and the mass eigenstates are not the same, we will label the mass eigenstates as ν_1 and ν_2, and the weak eigenstates as ν_e and ν_μ. We want to study the time dependence of the states, so we write $\nu_i(t)$ for any of them. We can guarantee that a beam of ν_μ is produced at $t = 0$ by producing charged pions, which decay mainly by $\pi^+ \to \mu^+ \nu_\mu$ or $\pi^- \to \mu^- \overline{\nu}_\mu$. In principle we can detect the μ^+ so we know a ν_μ was emitted.

The weak eigenstates must be some linear combination of the mass eigenstates; at $t = 0$,

$$\nu_\mu(0) = \nu_1(0) \cos \alpha + \nu_2(0) \sin \alpha$$
$$\nu_e(0) = -\nu_1(0) \sin \alpha + \nu_2(0) \cos \alpha$$

$$(29.1)$$

where α is an angle that parameterizes the mixing. If the interaction that gave rise to the masses was known, α could be calculated; we will assume α is to be measured. If $m_{\nu_i} = 0$ there is no way to distinguish weak eigenstates from mass eigenstates, so the states could always be expanded in a new set and the angle α rotated to zero. A non-zero α implies that some neutrino masses are nonzero and that the mass eigenstates are not degenerate. Since α expresses a relation between various eigenstates, it is not time dependent.

The mass eigenstates vary with time as

$$\nu_i(t) = e^{-iE_i t}\nu_i(0) \tag{29.2}$$

since they are free particles after they are produced, each with energy E_i. Thus

$$\nu_\mu(t) = e^{-iE_1 t}\nu_1(0)\cos\alpha + e^{-iE_2 t}\nu_2(0)\sin\alpha \ . \tag{29.3}$$

We can express $\nu_1(0)$ and $\nu_2(0)$ in terms of $\nu_e(0)$ and $\nu_\mu(0)$ from equation (29.1). That gives $\nu_1(0) = \nu_\mu(0)\cos\alpha - \nu_e(0)\sin\alpha$, and $\nu_2(0) = \nu_\mu(0)\sin\alpha + \nu_e(0)\cos\alpha$. Using these in equation (29.3) gives

$$\nu_\mu(t) = e^{-iE_1 t}\left(\nu_\mu(0)\cos^2\alpha - \nu_e(0)\cos\alpha\sin\alpha\right)$$
$$+ e^{-iE_2 t}\left(\nu_\mu(0)\sin^2\alpha + \nu_e(0)\sin\alpha\cos\alpha\right) \tag{29.4}$$

$$= \left(e^{-iE_1 t}\cos^2\alpha + e^{-iE_2 t}\sin^2\alpha\right)\nu_\mu(0)$$
$$+ \sin\alpha\cos\alpha\left(e^{-iE_2 t} - e^{-iE_1 t}\right)\nu_e(0) \ . \tag{29.5}$$

Since $E_1 = \sqrt{m_1^2 + p^2}$ and $E_2 = \sqrt{m_2^2 + p^2}$, where p is the momentum of the state, the second term is non-zero if $m_1 \neq m_2$. Thus a state that begins as a pure ν_μ has, at later times, some ν_e mixed in!

To calculate the probability that an initial beam of ν_μ later contains some ν_e, we just square the overlap. [We can think of the states we have been writing as kets, $|\nu_\mu(t)\rangle$, $|\nu_e(0)\rangle$, etc.]. Since $\nu_\mu(0)$ is orthogonal to $\nu_e(0)$,

$$P(\nu_\mu \to \nu_e) = |\langle \nu_e(0)|\nu_\mu(t)\rangle|^2$$
$$= \sin^2\alpha\cos^2\alpha|e^{-iE_2 t} - e^{-iE_1 t}|^2 \ . \tag{29.6}$$

Using $\sin 2\alpha = 2\sin\alpha\cos\alpha$, and

$$|e^{-iE_2 t} - e^{-iE_1 t}|^2 = 2\left(1 - \cos(E_2 - E_1)t\right) \ , \tag{29.7}$$

gives

$$P(\nu_\mu \to \nu_e) = \frac{1}{2}\sin^2 2\alpha\left[1 - \cos(E_2 - E_1)t\right] \ . \tag{29.8}$$

Remarkably, the probability that an initial beam of ν_μ will contain some ν_e is an oscillatory function of time! In principle it is a rather easy effect to look for experimentally. If the beam is composed purely of ν_μ , then all collisions will produce muons, as in Figure 29.1.

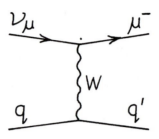

Figure 29.1
A pure ν_μ beam
produces muons.

If some of the beam has converted to ν_e, then some electrons will be produced, as in Figure 29.2.

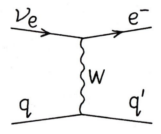

Figure 29.2
A ν_e beam will produce
electrons.

So any electrons produced by interactions in a ν_μ beam could signal oscillations described by equation (29.8).

In practice there are problems at a certain level of sensitivity. Some stray ν_e could be in the beam; probably it would be very hard to guarantee a beam without a percent or so of ν_e. A muon, produced as in Figure 29.1, will sometimes decay before it enters the detector, $\mu \to e\nu\bar{\nu}$, so it will appear as an electron. Various other experimental problems occur. So there is a limit of order a few percent on how small an effect could be observed.

If an effect is observed, equation (29.8) itself allows very clean checking. By changing either the energy difference or the time of observation the size of the effect can be controlled, and to be considered a real effect it must vary as shown in equation (29.8). The argument of the $\cos(E_2 - E_1)t$ is often written in a simplified form to see explicitly how it varies. In a time t, the beam will go a distance $x \simeq ct$. The energy difference can be written

$$
\begin{aligned}
E_2 - E_1 &= p\sqrt{1 + m_2^2/p^2} - p\sqrt{1 + m_1^2/p^2} \\
&\simeq p\left\{ \left(1 + \frac{m_2^2}{2p^2}\right) - \left(1 + \frac{m_1^2}{2p^2}\right) + \ldots \right\} \\
&\simeq \frac{m_2^2 - m_1^2}{2p}.
\end{aligned}
\tag{29.9}
$$

Then

$$
\cos(E_2 - E_1)t \simeq \cos 2\pi x/L
\tag{29.10}
$$

where $L = 4\pi p/(m_2^2 - m_1^2)$. L is an effective length that determines the distance over which one might expect to see an effect.

Suppose, for example, that $\Delta m^2 \equiv m_2^2 - m_1^2$ is 10eV^2. If we want L to be of order 100 m to allow an effect to show up at an accelerator, then, since 100 m $= 5 \times 10^8 \text{eV}^{-1}$, we need a momentum p of about $4 \times 10^8 \text{eV} \simeq 400$ MeV. This is typical of neutrino energies in the secondary beams at Brookhaven National Laboratory, where such experiments have been underway for some time. To see larger or smaller Δm^2, a larger or smaller momentum is needed. Because of that, finding Δm^2 much less than 10 eV^2 is probably not possible at accelerators, since the required momenta are too small. Reactors produce neutrinos with energies of a few MeV, so they allow probing somewhat smaller Δm^2. To confirm an effect, the detector is moved from one x to another at fixed L (*i.e.*, fixed momentum), and the rate must vary as $\cos 2\pi x/L$. If Δm^2 is very large, L becomes small compared to detector size even for the largest momenta; the cosine oscillates and averages to zero so an effect can be observed.

Note that the detectability depends on $m_2^2 - m_1^2$, not on m_2 or m_1 themselves. In practice there is another complication because there are of course three neutrinos and mixing could occur among all three. Then the appropriate formulas are considerably more complicated, with three mixing angles and three Δm_i^2. Conceptually the situation is similar to the simple case we have discussed, so we will not go into more detail. Finally, note that the effect is always proportional to $\sin^2 2\alpha$, so if the mixing is too small no effect will be observed, even if it is present, whatever Δm^2 is.

29.2   Solar Neutrinos

The processes which cause the sun to shine include some in which neutrinos (ν_e) are emitted, such as

$$pp \rightarrow D + e^+ + \nu_e, \qquad (29.11)$$

or

$$p + {}^7\text{Be} \rightarrow {}^8\text{B} + \gamma$$
$$\phantom{p + {}^7\text{Be}} \hookrightarrow {}^8\text{Be} + e^+ + \nu_e. \qquad (29.12)$$

These neutrinos escape from the sun, and there is some flux of them at the earth. In a remarkable experiment, ν_e from reaction (29.12) have been detected.

The neutrinos from reaction (29.12) constitute only about one in 10^4 of all those from the sun, but they are the most energetic. They are detected using the process

$$\nu_e + {}^{37}\text{Cl} \rightarrow {}^{37}\text{Ar} + e^-. \qquad (29.13)$$

The apparatus is about a mile underground, to reduce background events where a cosmic ray muon strikes a proton or nucleus which produces ${}^{37}\text{Ar}$ in a collision. In a large tank (hundreds of tons) of chlorine, about one atom of ${}^{37}\text{Ar}$ should be produced every few days (!) if the theoretical predictions are correct. To extract this tiny number of atoms from the huge sample, ${}^4\text{He}$ gas is circulated through the tank every month or so (the half life of ${}^{37}\text{Ar}$ is 35 days, which sets the scale) to capture the ${}^{37}\text{Ar}$ atoms. Then they are counted by observing ${}^{37}\text{Ar}$ electron capture decay. The system is calibrated by controlled experiments. The scale and precision of the entire experiment are extraordinary.

The theoretical prediction has fluctuated a bit but is about 6 ± 1.5 SNU, where a "SNU" is a "solar neutrino unit";

$$1 \text{ SNU} = 10^{-36} \text{ captures per target particle per second.} \qquad (29.14)$$

The experiment reports a result of 2.2 ± 0.4 SNU. Interpreting the result is not easy. While the theoretical prediction has been carefully studied, it depends on factors such as solar models, the temperature in the sun, amounts of heavy metals in the sun, various nuclear reaction cross sections, etc. Probably additional experiments are required to determine whether there is really a disagreement between theory and data.

We have discussed solar neutrinos in this chapter because an intriguing possibity is that there is a discrepancy between the theoretical and experimental numbers, and that some of the ν_e oscillate into ν_μ or ν_τ on the way to the earth. Then they would not participate in the weak interaction of equation (29.13), which requires an electron neutrino, so ν_μ and ν_τ are not counted.

In Section 29.1, we implicity assumed that the ν_i were propagating in vacuum. If they propagate in matter, as in the sun, a subtle effect occurs that can enhance the probability of an oscillation, allowing large probabilities even for small mixing angles α. The effect can occur because ν_e can interact with electrons in two ways, while ν_μ can only interact one way (see problem 5 in Chapter 7). The interactions behave somewhat like masses and can modify the oscillation rate (though they can not cause oscillations if none are present in vacuum). It turns out that for Δm^2 of order 10^{-5} eV^2, the amount of matter traversed by solar neutrinos can give probabilities near unity even for small $\sin^2 2\alpha$.

More experiments and analyses are in progress to clarify this situation.

29.3   Measurement of m_ν in Decays

Another way to look for neutrino masses is to use the kinematics of explicit decays. Many decays emit neutrinos, such as

$$\pi^+ \to \mu^+ \nu_\mu \, , \tag{29.15}$$

or

$$n \to p e \nu_e \, , \tag{29.16}$$

or

$$\tau \to \pi \nu_\tau \, . \tag{29.17}$$

There is a maximum energy allowed for the other final particles, μ^+ or e^- or π above, depending on the masses of all particles. So far all processes studied have given results consistent with ν_μ and ν_τ being zero, with the accuracies mentioned at the beginning of this chapter. For ν_e the most sensitive measurements may come from tritium β-decay. A Soviet group has reported a non-zero lower limit on m_{ν_e} from studying β–decay, but other experiments have so far not succeeded in reproducing that result. Several experiments in progress will either find a non-zero value of order 10-30 eV for m_{ν_e} or set an upper limit of about 10 eV.

29.4   Expectations for m_ν

There are no arguments for the values of m_{ν_i} that are convincing to any group of people. Any result could emerge, including zero. However, one particular range of values has interesting cosmological implications; it is perhaps worthwhile to look at the argument.

The ratio for the number of baryons in the universe to the number of photons in the universe is measured to be about $n_B/n_\gamma \simeq 10^{-10}$. One would expect crudely that as the universe cooled, photons and neutrinos decoupled at similar temperatures (*i.e.* they were unable to initiate a reaction because they had lost too much energy due to the expansion of the universe) and therefore $n_\nu \sim n_\gamma$. However, most other species annihilate to photons, increasing n_γ but not n_ν, so finally we might guess $n_\nu \sim n_\gamma/10$. The fraction of the critical mass (*i.e.*, the amount of mass in the universe just necessary to close the universe by gravitational attraction) taken up by baryons is, to about a factor of 2, observed to be $\Omega_B \simeq 0.05$.

Using these numbers, if neutrinos have mass then the fraction of the critical mass taken by neutrinos can be estimated by multiplying

$$\Omega_\nu \simeq \frac{n_\nu m_\nu}{m_{\text{crit}}} \simeq \frac{n_\gamma m_\nu}{10 m_{\text{crit}}} \simeq 10^9 \left(\frac{n_B m_B}{m_{\text{crit}}} \right) \frac{m_\nu}{m_B}$$
$$= 10^9 \, \Omega_B \, m_\nu/m_B \, . \tag{29.18}$$

If we assume that neutrinos provide enough mass to close the universe, then $\Omega_\nu \approx 1$, so

$$m_\nu \sim m_B/10^9 \Omega_B \simeq 2 \times 10^{-8} m_{\text{proton}}$$
$$\simeq 20 \text{ eV}. \tag{29.19}$$

This argument is meant only to indicate the interest of this range, not to imply that neutrinos have such masses. The conclusion is that, fortuitously, precisely the range of masses $m_\nu \geq 1$ eV that accelerators and reactors are sensitive to is the range of great cosmological interest. On the other hand, if the solar neutrino results are due to oscillations it probably means neutrino masses are too small to have direct cosmological significance.

Suggestions for Further Study

Good technical treatments of all the subjects of this chapter can be found in Cummins and Buchsbaum. Additional aspects of the study of neutrino masses, such as the implications of neutrinoless double beta decay, are also covered there. The resonant effect of the matter in the sun on neutrino oscillations is described clearly by Bethe, with references to the original literature. The beautiful experiment that has detected some solar neutrinos is described further by Davis.

Angular Momentum
and Spin
and $SU(2)$

Although we have asssumed that most readers are familiar with the treatment of spin in nonrelativistic quantum theory, this appendix provides a brief pedagogical summary for those whose memory needs refreshing. A familiarity with spin is very important for two reasons. First, the fundamental matter particles, quarks and leptons, are spin-1/2 states, and their spin enters into the structure of the theory in absolutely essential ways. Second, it seems as if nature knows about group theory; the Hamiltonian (Lagrangian) of the theory is invariant under several transformations that form groups. Luckily all that is needed to understand the group structure is a familiarity with the structure of the theory in spin space, which is an example of an $SU(2)$ group of transformations. From there it is possible to proceed by analogies. Appendix B provides a summary of the group theory ideas.

Since $\overrightarrow{L} = \overrightarrow{r} \times \overrightarrow{p}$, in quantum theory the angular momentum operators satisfy commutation relations

$$[L_i, \ L_j] = i\epsilon_{ijk}L_k \tag{A.1}$$

where i, j, and $k = 1$, 2, and 3 or x, y, and z. Since $[L_i, \ L^2] = 0$, L^2 and one of the L_i (usually L_z) can be simultaneously diagonalized. The eigenfunctions of L^2 and L_z are the orthonormal spherical harmonics $Y_{\ell m}(\theta, \ \phi)$.

The structure of the angular momentum theory is one that we want to generalize to more abstract situations. With any Hermitian operator $\sum$, we can associate a Hermitian matrix

$$\Sigma_{\ell'm',\ell m} = \int d\Omega \; Y^*_{\ell'm'}(\theta, \phi) \sum Y_{\ell m}(\theta, \phi). \qquad (A.2)$$

One way to think about it is to specify ℓ and ℓ' and construct a $(2\ell+1)$ by $(2\ell'+1)$ matrix for each ℓ, ℓ'.

Suppose we choose $\sum$ to be angular momentum operators with $Y_{\ell, m}$ as eigenstates. Then $\ell' = \ell$ or the result is zero. For $\ell = 0$ the result is one number. For $\ell = 1$ we have m, $m' = 1$, 0, -1 so the result is a 3×3 matrix. For L^2 the result is (remember, $\hbar = 1$)

$$L^2 = 2 \begin{pmatrix} 1 & & \\ & 1 & \\ & & 1 \end{pmatrix}, \qquad (A.3)$$

with zeros in the blank spots, since $L^2 Y_{\ell m} = \ell(\ell + 1)Y_{\ell m}$ and $\ell = 1$; we have used the orthonormality of the $Y_{\ell m}$. Similarly, for L_z, the result obviously is

$$L_z = \begin{pmatrix} 1 & & \\ & 0 & \\ & & -1 \end{pmatrix}. \qquad (A.4)$$

For L_x and L_y, the commutation relations can be used to derive

$$L_x = \frac{1}{\sqrt{2}} \begin{pmatrix} & 1 & \\ 1 & & 1 \\ & 1 & \end{pmatrix}, \qquad (A.5)$$

$$L_y = \frac{i}{\sqrt{2}} \begin{pmatrix} & -1 & \\ 1 & & -1 \\ & 1 & \end{pmatrix}. \qquad (A.6)$$

It is easy to check that the commutation relations are satisfied for these matrices.

These are matrix representations of the angular momentum operators. If we had chosen $\ell = 2$ we would have had 5×5 matrices, etc. There are an infinite number of possible representations of a given operator.

It is interesting that the representations we have found all have an odd number of rows and columns. By using the classical correspondence $\vec{L} = \vec{r} \times \vec{p}$ we will never get a 2×2 representation. However, by just writing 2×2 Hermitian matrices and substituting into the commutation relations an acceptable representation is quickly obtained,

$$L_z = \frac{1}{2} \begin{pmatrix} 1 & 0 \\ 0 & -1 \end{pmatrix}, \tag{A.7}$$

$$L^2 = \frac{3}{4} \begin{pmatrix} 1 & 0 \\ 0 & 1 \end{pmatrix}, \tag{A.8}$$

$$L_x = \frac{1}{2} \begin{pmatrix} 0 & 1 \\ 1 & 0 \end{pmatrix}, \tag{A.9}$$

$$L_y = \frac{1}{2} \begin{pmatrix} 0 & -i \\ i & 0 \end{pmatrix}. \tag{A.10}$$

In one sense this kind of angular momentum, which is called "spin", is an internal property of particles, different from the rotational angular momentum. In another sense it is not, since both are quantized in units of $\hbar$. The above argument shows that this internal spin exists in the theory. Whether it exists in nature is an experimental question; the answer is of course "yes". The electron and all fermions have spin $1/2$.

The matrices for spin $1/2$ are called the Pauli matrices, defined by

$$L_i = \frac{1}{2}\sigma_i . \tag{A.11}$$

When we are discussing spin we will call them σ_i; when we are using the same matrices to discuss an internal symmetry that has the same mathematical description as spin we will call them τ_i. Note that

$$\{\sigma_i, \ \sigma_j\} = \sigma_i\sigma_j + \sigma_j\sigma_i = 2\delta_{ij} , \tag{A.12}$$

$$[\sigma_i, \ \sigma_j] = 2i\epsilon_{ijk}\sigma_k \tag{A.13}$$

where ϵ_{ijk} is totally antisymmetric and $\epsilon_{123} = 1$.

The eigenstates are two component spinors. For example, with a, b as numbers,

$$\sigma_z \begin{pmatrix} a \\ b \end{pmatrix} = \begin{pmatrix} 1 & 0 \\ 0 & -1 \end{pmatrix} \begin{pmatrix} a \\ b \end{pmatrix} = \lambda \begin{pmatrix} a \\ b \end{pmatrix} \tag{A.14}$$

so

$$\begin{aligned} a &= \lambda\, a \\ -b &= \lambda\, b \end{aligned} \tag{A.15}$$

and solutions are $\begin{pmatrix} 1 \\ 0 \end{pmatrix}$, $\begin{pmatrix} 0 \\ 1 \end{pmatrix}$ with eigenvalues ± 1. The eigenstate $\begin{pmatrix} 1 \\ 0 \end{pmatrix}$ is referred to as "spin up" and sometimes denoted $|\uparrow\rangle$ while $\begin{pmatrix} 0 \\ 1 \end{pmatrix}$ is "spin down", $|\downarrow\rangle$. An arbitrary state can be expanded in terms of these,

$$|\psi\rangle = C_1 |\uparrow\rangle + C_2 |\downarrow\rangle , \tag{A.16}$$

so $|C_1|^2$ is the probability for finding spin up in a measurement, $|C_2|^2$ is the probability of finding spin down, and $|C_1|^2 + |C_2|^2 = 1$.

For a general quantum mechanical problem, we can write

$$i \frac{\partial \psi}{\partial t} = H\psi \tag{A.17}$$

where

$$\psi = \begin{pmatrix} \psi_1(\overrightarrow{x}, t) \\ \psi_2(\overrightarrow{x}, t) \end{pmatrix} \tag{A.18}$$

and ψ_i are normal wave functions. The Hamiltonian can be an operator in spin space but it must be invariant under rotations in spin space if angular momentum is conserved.

To understand how to construct invariant Hamiltonians note the important result that, in the appropriate sense, $\overrightarrow{\sigma}$ is a vector. Consider a general spinor as in equation (A.16). The complex numbers C_i can be parameterized completely generally as

$$\begin{aligned} C_1 &= e^{i\delta} e^{-i\phi/2} \cos\theta/2 , \\ C_2 &= e^{i\delta} e^{i\phi/2} \sin\theta/2 . \end{aligned} \tag{A.19}$$

Then

$$\psi^\dagger \sigma_1 \psi = \left(e^{-i\delta} e^{i\phi/2} \cos\left(\tfrac{\theta}{2}\right) \quad e^{-i\delta} e^{-i\phi/2} \sin\left(\tfrac{\theta}{2}\right) \right) \begin{pmatrix} 0 & 1 \\ 1 & 0 \end{pmatrix} \begin{pmatrix} e^{i\delta} e^{-i\phi/2} \cos\left(\tfrac{\theta}{2}\right) \\ e^{i\delta} e^{i\phi/2} \sin\left(\tfrac{\theta}{2}\right) \end{pmatrix}$$

$$= e^{i\phi}\cos\left(\frac{\theta}{2}\right)\sin\left(\frac{\theta}{2}\right) + e^{-i\phi}\sin\left(\frac{\theta}{2}\right)\cos\left(\frac{\theta}{2}\right)$$

$$= \sin\theta\,\cos\phi\,. \tag{A.20}$$

Similarly,

$$\psi^\dagger \sigma_2 \psi = \sin\theta\,\sin\phi\,, \tag{A.21}$$

$$\psi^\dagger \sigma_3 \psi = \cos\theta\,. \tag{A.22}$$

These are just the normal components of a unit vector with polar angle θ and azimuthal angle ϕ, so $\psi^\dagger \overrightarrow{\sigma} \psi$ can be interpreted as a unit vector in spin space. This fact is widely used in writing rotationally invariant Hamiltonians and Lagrangians. Given a vector, possible scalars can be constructed with other available vectors. For example, if a problem involves one other momentum $\overrightarrow{p}$ we can include a term $\overrightarrow{\sigma} \cdot \overrightarrow{p}$ in the Hamiltonian, where it is understood that $\overrightarrow{\sigma} \cdot \overrightarrow{p}$ always means $(\psi^\dagger \overrightarrow{\sigma} \psi) \cdot \overrightarrow{p}$. If two vectors $\vec{p}_1$ and $\vec{p}_2$ are available we can write $\overrightarrow{\sigma} \cdot \vec{p}_1 \times \vec{p}_2$, and so on. We will use such techniques extensively to construct Lagrangians invariant under transformations in a variety of internal spaces.

Finally we note that functions of operators or matrices frequently occur. Suppose we do a phase transformation where the phase is an operator in spin space,

$$\psi' = e^{i\vec{\sigma}\cdot\vec{\epsilon}}\psi. \tag{A.23}$$

The vector $\vec{\epsilon}$ is a set of three parameters ϵ_1, ϵ_2, ϵ_3 that determine the transformation, or we can write it as $\overrightarrow{\epsilon} = \epsilon\hat{n}$ in terms of a magnitude and a unit vector. All functions of matrices can be thought of as defined by their power series expansions. Those of the form of equation (A.23) are particularly simple, as we can see by writing it out,

$$e^{i\vec{\sigma}\cdot\hat{n}\epsilon} = 1 + i\overrightarrow{\sigma}\cdot\hat{n}\epsilon + \frac{(i\overrightarrow{\sigma}\cdot\hat{n}\epsilon)^2}{2!} + \frac{(i\overrightarrow{\sigma}\cdot\hat{n}\epsilon)^3}{3!} + \cdots \tag{A.24}$$

Now note

$$(\vec{\sigma}\cdot\hat{n})^2 = (\sigma_1 n_1 + \sigma_2 n_2 + \sigma_3 n_3)^2$$
$$= n_1^2 + n_2^2 + n_3^2 + n_1 n_2(\sigma_1\sigma_2 + \sigma_2\sigma_1) + \cdots$$
$$= 1 \tag{A.25}$$

using equation (A.12) and the unit vector property $n^2 = n_1^2 + n_2^2 + n_3^2 = 1$. Thus

$$e^{i\vec{\sigma}\cdot\hat{n}\epsilon} = [1 - \epsilon^2/2! + \epsilon^4/4! + \ldots] + i\vec{\sigma}\cdot\hat{n}[\epsilon - \epsilon^3/3! + \ldots] \qquad (A.26)$$

$$= \cos\epsilon + i\vec{\sigma}\cdot\hat{n}\sin\epsilon. \qquad (A.27)$$

We can write it out in more detail, putting in the Pauli matrices and the unit matrix,

$$e^{i\vec{\sigma}\cdot\hat{n}\epsilon} = \begin{pmatrix} \cos\epsilon + in_3\sin\epsilon & (in_1 + n_2)\sin\epsilon \\ (in_1 - n_2)\sin\epsilon & \cos\epsilon - in_3\sin\epsilon \end{pmatrix}. \qquad (A.28)$$

This is just a rotation matrix, since the components of n are $n_1 = \sin\theta\cos\phi$, $n_2 = \sin\theta\sin\phi$, $n_3 = \cos\theta$ if $\vec{\epsilon}$ has polar angle θ and azimuthal angle ϕ. For example, if $\theta = \phi = \pi/2$, $n_1 = n_3 = 0$ and $n_2 = 1$ and

$$e^{i\vec{\sigma}\cdot\hat{n}\epsilon} = \begin{pmatrix} \cos\epsilon & \sin\epsilon \\ -\sin\epsilon & \cos\epsilon \end{pmatrix}, \qquad (A.29)$$

a familiar rotation in the x-z plane by an angle ϵ.

The Pauli matrices plus the unit matrix form a group, as discussed in Appendix B. That is, they provide a set of transformations that connect one spin state to another, and the transformations generate a group (called $SU(2)$). Particles have, in addition to spin, other properties such that the same mathematical transformation structure connects different particle states. If one is familiar with the way spin and the Pauli matrices work, then one can by analogy work with all of the other properties of particles that transform similarly. In particular, one can write Hamiltonians (interaction Lagrangians) that are invariant under such transformations. The techniques are just the ones reviewed in this Appendix. In addition to $SU(2)$ transformations that are identical to the spin transformations, particles also know about $SU(3)$ transformations; working by analogy with $SU(2)$ we can easily construct all we need for using $SU(3)$, as is done in Appendix B.

Some
Group Theory

Since nature certainly knows about groups, it is helpful to know a little group theory when the laws of particle physics are being discussed. Here we will summarize some of the simple ideas that are relevant for this book.

A group is a set of elements plus a composition rule, such that:

(a) combining two elements under the rule gives another of the elements;

(b) there is an identity element I so that for any element E in the group of elements (where $\cdot$ represents the composition rule)

$$E \cdot I = I \cdot E = E \; ;$$

(c) every element E has a unique inverse E^{-1}, with

$$E \cdot E^{-1} = E^{-1} \cdot E = I \; ;$$

(d) and the composition rule is associative,

$$A \cdot (B \cdot C) = (A \cdot B) \cdot C \; .$$

For example, the set of all complex phase factors of a wave function, $U(\theta) = e^{i\theta}$ where θ is a real parameter, form a (continuous) group with multiplication for the composition rule. Checking (a),

$$U(\theta)U(\theta') = e^{i(\theta+\theta')} = U(\theta + \theta') \tag{B.1}$$

and $U(\theta + \theta')$ is an element of the group. $U(0)$ is an identity element. Since $U(\theta)U(-\theta) = U(-\theta)U(\theta) = U(0) = I$ for any θ, there is an inverse $U^{-1}(\theta) = U(-\theta)$. The associative law is satisfied since

$$\begin{aligned}
[U(\theta_1)U(\theta_2)]U(\theta_3) &= e^{i(\theta_1+\theta_2)}e^{i\theta_3} \\
&= e^{i(\theta_1+\theta_2+\theta_3)} \\
&= e^{i\theta_1}e^{i(\theta_2+\theta_3)} \\
&= U(\theta_1)[U(\theta_2)U(\theta_3)] \, .
\end{aligned} \tag{B.2}$$

In particle physics, groups enter because we can carry out transformations on physical systems and the physical systems often are invariant under the transformations. We will only consider continuous groups of transformations, where the parameters that describe the transformations are continuous variables, and can take on an infinite number of values (infinite continuous groups).

The group used as an example above is a one dimensional unitary group; it is given the name "$U(1)$". Each element is characterized by a continuous parameter θ, $0 \leq \theta \leq 2\pi$; θ can take on an infinite number of values. In addition, we can define

$$\begin{aligned}
dU = U(\theta + d\theta) - U(\theta) &= e^{i(\theta+d\theta)} - e^{i\theta} \\
&= e^{i\theta}(1 + id\theta) - e^{i\theta} \\
&= ie^{i\theta}\,d\theta \\
&= iU\,d\theta
\end{aligned} \tag{B.3}$$

so the elements are differentiable.

A Lie group is one where the elements E are differentiable functions of their parameters, as for the case we have just seen.

It can be shown that for a Lie group any element can be written in the form

$$E(\theta_1,\ \theta_2,\ \ldots\theta_n) = \exp\left(\sum_{i=1}^{n} i\theta_i F_i\right). \tag{B.4}$$

For n parameters there are n of the quantities F_i. They are called the generators of the Lie group. Physically they can be thought of as generating the transformations.

B.1   The $SO(n)$ Groups

The rotations in an n-dimensional Euclidean space form a group, called $O(n)$. The elements of $O(n)$ can be represented by $n \times n$ matrices, each with $n(n-1)/2$ independent elements. To derive that number, start with an $n \times n$ matrix R of real numbers, with n^2 elements. Since the rotations will be represented by an orthogonal matrix, $R^T R = 1$. $R^T R$ has n diagonal elements so there are n conditions of the form $d = 1$. The off-diagonal elements of $R^T R$ have to be zero; there are $(n^2 - n)/2$ of them, with the $1/2$ because those below the diagonal are not independent of those above it. Thus the number of independent elements is

$$n^2 - n - (n^2 - n)/2 = n(n-1)/2 . \tag{B.5}$$

If in addition the determinant of R is set to be $+1$ then the group is called $SO(n)$; the S is for "special." A determinant of -1 could have represented a transformation which also involved an inversion of the coordinate system.

The elements of $SO(2)$ are familiar; they are the rotations in a plane. Coordinates x and y rotate into x' and y' according to

$$\begin{pmatrix} x' \\ y' \end{pmatrix} = R \begin{pmatrix} x \\ y \end{pmatrix} , \tag{B.6}$$

$$R = \begin{pmatrix} \cos\theta & \sin\theta \\ -\sin\theta & \cos\theta \end{pmatrix} . \tag{B.7}$$

This group has one parameter. Note $x^2 + y^2$ is left invariant.

$SO(3)$ is a three parameter group, to describe rotations in three dimensions. One way to write the transformations is in terms of Euler angles,

$$R(\alpha, \ \beta, \ \gamma) = R_{z'}(0, \ 0, \ \alpha) \, R_y(0, \ \beta, \ 0) \, R_z(0, \ 0, \ \gamma) \tag{B.8}$$

where R_z is a rotation about the z axis by an angle γ, R_y a rotation about the y axis by an angle β, and then $R_{z'}$ is a rotation about the new z axis, the z' axis, by the angle α. This sequence can perform a general rotation. The separate rotations can be written in a simple way,

$$R_y(0,\ \beta,\ 0) = \begin{pmatrix} \cos\beta & 0 & -\sin\beta \\ 0 & 1 & 0 \\ \sin\beta & 0 & \cos\beta \end{pmatrix} \tag{B.9}$$

$$R_z(0,\ 0,\ \gamma) = \begin{pmatrix} \cos\gamma & \sin\gamma & 0 \\ -\sin\gamma & \cos\gamma & 0 \\ 0 & 0 & 1 \end{pmatrix}. \tag{B.10}$$

Note the $SO(3)$ rotations leave $x^2 + y^2 + z^2$ invariant.

In Appendix A it is shown that

$$e^{i\theta\sigma_2} = \cos\theta + i\sigma_2\sin\theta. \tag{B.11}$$

Using $\sigma_2 = \begin{pmatrix} 0 & -i \\ i & 0 \end{pmatrix}$, this gives

$$e^{i\theta\sigma_2} = \begin{pmatrix} \cos\theta & \sin\theta \\ -\sin\theta & \cos\theta \end{pmatrix}, \tag{B.12}$$

which according to equation (B.7) represents a rotation in two dimensions. Equation (B.12) is of the form of equation (B.4), and we see that σ_2 is the generator of rotations for $SO(2)$.

B.2   The $SU(n)$ Groups

Elements of the $SU(n)$ groups are represented by $n \times n$ unitary matrices, $U^\dagger U = 1$, with det $U = +1$.

They have $n^2 - 1$ independent parameters. Since they are $n \times n$, there are n^2 elements, each one being complex, so $2n^2$ parameters. $U^\dagger U = 1$ imposes n conditions for the diagonal elements and $2[(n^2 - n)/2]$ conditions for the independent off-diagonal complex elements. This gives $2n^2 - n - (n^2 - n) = n^2$ independent parameters, finally reduced by one since det U is fixed.

In the 2×2 case possible ways to parameterize U are

$$\begin{pmatrix} a & b \\ -b^* & a^* \end{pmatrix},$$

(B.13)

where a, b are arbitrary complex numbers and $|a|^2 + |b|^2 = 1$, or

$$\begin{pmatrix} \cos\theta e^{i\alpha} & \sin\theta e^{i\gamma} \\ -\sin\theta e^{-i\gamma} & \cos\theta e^{-i\alpha} \end{pmatrix}.$$

(B.14)

If H is a Hermitian matrix, then e^{iH} is unitary,

$$(e^{iH})^{\dagger}(e^{iH}) = e^{-iH^{\dagger}} e^{iH} = e^{-i(H-H^{\dagger})} = 1 \;.$$

(B.15)

Note that the second step is only correct for commuting matrices, which is all right here since H commutes with $H^{\dagger}$. Thus we can always write

$$U = e^{iH}.$$

(B.16)

The counting is consistent since there are n^2 Hermitian $n \times n$ matrices.

Then we can pick a particular set of n^2 Hermitian matrices, H_j, so any $n \times n$ unitary matrix can be written

$$U = \exp\left(\sum_{j=1}^{n^2} \theta_j H_j\right)$$

(B.17)

where the θ_j are real parameters. The n^2 H_j are the generators of the group $U(n)$. To specialize to $SU(n)$ one more step is needed, since in the present form the condition $\det U = 1$ is not satisfied.

To impose this condition we need the identity

$$\det e^A = e^{Tr A}$$

for any square matrix A. [To prove this, suppose A has been diagonalized, which is always possible. The diagonal elements are the eigenvalues λ_i of A. Then

$$e^A = 1 + A + A^2/2! + \ldots$$

$$= \begin{pmatrix} 1 + \lambda_1 + \lambda_1^2/2! \ldots & 0 & 0 & \ldots \\ 0 & 1 + \lambda_2 + \lambda_2^2/2! + \ldots & 0 & \ldots \\ 0 & 0 & & \\ \vdots & \vdots & & \end{pmatrix}$$

$$= \begin{pmatrix} e^{\lambda_1} & 0 & 0 & \ldots \\ 0 & e^{\lambda_2} & 0 & \ldots \\ \vdots & 0 & \ddots & \\ \vdots & \vdots & & e^{\lambda_n} \end{pmatrix}$$

so

$$\det e^A = e^{\lambda_1} e^{\lambda_2} \ldots e^{\lambda_n}$$
$$= e^{\lambda_1 + \lambda_2 + \ldots + \lambda_n}$$
$$= e^{TrA}$$

since the trace of A is the sum of its diagonal elements.]

Thus $\det U = 1$ requires $TrH = 0$ for every H, *i.e.* the matrices H must be Hermitian and traceless. The generators of $SU(n)$ are any set of $n \times n$ traceless Hermitian matrices.

B.3 $SU(2)$ and Physics

In the text we study the fact that quarks and leptons can be places in spinors in analogy to spin states $\begin{pmatrix} \uparrow \\ \downarrow \end{pmatrix}$. What do quark or lepton charge states and spin have to do with each other?

The set of all 2×2 unitary, unimodular, matrices form a group $SU(2)$. It depends on three continuous parameters. We have already expressed this in equations (B.13), (B.14) above. Further, this group is locally identical to the rotations in three dimensions, which form the group $SO(3)$. [They differ globally, *i.e.* for rotations that are not infinitesimal.] The two groups have the same Lie algebra, *i.e.* their generators have the same commutation relations.

What has happened is a fortunate simplification. The set of transformations on two objects has a mathematical structure that we can become familiar with in the case of spin. Since we do not have a deep understanding of why the Standard Model has an $SU(2)$ internal symmetry, we cannot know whether the presence of another $SU(2)$ is accidental or profound. In either case it is very helpful. We already have an intuition and some experience to guide us in treating the fermion states and the internal symmetries. In the next section we will look at $SU(3)$, which we need to describe the color symmetry. Our experience with $SU(2)$ helps greatly in understanding what we need of $SU(3)$.

B.4 $SU(3)$

The Pauli matrices have a simple generalization to $SU(3)$. For $SU(3)$ there are $3^2 - 1 = 8$ generators. They are called λ_a, $a = 1$, ..., 8. The commutation relations are

$$[\lambda_a, \lambda_b] = 2i f_{abc} \lambda_c$$

where

$$f_{123} = 1 \, ,$$
$$f_{458} = f_{678} = \sqrt{3}/2 \, ,$$
$$f_{147} = f_{516} = f_{246} = f_{257} = f_{345} = f_{637} = 1/2 \, ,$$

f_{abc} is totally antisymmetric, and the rest of the f's are zero. A simple representation for the λ's comes from embedding the Pauli matrices appropriately,

$$\lambda_1 = \begin{pmatrix} & 1 & \\ 1 & & \\ & & \end{pmatrix}, \; \lambda_2 = \begin{pmatrix} & -i & \\ i & & \\ & & \end{pmatrix}, \; \lambda_3 = \begin{pmatrix} 1 & & \\ & -1 & \\ & & \end{pmatrix},$$

$$\lambda_4 = \begin{pmatrix} & & 1 \\ & & \\ 1 & & \end{pmatrix}, \; \lambda_5 = \begin{pmatrix} & & -i \\ & & \\ i & & \end{pmatrix}, \; \lambda_6 = \begin{pmatrix} & & \\ & & 1 \\ & 1 & \end{pmatrix},$$

$$\lambda_7 = \begin{pmatrix} & & \\ & & -i \\ & i & \end{pmatrix}, \; \lambda_8 = \frac{1}{\sqrt{3}} \begin{pmatrix} 1 & & \\ & 1 & \\ & & -2 \end{pmatrix}.$$

Although we will rarely use the $SU(3)$ matrices, we will write them formally in a number of places; it may help if one keeps their properties and their simple forms in mind.

B.5   Abelian and Non-Abelian Groups

Finally we note the important distinction between groups where the transformations commute, $U_1 U_2 = U_2 U_1$, called Abelian groups, and the non-Abelian groups where the tranformations do not commute.

For example, if we have two transformations

$$R_1 = e^{\vec{\sigma} \cdot \hat{n} \epsilon_1} = \cos \epsilon_1 + i \vec{\sigma} \cdot \hat{n}_1 \sin \epsilon_1$$

and

$$R_2 = e^{i \vec{\sigma} \cdot \hat{n} \epsilon_2} = \cos \epsilon_2 + i \vec{\sigma} \cdot \hat{n}_2 \sin \epsilon_2$$

then using the commutation relations,

$$R_1 R_2 - R_2 R_1 = \sin \epsilon_1 \sin \epsilon_2 (\vec{\sigma} \cdot \hat{n}_2 \vec{\sigma} \cdot \hat{n}_1 - \vec{\sigma} \cdot \hat{n}_1 \vec{\sigma} \cdot \hat{n}_2)$$
$$= - \sin \epsilon_1 \sin \epsilon_2 \; n_{1i} n_{2j} (\sigma_i \sigma_j - \sigma_j \sigma_i)$$
$$= -2i \sin \epsilon_1 \sin \epsilon_2 \epsilon_{ijk} \; n_{1i} n_{2j} \sigma_k$$

which is another rotation. In general, when non-commuting operators are involved, the results depend on the order in which the operations are performed.

Some
Relativistic
Kinematics

Since the particles we consider are often moving at relativistic velocities, it is necessary to use relativistic kinematics to describe their motion. It is a basic point because finally much of what is really measured are the momenta and energies of some electrically charged particles. The rest is interpretation, and the kinematics is the first stage. In addition, relativistic kinematics is usually simpler than non–relativistic kinematics.

Consider a process $a + b \rightarrow c + d$. The particles have mass M_a, M_b, M_c, and M_d and four–momenta

$$p_a = (E_a; \vec{p}_a) , \tag{C.1}$$

$$p_b = (E_b; \vec{p}_b) , \tag{C.2}$$

$$etc.$$

Each four–momentum satisfies $p_i^2 = M_i^2$. Conservation of four–momentum gives $p_a + p_b = p_c + p_d$, four equations. Together, these give eight conditions. From the four–momenta, there are ten scalars that can be formed $[p_a^2, \ p_a \cdot p_b, \ p_a \cdot p_c, \ \ldots]$. Thus, there are two independent variables that describe the process. In non–relativistic quantum theory, they are often chosen to be the energy and the scattering angle. We will use instead Lorentz scalar variables, as is conventional in particle physics. Recall any scalar product is $a \cdot b = a^0 b^0 - \vec{a} \cdot \vec{b}$.

Define

$$s = (p_a + p_b)^2, \tag{C.3}$$
$$t = (p_c - p_a)^2, \tag{C.4}$$
$$u = (p_d - p_a)^2. \tag{C.5}$$

Only two of these can be independent, but it is normal to define the three variables for symmetry. A little algebra shows the relation is

$$s + t + u = M_a^2 + M_b^2 + M_c^2 + M_c^2. \tag{C.6}$$

Since s, t, and u are Lorentz scalars, they do not change from frame to frame. They can be evaluated in any frame that is convenient. We will work mainly in the center of mass system because we will mainly consider colliders.

We set

$$p_a = (E_a; 0, 0, p) \tag{C.7}$$
$$p_b = (E_b; 0, 0, -p) \tag{C.8}$$

so the initial particles move in the z–direction. That a and b have equal and opposite momenta defines the center of mass. The final particles can scatter at some angle, so

$$p_c = \left(E_c; \vec{p}'\right), \tag{C.9}$$
$$p_d = \left(E_d; -\vec{p}'\right), \tag{C.10}$$

where $\vec{p}'$ can be taken to be in the x–z plane,

$$\vec{p}' = (p' \sin\theta, 0, p' \cos\theta). \tag{C.11}$$

The angle θ is the normal scattering angle, $\cos\theta = \hat{p} \cdot \hat{p}'$.

Then, in the center of mass system,

$$s = (E_a + E_b)^2 = \left[\sqrt{M_a^2 + p^2} + \sqrt{M_b^2 + p^2} \right]^2 \qquad \text{(C.12)}$$

which can be solved for p,

$$p^2 = \frac{\left[s - (M_a + M_b)^2 \right] \left[s - (M_a - M_b)^2 \right]}{4s} . \qquad \text{(C.13)}$$

Similarly,

$$s = (E_c + E_d)^2 \qquad \text{(C.14)}$$

so

$$p'^2 = \frac{\left[s - (M_c + M_d)^2 \right] \left[s - (M_c - M_d)^2 \right]}{4s} . \qquad \text{(C.15)}$$

A little more algebra gives

$$E_a = \frac{(s + M_a^2 - M_b^2)}{2\sqrt{s}} \qquad \text{(C.16)}$$

$$E_c = \frac{(s + M_c^2 - M_d^2)}{2\sqrt{s}} , \qquad \text{(C.17)}$$

and E_b and E_d are given by the obvious modifications as can be confirmed from equations (C.12) and (C.14).

To bring in the scattering angle,

$$\begin{aligned} t &= M_c^2 + M_a^2 - 2p_c \cdot p_a \\ &= M_c^2 + M_a^2 - 2E_c E_a - 2\vec{p}' \cdot \vec{p} \\ &= M_c^2 + M_a^2 - 2E_c E_a + 2p'p \cos\theta \end{aligned} \qquad \text{(C.18)}$$

and so

$$u = M_d^2 + M_a^2 - 2E_d E_a - 2p'p \cos\theta . \qquad \text{(C.19)}$$

For a process where the masses are negligible,

$$E_a = E_b = p = E_c = E_d = p' = \frac{\sqrt{s}}{2} \qquad \text{(C.20)}$$

$$t = -\frac{s}{2}(1 - \cos\theta) , \qquad \text{(C.21)}$$

$$u = -\frac{s}{2}(1 + \cos\theta) . \qquad \text{(C.22)}$$

Differential cross sections are often written as $d\sigma/d\Omega$ where $d\Omega = d\phi\ d\cos\theta$. Sometimes it is convenient to write $d\sigma/dt$. At fixed energy, the relation is given by equation (C.18); for a process where the masses are negligible, there is just a factor $s/2$ relating the two cross sections.

The laboratory frame is the one where target particle b is at rest. Then $p_b = \left(M_b;\ \overrightarrow{0}\right)$ so

$$s = M_a^2 + M_b^2 + 2E_a E_b\ ,\tag{C.23}$$

and is determined by the energy E_a of the beam. The lab angle can be evaluated, for example, by equating values of t in the lab and center of mass systems.

The Point
Cross Section

Perhaps the most directly compelling evidence for quarks and their behavior as described by QCD is the size and angular distribution of the point cross section for $e^+e^- \rightarrow q\bar{q}$. As described in Chapter 15, the detector shows two cones of hadrons. The size of the cross section, and the distribution in the polar angle, θ, are exactly as expected for point–like particles of spin one–half.

Because of the presence of spin and relativistic kinematics, there does not appear to be any simple way to derive the angular distribution of the point cross section; either the full apparatus of the solutions of the Dirac equation, or helicity amplitudes, or even clumsier methods are needed. Because of the importance of the result, a derivation is given here.

Consider $e^+e^- \rightarrow \mu^+\mu^-$, with four momenta p, q, p', and q' respectively. The matrix element for figure (19.1) is

$$M = \frac{e^2}{s} \left(\overline{u}(p')\gamma^\mu v(q')\right)\left(\overline{v}(p)\gamma_\mu u(q)\right) \tag{D.1}$$

so, neglecting all masses, and using standard trace theorems,

$$\begin{aligned}
\overline{|M|^2} &= \frac{e^4}{4s^2}\mathrm{Tr}\ (\gamma \cdot p'\, \gamma_\mu\, \gamma \cdot q'\gamma_\nu)\ \mathrm{Tr}\ (\gamma \cdot p\gamma^\mu\, \gamma \cdot q\gamma^\nu) \\
&= \frac{4e^4}{s^2}\left[p'_\mu q'_\nu + p'_\nu q'_\mu - p' \cdot q'\, g_{\mu\nu}\right]\left[p^\mu q^\nu + p^\nu q^\mu - p \cdot q\, g^{\mu\nu}\right] \\
&= \frac{8e^4}{s^2}\left(p' \cdot p\, q' \cdot q + p' \cdot q\, p \cdot q'\right).
\end{aligned} \tag{D.2}$$

In the center of mass, we can choose four momenta

$$p = \frac{\sqrt{s}}{2}\,(1;\, 0,0,1)$$

$$q = \frac{\sqrt{s}}{2}\,(1;\, 0,0,-1)$$

$$p' = \frac{\sqrt{s}}{2}\,(1;\, \sin\theta,0,\cos\theta)$$
(D.3)

$$q' = \frac{\sqrt{s}}{2}\,(1;\, -\sin\theta,0,-\cos\theta)$$

with the initial particles incident along the z axis and the final particles scattered at a polar angle θ in the x–z plane. Then

$$p'\cdot p = \frac{s}{4}\,(1-\cos\theta),$$

$$q'\cdot q = \frac{s}{4}\,(1-\cos\theta),$$

$$p'\cdot q = \frac{s}{4}\,(1+\cos\theta),$$
(D.4)

$$p\cdot q' = \frac{s}{4}\,(1+\cos\theta),$$

$$\overline{|M|^2} = e^4\,(1+\cos^2\theta),$$
(D.5)

and

$$\frac{d\sigma}{d\Omega} = \frac{1}{64\pi^2 s}\overline{|M|^2} = \frac{\alpha^2}{4s}\,(1+\cos^2\theta),$$
(D.6)

which is the angular distribution we wanted to derive. Integrating over angles, the total point cross section is

$$\sigma = \frac{4\pi\alpha^2}{3s}\,.$$
(D.7)

Since we only used the spin (and the approximation of masslessness), these results hold for any case where spin one–half particles with masses small compared to $\sqrt{s}$ scatter through an s–channel vector current. If the final fermion has electric charge eQ_f, a factor of Q_f^2 should be inserted in equations (D.6) and (D.7). If a final fermion is a colored quark, the three colors are all produced equally so $d\sigma/d\Omega$ and σ are also increased by a factor of three.

When Are Our
Approximations
Not Valid?

In various places, we have seen that calculations for decay widths and lifetimes can be done very simply, giving results correct to a factor of two or better. This procedure is entirely satisfactory for understanding many of the tests and successes of the Standard Model. Indeed, it means the reader can understand many of the tests of the Standard Model without having to learn sophisticated calculational techniques.

The approximate methods are not always good, and here we summarize their limits of validity. One reason to do that is to insure that people will not get misled by the use of approximate methods.

(1) The approximation techniques are designed to work for processes involving quarks, leptons, and gauge bosons. They will not work without some modification for processes where hadrons are involved, such as $\pi^0 \to \gamma\gamma$, $\pi^\pm \to \mu^\pm \nu_\mu$, $\psi \to \mu^+ \mu^-$, *etc.* All of these need a factor for the hadronic wave function and may also involve interferences.

(2) They will not give angular distributions correctly in general, since often there are angular factors either in $u\bar{u}$ or in kinematical quantities that are replaced by maximum values.

(3) They will not give correctly the differences between processes that involve particles and antiparticles. For example, the distinction between the cross sections for $\nu_\mu e \to \nu_\mu e$ and $\bar{\nu}_\mu e \to \bar{\nu}_\mu e$ is lost by the approximation methods. The

total cross sections for both these would be given approximately correctly, but
the factor depending on $\sin^2 \theta_w$ that distinguishes them would not be.

(4) When there is a t–channel pole, such as a photon or neutrino exchange,
the results will include the effects of the pole only if the kinematics is done very
carefully.

(5) When polarization vectors of longitudinal gauge bosons are involved,
they will have to be included explicitly to get the effects of the $1/M_W$ dependence.
That is done in Chapter 21.

(6) Processes for which there is no tree level decay, such as $H \rightarrow \gamma\gamma$ or
$H \rightarrow Z\gamma$ or $b \rightarrow s\gamma$, will be zero in our approximation. Since they are typically
two to three orders of magnitude smaller than other rates, the approximation is
not bad, but the reader should be aware that some rare but interesting processes
are not included.

(7) An important set of processes that are not treated correctly are helicity–
suppressed decays. These are decays where conservation of angular momentum
and the left–handed couplings of the weak interactions provide opposite requirements on helicities of the decay products, so the decay can only occur if there is
a spin flip. That is only possible for particles with mass, which means that the
decay amplitude has an extra factor of the mass of the particle in question. An
example is $\pi^- \rightarrow \ell^- \overline{\nu}_\ell$, with $\ell = \mu$ or e. The ℓ^- must be left–handed, and the $\overline{\nu}_\ell$
right–handed, so the final spins point in the same direction. The π^-, however,
has spin zero, so the final helicities must point in opposite directions. The rate
then has an extra factor of M_μ^2 or M_e^2, which is why the $\mu\nu_\mu$ channel dominates.

It would be possible to extend the approximate techniques to include most
of these exceptions, but the method would begin to lose its simplicity. Probably
it is best to use a very simple method for the majority of the basic Standard
Model calculations as we have done; anyone who wants to go further should learn
the necessary techniques.

Lagrangians
and Symmetries;
The Euler–Lagrange
Equations

Here, for interested readers, we extend the treatment of Lagrangians and symmetries a little. First we write the Euler–Lagrange equations; then we briefly discuss Noether's theorem. The example of the electromagnetic field, begun in Chapter 2, is extended in a problem.

We did not need the Euler–Lagrange equations in Chapter 2, though we implicitly used them at one stage. They are the conditions on the Lagrangian which guarantee that the action is an extremum. We only consider the case where $\mathcal{L}$ depends explicitly on the fields and their derivatives but not on x^μ, $\mathcal{L} = \mathcal{L}(\phi, \partial_\mu \phi)$. We imagine a variation

$$\phi(x) \to \phi'(x) = \phi(x) + \delta\phi(x) \ . \tag{F.1}$$

Then

$$\begin{aligned} \delta S &= \int d^4x \, \delta\mathcal{L} \\ &= \int d^4x \left[\frac{\partial\mathcal{L}}{\partial\phi} \delta\phi + \frac{\partial\mathcal{L}}{\partial(\partial_\mu \phi)} \delta(\partial_\mu \phi) \right] . \end{aligned} \tag{F.2}$$

Using $\delta(\partial_\mu \phi) = \partial_\mu \delta\phi$ and integrating the second term by parts gives

$$\delta S = \int d^4x \left[\frac{\partial \mathcal{L}}{\partial \phi} - \partial_\mu \frac{\partial \mathcal{L}}{\partial(\partial_\mu \phi)} \right] \delta\phi \qquad (\text{F.3})$$

provided the contributions from the surface of space–time may be dropped. Thus the condition for the action to be stationary ($\delta S = 0$) is

$$\frac{\partial \mathcal{L}}{\partial \phi} - \partial_\mu \frac{\partial \mathcal{L}}{\partial(\partial_\mu \phi)} = 0 \,. \qquad (\text{F.4})$$

This is the Euler–Lagrange equation for ϕ. A similar equation holds for any field in $\mathcal{L}$. As we saw in Chapter 2, for a real scalar field with the Lagrangian $2\mathcal{L} = \partial^\mu \phi \partial_\mu \phi - m^2 \phi^2$, the Euler–Lagrange equation gives the wave equation or Klein–Gordon equation, $\partial_\mu \partial^\mu \phi + m^2 \phi = 0$.

Whenever the Lagrangian or the action is invariant under a set of continuous transformations, a divergenceless current arises. This leads to an explicitly conserved charge. Consider the case $\partial_\mu J^\mu = 0$. Integrating over d^3x gives

$$\int \partial_0 J^0 d^3x + \int \partial_i J^i d^3x = 0 \,.$$

The second term can, by Gauss' theorem, be transformed to an integral over the surface of space, and is assumed to vanish. The first term gives

$$\frac{\partial}{\partial t} \int J^0 d^3x = 0 \,,$$

so the charge $Q = \int J^0 d^3x$ does not change with time.

Whenever there is a conserved current, there is also a conserved charge, and vice–versa. Whenever there is an invariance of the theory under some transformation, a conserved current arises, as we saw in Chapter 2, and can be written down in terms of appropriate derivatives of the Lagrangian. Thus, whenever there is an invariance, there is a corresponding conserved quantity. It can be any kind of "charge" defined by the associated current. All of this is called "Noether's Theorem."

The importance of Noether's theorem is that it tells us when conserved quantities will exist, and how to define them. If there is a conserved quantity observed, we know there is an associated symmetry, and if there is a symmmetry, we know there is a conserved quantity. No guessing is needed. Since writing the Lagrangian is the goal, and knowing what symmetries to build into the Lagrangian is of great value in writing it, Noether's theorem plays a major role in relating observed symmetries and conservation laws to the structure of the theory.

Problems

F.1:   Show for the electromagnetic field that, given the Lagrangian of Chapter 2,

(a) $\dfrac{\partial \mathcal{L}}{\partial V} = -\rho$

(e) $\dfrac{\partial \mathcal{L}}{\partial A_x} = J_x$

(b) $\dfrac{\partial \mathcal{L}}{\partial \left(\frac{\partial V}{\partial x}\right)} = -E_x$

(f) $\dfrac{\partial \mathcal{L}}{\partial \left(\frac{\partial A_x}{\partial x}\right)} = 0$

(c) $\dfrac{\partial \mathcal{L}}{\partial \frac{\partial V}{\partial t}} = 0$

(g) $\dfrac{\partial \mathcal{L}}{\partial \left(\frac{\partial A_x}{\partial t}\right)} = -E_x$

(d) $\partial_k \dfrac{\partial \mathcal{L}}{\partial (\partial_k V)} = -\nabla \cdot \vec{E}$

(h) $\dfrac{\partial \mathcal{L}}{\partial \left(\frac{\partial A_x}{\partial y}\right)} = B_z$.

By using these (and the obvious extensions for $x \to y$, z) in the Euler–Lagrange equations, show that Maxwell's equations follow. Show that $\partial_\mu F^{\mu\nu} = J^\nu$ gives $\nabla \cdot \vec{E} = \rho$.

F.2:   Consider the Lagrangian

$$\mathcal{L} = \frac{1}{2} \int d^3x \left[i\left(\psi^*\dot{\psi} - \dot{\psi}^*\psi\right) - \frac{1}{m}\nabla\psi^* \cdot \nabla\psi \right] .$$

Find the equation that ψ satisfies, and interpret it. How would $\mathcal{L}$ have to be changed to add a potential $V(x)$ into the equation that ψ satisfies?

F.3:   (a) Suppose a theory is invariant under a transformation where $t' = t$, $x' = x - \delta x$, $y' = y$, $z' = z$, $\dot{x}' = \dot{x}$, $\dot{y}' = \dot{y}$ and $\dot{z}' = \dot{z}$. What conservation law can be deduced? (b) What if the transformation is $t' = t$, $x' = x + y\delta\theta$, $y' = y - x\delta\theta$, $z' = z$, $\dot{x}' = \dot{x} + \dot{y}\delta\theta$, $\dot{y}' = \dot{y} - \dot{x}\delta\theta$, and $\dot{z}' = \dot{z}$; then what conservation law do you expect?

Bibliography

Aitchison, I. J. R., and A. J. G. Hey, *"Gauge Theories in Particle Physics"*, Adam Hilger Ltd., Bristol, (1982).

Arnison, G., *et. al.*, Physics Letters, **B177** (1986), 244.

Bethe, H. A., Physical Review Letters, **56** (1986), 1305.

Bloom, Elliot D., and Gary J. Feldman, *"Quarkonium"*, Scientific American, May 1982, p. 66.

Cahn, R. and G. Goldhaber, "Experimental Foundations of Particle Physics", Cambridge University Press, to be published.

Cheng, Ta-Pei, and Ling-Fong Li, *"Gauge Theory of Elementary Particle Physics"*, Clarendon Press, Oxford, (1984).

Close, F. E., *"An Introduction to Quarks and Partons"*, Academic Press, (1979).

Collins, P. D. B. and A. D. Martin, *"Hadron Interactions"*, Adam Hilger Ltd., Bristol, (1984).

Commins, Eugene D. and Phillip H. Bucksbaum, *"Weak Interactions of Leptons and Quarks"*, Cambridge University Press, (1983).

Crease, Robert P. and Charles C. Mann, *"The Second Creation"*, Macmillan Publishing Co., (1986).

Cronin, James, Reviews of Modern Physics, **53** (1981), 367.

Davis, R., Science, **191** (1976), 264.

de Wit, B., and J. Smith, *"Field Theory in Particle Physics"*, Volume 1, North–Holland, (1986).

Dodd, J. E., *"The Ideas of Particle Physics"*, Cambridge University Press, (1984).

Drell, Sidney D., *"Electron-Positron Annihilation and the New Particles"*, Scientific American, June 1975, p. 50.

Duke, D. W., and J. F. Owens, Physical Review, **D30** (1984), 49.

Eichten, E., I. Hinchliffe, K. Lane, and C. Quigg, Reviews of Modern Physics, **56** (1984), 579.

Fitch, V., Reviews of Modern Physics, **53** (1981), 367.

Fraunfelder, Hans and Ernest M. Henley, *"Subatomic Physics",* Prentice–Hall, Inc., (1974).

Freedman, Daniel Z., and Peter van Nieuwenhuizen, *"Supergravity and the Unification of the Laws of Physics",* Scientific American, February 1978, p. 126.

Gasser, J. and H. Leutwyler, Physics Reports, **87c** (1982), 77.

Georgi, Howard, *"Unified Theory of Elementary Particles",* Scientific American, April 1981, p. 48.

Georgi, H., H. R. Quinn, and S. Weinberg, Physical Review Letters, **33** (1974), 451.

Glashow, S., Reviews of Modern Physics, **52** (1980), 539.

Gottfried, K., *"Quantum Mechanics",* Volume 1, W. A. Benjamin Inc., (1966).

Gottfried, Kurt and Victor F. Weisskopf, *"Concepts of Patricle Physics",* Volume I, Clarendon Press, Oxford, (1984).

Halzen, Francis and Alan D. Martin, *"Quarks and Leptons",* John Wiley & Sons, (1984).

Haber, Howard E., and Gordon L. Kane, *"Is Nature Supersymmetric",* Scientific American, June 1986, p. 52.

Harris, Edward G., *"A Pedestrian Approach to Quantum Field Theory",* John Wiley & Sons, (1972).

Hill, E. L., Reviews of Modern Physics, **23** (1951), 23.

Hughes, I. S., *"Elementary Particles"*, Cambridge University Press, Second Edition, (1985).

Ishikawa, Kenzo, *"Glueballs"*, Scientific American, November 1982, p. 142.

Jackson, J. D., M. Tigner, and S. Wojcicki, *"The Superconducting Super Collider"*, Scientific American, March 1986, p. 66.

Jacob, Maurice, and Peter Landshoff, *"The Inner Structure of the Proton"*, Scientific American, March 1979, p. 66.

Johnson, Kenneth A., *"The Bag Model of Quark Confinement"*, Scientific American, July 1979, p. 112.

Leader, Elliot and Enrico Predazzi, *"An Introduction to Gauge Theories and the New Physics"*, Cambridge University Press, (1982).

Lederman, Leon, *"The Upsilon Particle"*, Scientific American, October 1978, p. 72.

Lee, T. D., *"Particle Physics and Introduction to Field Theory"*, Harwood Academic Publishers, (1984).

LoSecco, J. M., Frederic Reines, and Daniel Sinclair, *"The Search For Proton Decay"*, Scientific American, June 1985, p. 54.

Mandl F. and G. Shaw, *"Quantum Field Theory"*, John Wiley & Sons, (1984).

Moriyasu, K., *"An Elementary Primer for Gauge Theories"*, World Scientific, (1983).

Nambu, Yoichiro, *"The Confinement of Quarks"*, Scientific American, October 1976, p. 48.

Ne'eman, Yuval, and Yoram Kirsh, *"The Particle Hunters"*, Cambridge University Press, (1986).

Okun, L. B., *"Leptons and Quarks"*, North–Holland Publishing Co., (1982).

Okun, L. B., *"Particle Physics, The Quest for the Substance of Substance"*, Harwood Academic Publishers, (1985).

Pais, Abraham, "*Inward Bound*" Clarendon Press, Oxford, (1986).

Particle Data Tables, "*Review of Particle Properties*" Physics Letters **170B**, 10 April, (1986).

Perkins, Donald H., "*Introduction to High Energy Physics*" Third Edition, Addison Wesley Publishing Co., (1986).

Perl, Martin L., "*High Energy Hadron Physics*" John Wiley & Sons, (1974).

Perl, Martin L., "*The Tau Heavy Lepton*", Nature, **275** (1978), 273.

Perl, Martin L., and William J. Kirk, "*Heavy Leptons*" Scientific American, March 1979, p. 50.

Pickering, Andrew, "*Constructing Quarks*" University of Chicago Press, (1984).

Quigg, C, "*Gauge Theories in High Energy Physics*" 1981 Les Houches Summer School, edited by M. K. Gaillard and R. Stora, North-Holland, Amsterdam, (1983).

Quigg, Chris, "*Gauge Theories of the Strong, Weak and Electromagnetic Interactions*" Benjamin/Cummings Publishing Co., Inc., (1983).

Rebbi, Claudio, "*The Lattice Theory of Quark Confinement*" Scientific American, February 1983, p. 54.

Renard, Fernand M., "*Basics of Electron Positron Collisions*" Editions Frontières, (1981).

Richter, B., Adventures in Experimental Physics, **5** (1976), 143.

Ross, Graham G., "*Grand Unified Theories*" Benjamin/Cummings Publishing Co., Inc., (1985).

Rubbia, Carlo, Reviews of Modern Physics, **57** (1981), 699.

Rubin, Vera C., "*Dark Matter in Spiral Galaxies*" Scientific American, June 1983, p. 96.

Ryder, Lewis H., "*Quantum Field Theories*" Cambridge University Press, (1985).

Sakharov, A. D., "*Violation of CP Invariance, C Asymmetry, and Baryon Asymmetry of the Universe*," JETP lett. , **5** (1967), 24.

Sakurai, J. J., "*Invariance Principles and Elementary Particles*," Princeton University Press, (1964).

Sakurai, J. J., "*Advanced Quantum Mechanics*," Addison–Wesley Publishing Co., (1967).

Sakurai, J. J., "*Modern Quantum Mechanics*," Benjamin/Cummings Publishing Co., (1985).

Salam, A., Reviews of Modern Physics, **52** (1980), 525.

Saxon, D. H. "*Measurement of Electroweak Effects in e^+e^- Annihilation*," Proceedings of the IX Warsaw Symposium on Elementary Particles, Kazimierz, Poland, May 1986.

Scadron, Michael D., "*Advanced Quantum Theory*," Springer–Verlag, (1979).

Schwitters, Roy F., "*Fundamental Particles with Charm*," Scientific American, October 1977, p. 56.

Söding, P. and G. Wolf, "*Experimental Evidence on QCD*," Annual Reviews of Nuclear Science, **5** 1981, p. 231.

't Hooft, Gerard, "*Gauge Theories of the Forces Between Elementary Particles*," Scientific American, June 1980, p. 104.

TASI Lectures..., see Williams, David N.

Taubes, Gary, "*Nobel Dreams*," Random House, (1986).

Ting, S. C. C., Adventures in Experimental Physics, **5** (1976), 115.

Veltman, Martinus, "*The Higgs Boson*," Scientific American, November 1986, p. 76.

Watkins, Peter, "*Story of the W and Z*," Cambridge University Press, (1986).

Weinberg, Steven, Reviews of Modern Physics, **52** (1980), 515.

Weinberg, Steven, *"The Decay of the Proton",* Scientific American, June 1981, p. 64.

Weinberg, Steven, *"The Problem of Mass"* in *"A Festschrift for I. I. Rabi",* Transactions of the New York Academy of Sciences, Series II Vol 38, p. 185.

Wentzel, Gregor, *"Quantum Theory of Fields",* Interscience Publishers, (1949).

Wilczek, Frank, *"The Cosmic Asymmetry Between Matter and Antimatter",* Scientific American, December 1980, p. 82.

Williams, David N., editor, *"TASI Lectures in Elementary Particle Physics",* World Scientific, (1984).

Zee, A., *"Fearful Symmetry",* Macmillan, (1986).

Index

A

Abelian gauge invariance 35-40, 78, 103-04, 318.
Abelian vector field 31.
accelerators 155-62, 168-70.
α_i couplings
    measurement of α_3 150-51.
    measurement of α and α_2 143-44.
    variation of 223, 226-29, 231, 273.
    values 93.
angular momentum and $SU(2)$ 306.
antiparticle 10, 26, 33, 64, 258.
antiscreening 231.
approximate calculation 146, 214, 237, 326.
asymptotic freedom 205, 231, 232.
atoms 6, 10, 12.

B

b (bottom) quark 7, 92, 199, 217-18.
    (see also quarks).
    decay 253, 263-64.
    decay width Γ_b 253.
    electric charge 7.
    interactions 261.
    lifetime 167, 254.
    mass 8.
baryon 175, 178, 180.
    $L = 0$ states 187.
    baryon number 7, 280, 283.
    baryon number of the universe 257, 283.
    conservation of baryon number 280.

baryon asymmetry 283-4.
beams 132, 155-160, 164.
β decay 89.
big bang theory 283.
Bjorken scaling 203.
boson 285.
    vector bosons 33.
branching ratio 128, 129.
Breit–Wigner resonance 119.
    figure 120.
Brookhaven National Laboratory 160, 191, 299.

C

c (charm) quark 7, 92, 193, 217-18.
    (see also quark and D mesons).
    charmed mesons 195, 196-98.
    charmonium 193-96.
    electric charge 7.
    lifetime 167.
    mass 8.
Cabibbo angle 250.
Čerenkov cone 283.
CERN 10, 132, 159, 168, 220, 290.
charge 4, 26, 27, 38.
charge conjugation 186, 255-260.
charge conjugation invariance 257-260.
    CP 258.
    CP violation 257, 283.
    CPT 258.
    decay of kaons 259-60.
charge states of isovector 44, 72.
charged current 88-90, 92, 146, 147, 249, 252, 257.
    Kobayashi-Maskawa matrix 253.

charmed mesons 195, 196-98.
    (see c quark and D mesons).
charmonium
    properties 193-94.
    spectrum 195-96.
classical electrodynamics 4-5, 17, 35-36, 330.
color 47, 51, 72, 75, 90, 91.
    color charge 6, 7, 10, 72, 91, 230-31.
    color force 6, 72, 178.
    confinement of color 176-79.
    number of colors 216-20, 271-72.
    space 51.
color singlets 75.
color triplets 75.
color–singlet states 179-81, 185.
complex scalar field 23, 30, 101-02, 114.
conserved current 25, 26, 60-62, 327-30.
constituent cross section $\hat{\sigma}$ 133, 203, 210, 289.
constituent mass 8, 185, 255.
constituents of matter 7, 10, 76.
cosmological constant 113.
Coulomb's law 5, 223.
covariant derivative 38-40, 48, 76, 103, 108, 276.
critical mass to close the universe 292, 302.
cross section 12, 116, 121, 133, 136, 157, 193, 202, 210, 214, 216-18, 289.
current 22-27, 60-62, 67, 328-29.

D

d (down) quark 7, 10, 75-77, 217-18.
    (see also quarks).
    electric charge 7.
    mass 8.

D meson
    composition of 196.
    decay 198, 200.
    mass 197.
    produced in charmonium decay 195-96.
dark matter 174, 291-93, 302.
decay rates and modes
    charmonium 193-96, 240.
    electromagnetic 188.
    fermion 146.
    forbidden and rare 174.
    general 116-18.
    Higgs boson 237-39, 247.
    muon 146.
    neutrino 301.
    proton 280.
    strong 188.
    τ 219.
    toponium 199.
    weak 188, 131.
    b quark 253, 263-64.
    c quark 197, 251.
    D meson 198, 200.
    K_L 259.
    s quark 251.
    t quark 199.
    W boson 124, 136-37, 140-41.
    Z boson 125, 131, 265.
deep inelastic scattering 201-05.
detectors 163-172.
    4π detectors 163.
    calorimeters 167.
    proton decay 282-83.
    solar neutrinos 300.
    triggering 166-67.
    vertex detector 167.

Dirac current 60.
Dirac equation 60.
Dirac mass 295.
doublets
    of leptons 8, 73, 81.
    of nucleons 44.
    of quarks 7, 73, 90.
Drell–Yan cross section 210.

E

eigenvalue of T_3 87.
    T_3^b 262.
    T_3^τ 264.
electric charge 43, 26.
electric charge operator 87, 90, 92, 96,
    107, 272.
electromagnetic current 61, 84, 226.
electromagnetic decays 188.
    (see also decays).
electromagnetic interaction 38, 67, 83.
electromagnetism 5, 6, 9, 35, 82-87.
electron e 10, 12, 74.
    charge 8.
    conservation of electron number
    174.
    fermion field for 19.
    mass 8.
electron neutrino ν_e 10, 74.
electroweak charge 86, 87.
electroweak force 3.
electroweak mixing angle θ_w 85.
    $\sin^2\theta_w$ calculation 276.
    $\sin^2\theta_w$ figure 145.
    $\sin^2\theta_w$ value 144.
electroweak theory 6, 10, 81-90.
Euler–Lagrange equations 17, 18, 25, 27,
    327-30.
expanding universe 284.

F

families 76, 92, 268.
    additional 127.
Fermi coupling 147, 93.
fermion f 33, 285.
    decay 146.
    fermion–Higgs vertex
    $H \to f\bar{f}$ 237.
    masses 8-9, 152.
Feynman rules 3, 5, 19, 32-33, 61, 92,
    249.
fine structure constant 144, 223.
flavor changing neutral current 174, 252.
flavors (see families)
    of leptons 8.
    of quarks 7.
force 3, 4, 5, 7, 12, 17.
    as an interaction 29.
form factor 202.
forward–backward asymmetry 262, 266.
foundations of the Standard Model 267.
four–vector 16, 17.
fundamental scalars 105, 267.

G

gauge boson 1, 7, 9, 10, 37-40, 82-91,
    105-08, 238, 244-45, 279.
    mass 104.
gauge transformation 35, 46-47.
    global 26, 36.
    local 26, 36, 103.
    of the first kind 26.
    of the second kind 26.
generators 317.
    $SU(2)$ 46, 311, 316-17.
    $SU(3)$ 47, 317.
    $SU(5)$ 276.
    $SU(n)$ 272, 316.
    hypercharge generator 51, 77, 81,
    83-85, 273.
    trace theorem for generators 277.

GIM mechanism 252.

glueball states 180, 184.

gluinos $\tilde{g}$ 288.

    (see also supersymmetry).

gluon g 1, 9, 75, 201.

    emission 206.

    in the structure of hadrons 132.

    interactions 189.

    observation 220-21.

    radiation 217, 219.

Goldstone boson 102, 105, 238, 247.

Goldstone theorem 102.

grand unified theory (GUT) 5, 6, 268, 271.

graviton 9.

gravity 5, 9, 113.

group theory 311-318.

    rotation in n-dimensional Euclidian space $SO(n)$ 313.

    rules for a group 311.

    unitary groups $SU(n)$ 314.

    $SU(n)$ invariance 50, 314.

    $SU(2)$ 44, 50, 52, 72, 107, 310.

    $SU(3)$ 52, 72.

    $U(1)$ 51, 71, 72, 107.

H

hadron 7, 10, 185, 232.

    color singlets 179-81.

    definition 6.

    gluon and quark structure of 132.

    mass 255-56.

hadronization 177, 217.

Hamiltonian 3, 12, 20.

hard collisions 165.

Higgs boson H 2, 10, 105, 235.

    fermion–Higgs boson vertex $H \rightarrow f\bar{f}$ 237, 110.

    Higgs boson width Γ_H 237-39.

    Higgs boson–gauge boson vertex $H \rightarrow WW$ 113, 237, 238.

    limits on M_H 236.

Higgs boson H (*continued*)

    mass 243, 246.

    rare decays 243.

    two doublets of Higgs fields 247.

Higgs mechanism 2, 94, 105-09.

Higgs boson searches 240-247.

higher order corrections 145, 223, 236, 256.

Horizontal gauge symmetries 268.

hypercharge generator 51, 77, 81, 83-85, 273.

hypercharge (weak), see weak hypercharge

I

interaction 27-30.

    as a force 4, 29.

interaction Lagrangian 18, 32, 45, 92.

invariance and conserved quantities 26, 329.

invariance under transformations 24, 239.

invariant Lagrangians 19, 45.

invariant rotations in isospin space 45.

isospin space 43-46.

    (see also strong isospin and weak isospin).

J

Jacobian Peak technique 138-39.

jets 177, 215, 218, 220.

    figures 215, 221.

K

kaons K 27.

    neutral kaon system 259.

    rare and forbidden decays 173, 174.

kinematics 134, 319-322.

Klein–Gordon equation 328.

Kobayashi–Maskawa matrix 253.

L

Lagrangian 3, 19, 20, 24, 26, 27, 35.
    mass term 67.
    source term 27.
    summary of 30-32.
Lagrangian density 20.
    for electromagnetism 18.
lattice gauge theory 179.
left–handed fermions 66, 73-75, 81-82.
left–handed interaction 67.
left–handed neutrino 86.
left–right symmetric theories 268.
lepton ℓ 1.
    basic particle of nature 9.
    in matter 7.
    lepton number 8.
    masses 9, 12.
    mixing angles 254.
    universality 92.
lifetime 117-18, 167.
lightest supersymmetric particle 291.
like–sign dileptons 171.
longitudinal polarization state of the
    gauge boson 105, 238, 245.
longitudinal vector bosons 238.
Lorentz invariant 15-16, 19.
luminosity 137, 157, 171.

M

Majorana fermion 296.
Majorana neutrino 171.
matter 7.
Maxwell's equations 18, 330.
meson field
    range of force 29.
mesons 29.
    D mesons 195-98.
    $L = 0$ states 186.
missing energy 167, 219.
missing momentum 138, 290.
molecules 6, 10.

muon μ 8, 92.
    charge 8.
    conservation of muon number 174.
    decay (figure) 146.
    decay 146-49.
    decay width Γ_μ 148.
    mass 9.
    rare decays 173.

N

narrow width approximation 134.
natural units 11-12, 118, 136, 179.
neutral current 82-88, 251.
    flavor changing 252.
neutral kaon system 259-60.
neutrino ν 8, 295.
    electric charge 8.
    mass 8, 111, 173, 295-303.
    mass ν_τ 264-65.
neutrino oscillations 173, 296-99.
neutrinoless double β–decay 173.
neutron n 43.
neutron "β–decay" 89.
neutron–proton mass difference 43, 256.
Newton's constant G_N 12.
Newton's laws 3, 17.
Noether's theorem 26, 328.
Non–Abelian vector field 31-32.
non–perturbative phenomena 153, 179,
    188, 255.
nuclear force 6, 10.
nuclei 6, 10, 12.
nucleon decay 174, 280-83.
number of colors
    from e^+e^- annihilation 218.
    in τ decay 220.
    linked to the quark charge 272.
    from W and Z decay 128.
numerical values for couplings 93.

P

parity 66, 76.
   of pseudo scalar mesons 186.
   violation 66, 88, 257-60.
particles
   (see also leptons, boson,
   and quarks).
   $\overline{D}^{(-,0)}$ 196.
   $D_s^{\pm}$ 197.
   $D^{(+,0)}$ 196.
   $F^{\pm}$ 197.
   J/ψ 191.
   ω 191.
   ϕ 191.
   ψ 191.
   ψ' 192.
   ρ^0 191.
   baryon $L = 0$ states 187.
   meson $L = 0$ states 186.
parton model 133, 204-05, 206-07.
   figure 209.
Pauli principle 182.
Pauli spin matrices 45, 57, 77, 307.
PETRA 10, 159, 217, 220, 261, 290.
phase space integration for μ decay 147.
phase transformation 36, 46.
   (see also gauge transformations).
   as gauge transformations (global or
   local) 26.
photinos $\widetilde{\gamma}$ 288.
   (see also supersymmetry).
photon γ 9, 19, 21, 29, 35, 37, 84, 108-
   09, 164, 202.
   mixing with J/ψ 194.
pions 43.
   charge states 45.
   mass m_π 187.
point cross section 213, 214, 323-24.
point–like constituent 203, 213.
point–like objects 201.
point-like scalars 247.

propagator 30, 33, 146.
proton 43.
proton decay 280-83.

Q

quantum chromodynamics (QCD) 1, 6,
   72, 91, 150, 175, 205, 220, 230-36,
   255, 273.
quantum electrodynamics (QED) 224-30.
quantum field theory 1, 3, 6, 9, 19, 22,
   29.
quark q 1, 199, 201.
   (see also u, d, c, s, t, and b quarks)
   basic particle of matter 9.
   bottom 7.
   charmed 7.
   chemistry 178.
   down 7.
   electroweak eigenstate 249.
   fractional charge 7, 272.
   "free" mass 8, 185, 255.
   in matter 7.
   mass eigenstate 249.
   masses 8, 12, 255-56.
   quarkonium 240.
   sea quarks 204.
   strange 7.
   structure of hadrons 132.
   top 7.
   up 7.
   valence quarks 204.
quark spin-flip transition 188.

R

$R = \sigma/\sigma_{\text{point}}$ 216.
radiated gluon 230.
radiative corrections 145, 223, 236, 256.
real scalar field 19-21, 30.
renormalizable theory 53, 244.
representation of the rotation matrices
   for spin–one 50.
rho (ρ) parameter 109.

right–handed current 67, 173.
right–handed fermion 66, 74, 76, 81.
right–handed neutrino 75, 111, 295.
running coupling strength 223-33, 273.

S

s (strange) quark 7, 92, 217-18.
    (see also quarks).
    electric charge 7.
    mass 8.
    strange particles 185-88.
scalar bosons 33, 104, 235-47, 267.
scalar fields.
    (see real scalar field and complex
    scalar field).
scaling violations 206.
Schrödinger equation 3, 22, 37.
screening 229.
sea quark 204.
$\sin^2 \theta_w$
    (see electroweak mixing angle).
SNU 300.
soft collisions 165.
solar neutrino 173, 300-01.
solar neutrino unit (SNU) 300.
spin flip transition 194.
spontaneously broken symmetry 98-113.
squarks $\tilde{q}$ 288.
    (see also supersymmetry).
Standard Model tests 109, 124, 128, 137,
    140, 144, 174, 183, 190, 205, 215-18,
    220, 262
strange particles 185-8.
    (see also s quark and quarks).
strange quark 7.
    (see also s quark and quarks).
    decay 249-52.
Strong decays 188.
    (see also decays).
strong force 1, 3, 6, 9.
    see also quantum chromodynamics.

strong isospin 43-46.
    (see also isospin).
    invariance 189-90.
    space 44.
    symmetry 43.
structure function 133, 135, 204-09, 211,
    290.
    $F_2(x)$ 206, 290.
    figure 208.
structure of quarks and leptons 201, 216.
sun 6.
supersymmetry 267, 268, 285-93.
    particles 286.

T

t (top) quark 7, 92, 199, 217-18, 261.
    (see also quarks).
    electric charge 7.
    mass 8, 265.
    toponium 242.
tau lepton τ 8, 92, 199.
    τ cross section 219.
    τ signatures 218.
    conservatiuon of tau number 174.
    decay (figure) 219.
    lifetime 167.
    mass 9.
tau neutrino ν_τ 9, 199, 261.
    (see also tau lepton).
    detection of ν_τ 265.
theta (θ) parameter 153.
time reversal 258.
    (see also charge and parity conjuga-
    tion, CPT).
toponium 199, 242.
transverse mass 139.
two doublets of Higgs fields 247.

U

u (up) quark 7, 10, 217-18.
(see also quarks).
electric charge 7.
mass 8.
unification scale 278.
unitarity violation 245.

V

$V-A$ charged current interaction 67, 89, 140, 147.
vacuum 97-98, 107, 112.
vacuum energy density 112.
vacuum expectation value 99, 104, 107.
vertex factor 33, 87-88, 92.
violation of conservation of probability 245.
violation of unitarity 245.

W

W particle 9.
decay asymmetry 140-41.
decay branching ratios 124-25.
discovery 10, 137.
mass 108, 137-39, 256.
production 132-36.
WW scattering 244.
$W^+ \rightarrow e^+ \nu_e$ vertex 122.
$W^{\pm}$ width 124.

weak decays 136, 140, 146, 188, 197, 219, 237, 249-53, 259, 264.
(see decays).
weak force 3, 5, 6, 9.
weak hypercharge 51, 71, 77, 81-85, 107.
weak isospin 43, 48, 72.
(see also isospin and $SU(2)$).

X

X bosons 281.

Y

Y bosons 281.
Yang–Mills gauge theory 48.

Z

Z particle 9.
Z width 125-28, 265.
decay branching ratios 128.
discovery 10, 137.
mass M_Z 109, 137, 256.